Genetic Resources, Traditional Knowledge and the Law

Genetic Resources, Traditional Knowledge and the Law

Solutions for Access and Benefit Sharing

Edited by
Evanson C. Kamau and Gerd Winter

First published by Earthscan in the UK and USA in 2009

For a full list of publications please contact:

Earthscan
4 Park Square, Milton Park, Abingdon, Oxon OX14 4RN
605 Third Avenue, New York, NY 10017

First issued in paperback 2013

Earthscan is an imprint of the Taylor & Francis Group, an informa business

ISBN: 978-1-84407-793-9 (hbk)
ISBN: 978-0-41584-790-2 (pbk)

Typeset by 4word Ltd, Bristol
Cover design by Susanne Harris

Earthscan publishes in association with the International Institute for Environment and Development

A catalogue record for this book is available from the British Library

Library of Congress Cataloging-in-Publication Data

Genetic resources, traditional knowledge, and the law solutions for access and benefit sharing / edited by Evanson C. Kamau and Gerd Winter.
    p. cm.
  Includes bibliographical references and index.
  ISBN 978-1-84407-793-9 (hardback)
  1. Biotechnology--Law and legislation. 2. Biotechnology--Patents. 3. Indigenous peoples--Legal status, laws, etc. 4. Biotechnology--International cooperation. 5. Transgenic organisms. I. Kamau, Evanson C. II. Winter, Gerd.
  K3611.G46G465 2009
  343'.0786606--dc22
                                2009012695

Contents

List of Figures, Tables and Boxes

Figures

Tables

Boxes

About the Authors

Anne N. Angwenyi is an advocate of the High Court of Kenya. She holds an LLB (University of Nairobi), MSEL (Vermont Law School) and a MALD (Fletcher School of Law and Diplomacy, Tufts University), and is a Chevening Scholar in Governance and Environmental Democracy. She was the Head of the Legal Services Department with the National Environment Management Agency (NEMA), Kenya, until July 2008. She is currently a Programme Officer with the Royal Danish Embassy Nairobi, in charge of the agriculture and natural Resources portfolio. Her research interests include environmental governance, sustainable development and management of natural resources. Anne_Angwenyi@alumni.tufts.edu

Geoff Burton is a Visiting Senior Fellow of the United Nations University Institute of Advanced Studies and consultant to various governments on Access and Benefit Sharing (ABS). Prior to leaving the public sector he was heavily engaged on genetic resources as Australia's National Competent Authority, its lead negotiator and head of delegation at the Convention on Biological Diversity (CBD) negotiations on genetic resources. geoff@jeanshannon.com

Jorge Cabrera Medaglia is the National Biodiversity Institute Legal Adviser and Professor of environmental law at the University of Costa Rica. He was a negotiator on behalf of his country at the CBD Conference of the Parties (COP) and the Access and Benefit Sharing Working Group (ABS-WG). Between 1999 and 2001 he served as Co-chair, CBD ABS Expert Panel, and has been a member of the Technical Expert Group on Certificates (2007) and on Use of Terms (2008) of the CBD. He is an international consultant in the area of ABS, intellectual property rights (IPR) and biodiversity, biotechnology and biosafety and biotrade. jcabrera@cisdl.org

Andreas Drews is a biologist holding a PhD in natural sciences. He has been involved in the ABS process since 1995, for Deutsche Gesellschaft für Technische Zusammenarbeit (GTZ) GmbH, and is an adviser to the German Federal Ministry for Economic Cooperation and Development (BMZ). andreas.drews@gtz.de

Christiane Gerstetter is a lawyer by training, with a strong interdisciplinary background and a keen interest in issues of global justice. She currently works for the environmental think-tank Ecologic Institute on projects related to intellectual property, trade, development and sustainability. christiane.gerstetter@ecologic.eu

Jack Kaguu Githae is a medical herbalist and Director of the School of Alternative Medicine and Technology (SAMTECH). He studied veterinary medicine in the USA and has worked in the UK, Australia and America. Later, he returned to Africa where he has dedicated his time since then to developing traditional medicine and its practice, in which he has now more than 20 years' experience. He has used traditional African medicinal herbs and recipes in ten African clinics with a lot of success, and proved that many therapeutic processes are applicable and useful in other parts of the world as well. kaguugithae@yahoo.com

Christine Godt, Law Professor, is currently visiting at the University of Oldenburg, teaching European and international economic law, private law (especially property), comparative law and intellectual property law. PhD: 1995–1997 ('Haftung für Ökologische Schäden'); Habilitation: 2005–2007 ('Eigentum an Information'). Her research focus is on intellectual property and cross-cutting policies like competition and innovation, public health, environmental and agricultural policies and regulation theory. cgodt@zerp.uni-bremen.de

Fabian Haas is an entomologist by profession and was the National Focal Point for the Global Taxonomy Initiative in Germany. Since 2006 at ICIPE, he explores the consequences of the ABS regime for biocontrol and taxonomic research in developing countries. fhaas@icipe.org

Hiroji Isozaki is a Professor of International Law at the Meiji-Gakuin University, as well as a Visiting Professor at the United Nations University. He has mainly carried out academic researches on the North-South Problem, Law of the Sea Issues and Environmental Issues, including ABS from the international law perspectives. isozaki@law.meijigakuin.ac.jp

Suhel al-Janabi holds an MSc after studying geography, ecology and economics of transport. He is Executive Director of GeoMedia GbR – a consultancy advising at the interface of environment, development and communication. Mr al-Janabi is also a Member of the Informal Advisory Committee on Communication, Education and Public Awareness of the Convention on Biological Diversity. s.aljanabi@geo-media.de

María Julia Ochoa graduated in Law at the University of Los Andes (Venezuela). She obtained her Magister Iuris at the University of Göttingen (Germany), where she is a PhD candidate and works as an assistant researcher at the Institute of International Law. mariajulia85@yahoo.com

Bram De Jonge is a PhD candidate at Applied Philosophy, Wageningen University. His research project, 'The Ethics of Benefit-Sharing in the Field of Plant Genomics', is part of the Centre for Society and Genomics in the Netherlands, funded by the Netherlands Genomics Initiative. bram.dejonge@wur.nl

Evanson Chege Kamau is a Kenyan lawyer currently working in Germany as a senior research fellow at the Forschungsstelle für Europäisches Umweltrecht (FEU), University of Bremen. He studied law in the former USSR acquiring an LL.M in 1992. From 1995 to 1997 he did a masters course in international and European law at the University of Bremen and acquired an LL.M. Eur. From 1998 to 2003 he did his doctoral studies focusing on the impacts of patents (IPRs) on technological growth in developing countries, including on traditional rights on biological resources and traditional knowledge. He has worked in various environmental projects in FEU since 2000 and has focused on ABS issues since 2004. echege@uni-bremen.de

Sandra Akemi Shimada Kishi is Regional Federal Prosecutor for the Republic; she has an MA in environmental law, is guest professor in *lato sensu* graduate courses in environmental law at the Piracicaba Methodist University, professor in the Higher School for Federal Prosecuting Attorneys and coordinator of the journal 'Revista Internacional de Direito e Cidadania'. skishi@PRR3.MPF.GOV.BR

John Bernhard Kleba has a PhD in Social Sciences, at the University of Bielefeld. Since March 2005 he has been Professor of Political Science and Sociology at the Aeronautics Technological Institute (ITA), Brazil. Currently also working in a project on ABS funded by DFG and FAPESP. jbkleba@ita.br

Niels Louwaars is senior scientist at the Centre for Genetic Resources, the Netherlands, part of Wageningen University. Trained plant breeder, he worked in developing country seed systems, and now concentrates on policy and regulatory issues affecting the management of agricultural genetic resources. niels.louwaars@wur.nl

Peter Munyi holds degrees in law as well as specialized certificates in genetic resources policy and genetic engineering regulation. He is the Chief Legal Officer, ICIPE, where his work involves, among others, legal and policy issues in agriculture and environment. pmunyi@icipe.org

Alexander Proelss is Professor of public law (with particular focus on the law of the sea), Christian Albrechts University Kiel, and is a Member of the Excellence Cluster 'The Future Ocean' (www.ozean-der-zukunft.de/english/). aproelss@internat-recht.uni-kiel.de

Tianbao Qin, PhD, Professor of Law, is Assistant Director of the Research Institute of Environmental Law and a Senior Research Fellow of European Studies Centre, Wuhan University, China. tianbaoq@hotmail.com

Juliana Santilli, Prosecutor (Ministério Público do Distrito Federal), is co-founder of the Brazilian NGO Instituto Socioambiental and is a PhD candidate in Socioenvironmental Law. juliana.santilli@superig.com.br

Monika Ribadeneira Sarmiento is an Ecuadorian environmental lawyer. She studied in Ecuador, Spain, Sweden and Germany. She focuses on CBD issues, mainly access to genetic resources and benefit-sharing. Now she works at the German Research Foundation (Deutsche Forschungsgemeinschaft, DFG) in Bonn as a Programme Officer for CBD and ABS. Monica.Ribadeneirasarmiento@dfg.de

Peter-Tobias Stoll is a Professor of public law and public international law at the Georg-August-University of Göttingen. He focuses mainly on international economic and environmental law (http://inteurlaw.uni-goettingen.de/intecolaw/). PT.Stoll@jur.uni-goettingen.de

Mandy Taylor is a South African attorney and a UK solicitor. She is currently working for the national water regulator in the UK and previously worked in private legal practice in South Africa, specializing in public sector law and environmental law. She drafted the South African regulations governing bioprospecting, access and benefit-sharing. mandytaylor.rsa@gmail.com

Brendan Tobin (Ashoka Fellow) is a consultant on international environmental law and the rights of indigenous peoples. From 2003 to 2007 he was a research fellow and coordinator of the Biodiplomacy Initiative at UNU-IAS. Prior to this he was coordinator of the International and Biodiversity Program at the Peruvian Environmental Law Society. tobin@ias.unu.edu

Gerd Winter is professor of public law and the sociology of law at the University of Bremen. He is co-director of the Research Centre for European Environmental Law (FEU) (www.feu.uni-bremen.de). His focus of teaching, research and publications is on administrative and environmental law in comparative and international perspectives. He has also worked as legal consultant in various countries. For further information see his personal website www-user.uni-bremen.de/~gwinter/. gwinter@uni-bremen.de

Rachel Wynberg is an environmental policy analyst, academic and activist, based in Cape Town, South Africa and at the University of Strathclyde, where she does research on the commercialization of SA biodiversity and from where she completed a doctorate that explores pro-poor models of biodiversity commercialization in southern Africa. Her work is focused on issues relating to the commercialization and trade of biodiversity and the integration of social justice into biodiversity concerns. Rachel@iafrica.com

Preface

This book is the outcome of an international expert workshop on access and benefit sharing (ABS) and related traditional knowledge (TK) issues held in Bremen in February 2008, and hosted by the Forschungsstelle für Europäisches Umweltrecht (FEU) of the University of Bremen. It is gratefully acknowledged that the workshop was sponsored by the Deutsche Forschungsgemeinschaft (DFG) and the Bremen International Graduate School for Marine Sciences (GLOMAR).

It was clear for the workshop organizers from the onset that the effectiveness of measures to regulate ABS and TK will depend upon the capacity of decision makers to develop measures that build bridges between the needs, rights and interests of both providers and users of resources. This requires a true understanding of a range of issues including the economic, environmental, cultural and inherent value of genetic resources; the nature of TK innovation systems; the effectiveness of existing national ABS and TK related legislation; the challenges for development of functional prior informed consent (PIC) procedures; mechanisms for promoting fair and equitable benefit sharing, including modalities for promoting equitable contractual arrangement mechanisms for protection of sovereign rights over genetic resources and of the rights of indigenous peoples and local communities over their TK; and the potential and impediments to adoption of user measures. Therefore, the workshop aimed to produce practical and amicable solutions to some of these issues, hence its title, 'Undoing the Knot in A&BS Transactions. In Search of Amicable Solutions'. From the conclusions and recommendations of the workshop, it was noted with much satisfaction that the workshop had achieved most of its objectives. A large number of the contributions were selected for further research on the salient issues and, together with invited contributions, this book was born. Some of the contributions, and the workshop itself, were elaborated in the framework of a research project on access to genetic resources and benefit sharing of the FEU. We gratefully acknowledge the funders of the project, the DFG.

During the preparation for publication, all authors worked tirelessly and on most occasions were keen to respond in a timely fashion to comments, suggestions and revisions. We appreciate their remarkable work and reliable cooperation, which made the birth pains bearable. Last but not least we acknowledge the great assistance rendered in organizing the workshop and formatting the book manuscript by the secretary of the FEU, Ms Antje Spalink.

Evanson Chege Kamau and Gerd Winter
Bremen, March 2009

List of Acronyms and Abbreviations

ABS	access and benefit sharing
ABS-WG	Access and Benefit-Sharing Working Group
ADN	Asociacion para la Defensa de los Derechos Naturales
ADR	alternative dispute settlement
AFD	Agence Française du Développement (French Development Agency)
ANDES	Association for Nature and Sustainable Development
BCH	Biosafety Clearing House
BMZ	Federal Ministry for Economic Cooperation and Development
BSA	benefit-sharing agreement
CAMBIA	Center for the Application of Molecular Biology to International Agriculture
CAN	Andean Community of Nations
CBD	Convention on Biological Diversity
CCPR	International Covenant on Civil and Political Rights
CEA	Civil Execution Act
CEPA	Communication, Education and Public Awareness
CESCR	International Covenant on Economic, Social and Cultural Rights
CGEN	Genetic Patrimony Management Council
CGIAR	Consultative Group on International Agricultural Research
CISDL	Centre for International Sustainable Development and Law
CITES	Convention on International Trade in Endangered Species of Wild Fauna and Flora
COICA	Coordinating Body of Indigenous Peoples of the Amazon Basin
COMARU	Rio Iratapuru Producers and Extracting Cooperative
COMIFAC	Working Group on Biological Diversity of the Central African Forest Commission
COP	Convention/Conference of the Parties
CSG	Centre for Society & Genomics
CSIR	Council for Scientific and Industrial Research
DEFRA	Department for Environment, Food and Rural Affairs
DFG	Deutsche Forschungsgemeinschaft (German Research Foundation)
DGIS	Directorate General for International Cooperation

DNA	deoxyribonucleic acid
DTK	disseminated traditional knowledge
EC	European Community
ECOSOC	Economic and Social Council
EEZ	exclusive economic zone
EGBGB	German Conflict of Law Code
EIA	environmental impact assessment
EIAR	Environmental Impact Assessment Report
EMCA	Environment Management and Co-ordination Act
EU	European Union
EWGPB	Ecuadorian Working Group on Prevention of Biopiracy
EWI	ecosystem well-being index
FA	Forests Act
FAO	Food and Agriculture Organization
FAPESP	Fundação de Amparo à Pesquisa do Estado de São Paulo (State of São Paulo Research Foundation)
FECON	Federation for the Conservation of the Environment
FEEA	Federal Economic Espionage Act
FEU	Forschungsstelle für Europäisches Umweltrecht (Research Centre for European Environmental Law)
FSC	Forest Stewardship Council
FUDECI	Foundation for the Development of Physical, Mathematical and Natural Science
GAP	General Access Procedure
GEF	Global Environmental Facility
GLOMAR	International Graduate School for Marine Sciences
GMO	genetically modified organism
GR	genetic resource
GTZ	Deutsche Gesellschaft für Technische Zusammenarbeit (German Organization for Technical Cooperation)
HWI	human wellbeing index
IARCs	International Research Centres
IBAMA	Brazilian Institute of Environment and Natural Resources
ICBG	International Biodiversity Group
ICC	International Chamber of Commerce
ICIPE	International Centre of Insect Physiology and Ecology
ICSID	International Centre for Settlement of Investment Disputes
IDRC	International Development Research Centre of Canada
IEPF	Institut de l'énergie et de l'environnement de la Francophonie (Francophone Institute for Energy and Environment)
IFA	Institute of Fishing and Aquaculture
IGC	Intergovernmental Committee
IGOs	Inter-Governmental Organizations
ILO	International Labour Organization

INBio	Instituto Nacional de Biodiversidad (National Biodiversity Institute)
INPI	Brazilian patent office
IOs	international organizations
IPA	Industrial Property Act
IPEN	International Plant Exchange Network
IPF	Intergovernmental Panel on Forests
IPP	intellectual property protection
IPPC	Integrated Pollution Prevention and Control
IPRs	intellectual property rights
ISA	International Seabed Authority
ISO	International Organization for Standardization
ITA	Aeronautics Technological Institute
ITPGR	International Treaty on Plant Genetic Resources
ITPGRFA	International Treaty on Plant Genetic Resources for Food and Agriculture
IUCN	International Union for Conservation of Nature
IU-PGRFA	International Undertaking on Plant Genetic Resources for Food and Agriculture
KFS	Kenya Forest Service
LB	Law of Biodiversity
MAT	mutually agreed terms
MDGs	Millennium Development Goals
MEE	Ministry of Environment and Energy
MLS	multilateral system
MOA	Ministry of Agriculture
MP	provisional measure (Medida Provisória)
MPF	Federal Public Prosecutor's Office
MSC	Marine Stewardship Council
MTA	material transfer agreement
NACOMB	National Commission for the Management of Biodiversity
NASYCA	National System of Conservation Areas
NCP	non-compliance procedures
NCST	National Council for Science and Technology
NEAFC	North East Atlantic Fisheries Commission
NEMA	National Environment Management Authority
NGO	non-governmental organization
NWO	Netherlands Organisation for Scientific Research
NYBG	New York Botanical Garden
PCA	Permanent Court of Arbitration
PGRFA	plant genetic resources for food and agriculture
PIC	prior informed consent
PIPRA	Public Intellectual Property Resource for Agriculture
PPPs	public–private partnerships

R&D	research and development
RCPs	Regional Common Genetic Pools
SAMTECH	School of Alternative Medicine and Technology
SBSTTA	Subsidiary Body on Scientific, Technical and Technological Advice
SDR	Sustainable Development Reserve
SEARICE	Southeast Asia Regional Initiatives for Community Empowerment
SEMA	Amapá State Environmental Secretariat
SEPA	State Environmental Protection Administration
SMTA	standard material transfer agreement
TB	tuberculosis
TEK-PAD	Traditional Ecological Knowledge Prior Art Database
TK	traditional knowledge
TKDL	Traditional Knowledge Digital Library
TMK	traditional medicine knowledge
TMPs	traditional medicine practitioners
TO	Technical Office
TRIPS	Agreement on Trade-Related Aspects of Intellectual Property Rights
TRR	traditional resources rights
UFS	University of the Free State
UNCCD	United Nations Convention to Combat Desertification
UNCED	United Nations Conference on Environment and Development
UNCITRAL	United Nations Commission on International Trade Law
UNCLOS	UN Convention on the Law of the Sea
UNCTAD	United Nations Conference on Trade and Development
UNDRIP	United Nations Declaration on the Rights of Indigenous Peoples
UNFCCC	United Nations Framework Convention for Climate Change
UNIDROIT	International Institute for the Unification of Private Law
UNIFESP	Federal University of São Paulo
UNU-IAS	United Nations University – Institute of Advanced Studies
USPTO	US Patent and Trademark Office
UTSA	US Trade Secrets Act
WCMA	Wildlife (Conservation and Management) Act
WGAL	Working Group on ABS Legislation
WIPO	World Intellectual Property Organization
WIPOIGC	Intergovernmental Committee on Intellectual Property and Genetic Resources, Traditional Knowledge and Folklore of the World Intellectual Property Organization
WSSD	World Summit on Sustainable Development
WTO	World Trade Organization
ZPO	Zivilprozessordnung (German Civil Procedure Act)

Introduction

Evanson Chege Kamau and Gerd Winter

The Convention on Biological Diversity (CBD) of 1992 constituted genetic resources (GRs) as subject to sovereign rights of the states hosting them. The states have the right to control the access to their GRs and thus the power to set conditions relating to research, development of uses and the sharing of benefits. Likewise, the user states are asked to ensure the sharing of benefits. Where traditional knowledge (TK) that is vested in the GR or accumulated knowledge about its use is involved, both host states and user states are obliged to ensure that such knowledge is maintained and that benefits drawn from it are shared with the holders of the knowledge.

The rationale of this regime is an exchange relationship: providing GRs for a share in benefits. Simple as this may appear, the legal framing of the exchange has caused tremendous difficulties. These are due to the fact that, while in theory the exchanged goods must be precisely carved out, in reality they eschew definition. Both sides – GRs and benefits – are diffuse phenomena. Many steps, both on the international and national level, have been taken to establish a fair deal. The Bonn Guidelines have been regarded as soft law guidance but have left many salient questions unsolved. National laws have been enacted by more and more resource states requiring access permits and the conclusion of contracts. These laws are often very wide in scope, ambitious in instrumentation and vague in language. This is due to the said difficulties, but also the mistrust that exists about the user states' willingness to cooperate, for user states have failed to introduce corresponding legislation. The ambiguity and ambition of national requirements have had the perverse effect that basic research has been hindered and commercial research has turned to a reliance on ex situ collections.

In this situation of mistrust, amicable solutions are needed. This book wishes to contribute to this search. It contains both practical suggestions and theoretical reflections because, as Immanuel Kant said, 'theory without practice is empty and practice without theory is blind'.

The book is structured as follows: The first two parts contain theoretical reflections on the concept of justice underlying the access and benefit-sharing (ABS) regime, one focusing on GRs as such and the other on TK. The third part has a practical goal: to show how different countries struggle with improving their ABS legislation and to indicate what can be learnt from this by other countries. The fourth and fifth parts then concentrate on specific instruments of ABS legislation discussed and enacted in either provider or user states.

In the following the individual contributions will be summarized.

Part One Theorizing on ABS

Peter-Tobias Stoll explains how the conception of common heritage, which was for some time propagated for GRs, fell victim to a conception of various proprietary claims ranging from sovereign rights of resource states and sui generis rights of resource state communities to intellectual property (IP) in plants, genes and biotechnology. These claims enter into exchange relationships which aim at fair deals, called *iustitia commutativa*, but suffer from negotiation asymmetries disadvantaging resource states. In order to make this good, user states are obligated to ensure benefit sharing aiming at fair distribution, called *iustitia distributiva*. The author suggests that the ongoing international negotiations should find a way to accord both kinds of justice. This entails, on the one hand, more active involvement of user state measures and, on the other, the facilitation of access and research by resource states. ˙

Along a similar ABS line, Gerd Winter argues that neither provider state measures nor user state measures can provide an efficient and just ABS regime. Given the problems of defining the scope of the regime and the problems of enforcing regulatory and contractual obligations of users, provider states lack the means of efficiently controlling the downstream process of utilization of GRs. Moreover, the right of the provider state to take all of the shares in benefits appears to be unjust where GRs and TK are common to more than one country. On the other hand, user states, even if willing to enact regulations on benefit sharing, will encounter difficulties in tracing GRs back to provider states. Winter therefore suggests the introduction of regional common pools. They could bundle the negotiation and enforcement power of resource states, facilitate the user countries' own regulation of benefit sharing and ensure justice among multiple owners of GRs.

The question of common pools is also posed in relation to GRs in the high seas. Alexander Proelss addresses this question in his chapter on marine GRs. While GRs in the coastal zone and in the Exclusive Economic Zone and continental shelf are subject to ABS legislation by coastal states (which, however, is still widely lacking), GRs in the high seas are freely accessible to everybody. They are not covered by the CBD concept of sovereign rights of states over GRs, nor are they captured by the common heritage regime established for mineral resources by the UN Convention of the Law of the Sea. Proelss discusses whether the common heritage concept should be extended to them. He however rejects this option on the ground that experiences with the present regime concerning minerals are not satisfactory. While in any case environmental protection must be ensured, he rather advocates exemptions of IP rights (IPRs) as a means of redistribution of benefits from high-seas GRs.

While Stoll, Winter and Proelss discuss ABS within a framework of fair exchange between resource and user states, Bram De Jonge and Niels Louwaars explain that ABS can also be seen in a framework of substantial policies. Studying the perceptions of different stakeholders, they have identified six principles underlying ABS: while the first three (the South–North imbalance in resource allocation and exploitation, biopiracy and the imbalance in IPRs, and an imbalance between IP protection and the public interest) are driven by the perception of imbalance and a motivation to increase equity, the other three (the need to conserve biodiversity, a shared interest in food security and the protection of the cultural identity of traditional communities) concentrate on other aims such as nature conservation, food security and the preservation of traditional cultures. The authors show that not all of these goals point in the same direction but may conflict with each other. This explains why progress in solidifying the ABS regime has been so slow.

Part Two TK From New Perspectives

Focusing on TK about the medicinal use of GRs, Jack Githae shows that human health is fundamentally interlinked with the ecosystem, the spiritual culture and the individual mind into which a person is embedded. Human health is thus basically local. TK about the conditions of human well-being can therefore not easily be disconnected from its local roots. This warns against an unembedded transfer of TK to modern-life conditions. An inattentive transfer would not only be unfair to the developers of the knowledge, but also unworkable. Both traditional and modern medicine should be married into hybrid forms which mutually further each other. This should be guided by ethical reflections on the value of each others' origin and performance. Giving numerous examples, Githae demonstrates that many achievements of modern medicine are due to older African knowledge.

Brendan Tobin puts ABS for TK in the broader framework of human rights. Rights to life, food, health, culture, traditional territories, resources, work and development are all relevant to and variously dependent upon protection of TK. The international community and national authorities will therefore need to broaden their approach to TK protection to include not only control of third-party commercial use, but also the adoption of measures and mechanisms, including necessary funding, to strengthen and invigorate TK systems. In addition, means must be developed to cope with the problem of ownership of TK by several communities. Tobin suggests community protocols as a means to develop the relevant customary law further towards joint ownership and representation in order not to facilitate ABS in cases of multiple ownership.

TK is construed by the CBD to be hosted by local communities. However, it can be spread in various ways. John B. Kleba, in his chapter, explores how community-owned knowledge can be delineated from freely usable knowledge, or 'disseminated knowledge' as it is called by the draft Brazilian law on ABS. Kleba illustrates the problem with two case studies: a bioprospecting project about psychoactive effects of plants used by the Krahô Indians, where three communities were asked for consent but 14 more communities claimed co-ownership in the knowledge; and the use of knowledge about cosmetic uses of plants, ownership of which was claimed by sales women in a local market. The cases show that knowledge can be disseminated 'horizontally' over several communities and 'vertically' along the chain of commercialization. Proposing a fourfold typology, Kleba suggests that herbalists who are detached from local communities but practise TK are legitimate holders of TK. He advocates anthropological studies to identify the appropriate category in a given case.

Evanson Chege Kamau strikes a slightly different chord concerning disseminated TK. Echoing Githae, he points to the communal origin of TK. Even if it is disseminated locally or nationally, it would be unjust to ignore that the knowledge was a product of local common traditions, experience and efforts. Drawing benefits from it would be freeriding. Kamau therefore looks for ways of protecting TK from an easy slip into the public domain, and how to ensure BS once the knowledge has been disseminated. He considers trade secrets as an option, concluding that resource states might introduce national legislation based on that concept. Another possibility is to develop a sui generis form of IP. Concerning BS he proposes communal, multi-communal and national funds. As an auxiliary measure he advocates making patent law more attentive to TK.

Part Three Recent Developments in Exemplary Countries

Some countries were selected for a closer look at their recent development. Some are exemplary for the sophistication of their experience, others for the fact that they are still in the beginning stages of experimentation.

As Anne Angwenyi shows in her chapter, Kenya exemplifies a common law country which perceives GRs as part of the property of a landowner. Therefore, the landowners must give their consent for access. However, common law was superseded by a regulatory framework of 2006 which requires administrative authorization for access, prior informed consent (PIC) of a local community if TK is involved, a BS agreement with the landowner, be it an individual, a local community or the state, and a material transfer agreement if the resource is exported. Assessing this regulatory approach, Angwenyi identifies a number of shortcomings. She suggests that more guidance is necessary on how to understand TK and how to obtain PIC. Benefits shared should flow into a trust fund. She also sees a lack of differentiation between access for basic research and access for commercial purposes. Concerning procedures, she calls for better coordination of competent administrative agencies and broad capacity-development of the relevant personnel.

Brazil is both resource rich and highly active in research and development (R&D) activities. The balance between providers and users is therefore to be struck within the country itself, beside its being a provider country for foreign user countries. Juliana Santilli, in her chapter, summarizes the Brazilian legislation and enforcement administration as introduced by a law of 2001, and identifies shortcomings which will be removed by the pending draft of a new ABS law. Access to GRs for research purposes requires authorization but no BS agreement. The new draft will privilege certain kinds of research by allowing access even without authorization. Access for commercial purposes requires, besides authorization, a BS agreement. If TK is involved, PIC of the local community must be obtained. Santilli identifies as a critical point that the law does not mirror the fact that TK is often produced and shared by several communities. This should lead to payments into appropriate funds. Focusing on traditional seeds, she discusses how to extend the concept of the International Treaty on Plant Genetic Resources (ITPGR) to seeds not covered by the treaty, but also how to improve the concept in order to trigger more payments, which would allow support of traditional breeding.

Although South Africa is, as Rachel Wynberg and Mandy Taylor demonstrate, the country with possibly the most significant history of cases and debates about ABS, it adopted legislation only in 2004 and regulations only in 2008. Access to GRs requires a permit by the competent central or provincial administration, as well as consent of and a BS agreement with the 'stakeholder', which is the landowner and, in the case of access to TK, also the local community. One shortcoming Wynberg and Taylor see is that the law ties ownership in GRs to ownership in land. This is unjust because the landowner is entitled to the benefit share although the GR may also be found on many other properties. The authors argue in favour of making the state the owner of GRs and thus the recipient of shared benefits. Other concerns are that the law does not adequately distinguish between basic and commercial research, and that it does not provide sufficient guidance on how to cope with disseminated TK. The authors also discuss possibilities of integrating biochemicals and derivatives into an ABS regime.

While China belongs to the megadiverse countries, it has only of late taken steps to introduce legislation on ABS. As Tianbao Qin explains in his chapter, the Chinese constitution, following socialist ideas, establishes state ownership in most natural resources. Various laws on wild and domesticated animals and plants are in force overseeing and protecting the use of natural resources. They are administered by many different agencies with often overlapping competences. The focus of existing laws and institutions is, however, on biological resources, not GRs in the narrower sense. Since 2000 and the advent of the Bonn Guidelines, attempts have increased to regulate GRs. A working group has been set up and mandated to elaborate drafts for ABS legislation. It is slowly moving towards solutions to intricate problems such as the future competent national authority, the form of the legislation, the scope of regulated GRs, differentiated procedures for basic and commercial research, and the protection of TK. It is noteworthy that a large project is underway producing an inventory of all indigenous biological resources and TK.

Costa Rica probably is the country with the most experience in managing ABS. As Jorge Cabrera Medaglia explains in his chapter, the country started as early as 1991 with an agreement between the pharmaceutical company Merck and the National Biodiversity Institute (INBio), a nonprofit research institute recognized as being of public interest. The agreement helped INBio to accumulate considerable R&D expertise. Afterwards, the national legislation was enacted, establishing a regulatory framework for access to GRs. It requires a permit for access and the PIC of the landowner, which can be a private person, a local community entrusted with the land or a public entity. The PIC is normally given in the form of a BS contract. Monetary benefits flow to the landowner, that is, not necessarily towards the conservation of resources. Cabrera Medaglia adds a critical note on unsolved issues of TK, of facilitating access for basic research purposes, and of monitoring and enforcement of contract duties.

In practice, INBio seems to be the crucial player in ABS R&D. Cabrera Medaglia gives examples of how, in the framework of an agreement with the state, INBio conducts bioprospection jointly with partners from other countries. The projects are based on detailed contracts that ensure the sharing of knowledge and monetary benefits. The policy of Costa Rica can thus be described as aiming at joint development rather than the selling of resources for money.

Australia has one of the most sophisticated legal systems of ABS, which is analysed by Geoff Burton in his chapter. Federal Regulations of 2000 provide that access requires an administrative permit. With the application, the consent of the landowner (which can be a private person, a local community or the federation) must be submitted. A BS agreement must be concluded in the case of commercial research. In the case of basic research, a statutory declaration must be signed ensuring renegotiation if the research turns commercial. To assist applicants, two model agreements have been published, one for publicly owned areas and another for indigenous peoples' privately owned lands. As these are probably the most developed model texts, one of them is reproduced in the Annex to this book. The landowner has the power to determine the recipient of the share in benefits. This reflects the common law concept of land property. It is noteworthy that all permits are entered into a publicly accessible register. Assessing the Australian approach, Burton examines the implementation of the Bonn Guidelines on Access to Genetic Resources and Fair and Equitable Sharing of the Benefits to see to what extent the ABS law and its ABS administrative arrangements represent the development of a model ABS law and implementation.

Part Four Core Problems of Provider Country Measures

A number of contributions address questions and issues cutting across individual provider states. The CBD requires that, prior to access, the PIC of the party providing GRs and/or TK associated with it is sought. This is, in most cases, a very complex procedure due to the nature of TK, which is often common to or shared between several communities. Sandra A.S. Kishi examines the importance of self-representation of ethnic communities in PIC procedures, as well as in ABS agreements. Kishi uses a case study of the Krahô tribe to demonstrate that exclusion of pertinent communities from the PIC procedure can lead to a failed agreement. In the given case, the Federal University of São Paulo (UNIFESP) accessed plants used in medicinal rituals and traditional practices by several ethnic groups of the Krahô peoples in the State of Tocantins with the PIC of only three Krahô villages (represented by an organization called the Vyty-Cati), while ignoring the other 14. The latter, represented by the Kapéy organization, opposed the continuation of the research project and conditioned further discussions on the prior payment of an indemnification for moral damages, that is, collection of genetic material in a manner that violated their uses and customs, and an up-front prospecting fee. Kishi notes that the existence of an independent anthropological study could have helped to avoid the exclusion of any village from the PIC and to protect the shamanic practices in the collection and use of the plants. In addition, it would assist in identifying the legitimate holders of knowledge in cases of disseminated TK, an opinion that is likewise supported by Kleba. Kishi concludes that PIC can be used to include an entire range of instruments for the protection of TK, as well as the communities that hold it.

María J. Ochoa looks at the use of databases as an option for protecting TK associated with GRs. This could be done by regulating access to TK by using information stored in databases as prior art in patent registration procedures. According to Ochoa, databases offer preventive and positive protection: the former obviate unauthorized persons from obtaining IPRs on products based on TK, while the latter gives indigenous and local communities the right to restrict the use of their TK and to share benefits from its usage. Ochoa examines the practice of India, China, USA and Peru, pointing out the kind of protection databases grant. Concerning Venezuela she notes that, though useful, the creation of databases suffered failure due to lack of authorization of all pertinent indigenous peoples prior to the documentation of their TK. As a result, the indigenous peoples through their representatives demanded that the collection of biological resources on their lands should stop and the database given back to them. Ochoa notes that extreme suspicion towards researchers, scientists and government workers has developed among the indigenous and local communities, causing reluctance by the latter to give consent for access to or enter into any agreement involving their knowledge. She therefore concludes by highlighting the role of PIC in qualifying the legality of access, which she, like Kishi, believes should precede any access to avoid the failure of any access arrangements. In addition she stresses the leading role of capacity building.

With rising suspicion, many potential access applications are often tagged 'biopiracy'. From experience gathered while working with a working group on prevention of biopiracy in Ecuador, EWGPB, Monica Ribadeneira Sarmiento discusses ways of distinguishing legitimate cases of access from irregular and illegal ones. She states that, according to the working group, biopiracy is a violation of national access rules. Therefore, she adds, in cases where national rules are either non-existent or not determined, it would be difficult to identify biopiracy cases. However, Ribadeneira Sarmiento notes that there are international laws that recognize the sovereign rights of states over GRs, as well as international principles on access. She hence concludes that cases that do not observe these principles are misappropriation cases in which administrative felonies, biopiracy and wildlife trafficking cases, amongst others, are to be found. According to her, fighting biopiracy successfully entails eradication of irregularities in observation of national access rules and suggests an expert working group strategy.

How could unwarranted use of traditionally cultured GRs be ensured? Christiane Gerstetter examines the public and private law approaches of protecting GRs from unwarranted use, discussing their disadvantages and advantages. She notes that public law approaches have not been sufficient to effectively protect the rights of indigenous communities against violation, but underlines that classic IPRs, for a number of reasons, are not suitable for protecting traditional resource rights either. However, based on a hypothetical assumption, she asserts that sui generis IPR systems would be instrumental in protecting the rights of local communities against violation by others. Gerstetter, however, seems to suggest that their operationalization might be hindered by antagonistic interests of the national states, who the CBD accords sovereignty over GRs, as well as users who would still be interested in gaining easy access to GRs, lowering transaction costs and claiming IPRs over GRs and their components. Finally, Gerstetter looks at the issue of fairness and equitability in sharing benefits from utilized GRs and suggests some administrative rules that states should adopt to ensure fairness and equitability in ABS arrangements, which are entailed in procedural and substantive elements.

Evanson Chege Kamau and Gerd Winter look at Article 15 of the CBD in view of the obligation placed upon states to facilitate access to GRs. Whereas the CBD recognizes the sovereign rights of states over their natural resources and their authority to determine access to GRs subject to their national legislations, it also requires that contracting parties facilitate access to GRs and do not impose restrictions that run counter to the objectives of the CBD. Kamau and Winter note that many provider measures to date have rather hindered than facilitated access: most access regimes hope to extract tremendous gains from ABS deals and/or ward off illegal access. Taking the Kenya access regime as an exemplary provider measure, the authors analyse its impact on the access procedure. It is noted with concern that such an access regime can only discourage any potential research activity, be it basic or commercial – for example, it is long, cumbersome, entails high costs, contains vague and overlapping procedures, and creates uncertainty. As a remedy, they suggest that the ABS legislation should be

revised. Kamau and Winter make a concrete proposal that access procedures and conditions for permits should be overhauled either through a
substantive or a procedural approach.

With the challenges that face most provider countries in implementing
the ABS provisions of the CBD, capacity building is a key issue. With this
in mind, the Dutch–German ABS Capacity Development Building
Initiative for Africa was initiated to support African stakeholders. Peter
Munyi, Fabian Haas, Andreas Drews and Suhel al-Janabi discuss the
activities of this Initiative with particular focus on the third objective of
the CBD, the fair and equitable sharing of the benefits arising from the
utilization of GRs. From the onset, they note that the CBD was the first
multilateral environmental agreement to explicitly link biodiversity conservation with sustainable development by adopting what is now referred to as
the third objective of the CBD. This objective, they add, underscores the
right of owners and users to enjoy the benefits from sustainable use.
Likewise, it gives motivation to conserve biodiversity. In order to promote
both the interests of the providers and the users, proper mechanisms must
be in place. However, as the authors note, the conflicting interests of these
parties often breed complex and impeding regulations. The Initiative helps
to build human and institutional capacity in developing countries to deal
with the complex ABS issues. Munyi et al state that the objectives of the
Initiative are to increase awareness of African policy makers and legislators,
foster meaningful participation of all relevant stakeholders, improve
regional cooperation on ABS issues and support the development of partnerships for business opportunities. The authors devote most of the chapter
to a report on the capacity-building activities of the Initiative.

Part Five User Countries' Measures

User countries until now have taken lightly their obligation under Article
15.7 CBD to ensure BS. Voluntary and informatory means have been discussed so far, including research guidelines, disclosure of origin and
certificates of origin. But few states to date have introduced regulatory legislation requiring the monitoring of R&D activities and the sharing of
benefits derived from such activities.

Evanson Chege Kamau discusses disclosure of origin clauses in the
European Union (EU). He starts his chapter with insisting that Article 15.7
CBD does contain an international duty of law to introduce national legislation effectively ensuring BS. He even considers whether this Article is not
precise enough to qualify as self-executing, thus being directly applicable
within user states, at least if their conception of international and national

law is of the monist kind, but possibly even if it is dualistic. Analysing the EU and member state legislation on disclosure of origin requirements, he shows that where disclosure is required this is without any procedural and substantial bearing on the right of patenting. Disclosure is only required if the origin is known. Cheating is easy because only in very few member states is disclosure sanctioned. Those who do disclose the origin are not obliged to also disclose benefits and ensure the sharing of them. Kamau suggests that user states must take more stringent regulatory measures in order to fulfil their obligations.

In the absence of such regulations, it is interesting to consider whether BS can be achieved by liability under civil law. Christine Godt shows in her chapter that a claim for BS can indeed be raised on the basis of torts. Prospects for success are better with regard to immaterial property than to material property. With regard to material property, it is the economic value of the good as such which is taxed. In contrast, it is the rule to recover a share in profits or the equivalent to the licence fee for the violation of an immaterial right. This rationale is better suited for cases of illegal bio-prospecting and is applicable to IPRs and autonomy rights alike. Claiming a share in profits or an equivalent to a licence fee would presuppose that the resource state frames its GRs as a private immaterial right sui generis. Litigation of this sort before a user country's court would be possible. Even if this court applies the resource state or user state tort law, it would have to build on the fact that an immaterial right was violated.

Hiroji Izozaki broadens the view on user country measures. He discusses regulatory means such as import requirements checking compatibility of the taking of biological material with provider country legislation, such as in the US Lacey Act and other regulations based on the Washington Convention on Trade in Endangered Species (CITES). He also mentions the prosecution of biopiracy under criminal law. He then focuses on the enforcement of material transfer and BS agreements between providers and users of GRs. The Japanese rules on enforcement of judgements of foreign courts are explained. Hinting at problems of trust and the time these formal procedures entail, Izozaki rather recommends consideration of alternative dispute resolutions such as the mechanism and rules of arbitration of disputes relating to natural resources established by the Permanent Court of Arbitration. He also reflects on means how to avoid disputes. One possibility is the founding of regional ABS centres representing shared resources; another one is the privileging of basic research combined with the obligation to renegotiate a shift of purpose to commercialization.

PART ONE

THEORIZING ON ABS

Chapter 1

Access to GRs and Benefit Sharing – Underlying Concepts and the Idea of Justice

Peter-Tobias Stoll

Introduction

The provisions of the CBD on the fair and equitable use of GRs, including access and benefit sharing (ABS), have often been considered as a model case of sustainable development. Providing GRs for use in research and development, including commercialization and sharing the benefits of such activity, appeared to be beneficial both for the environment and for social and economic development. A few prominent cases of cooperation between resource institutions, research agencies and business have inspired this vision. However, today, business people, researchers, diplomats and officials mainly agree that this vision has not translated into reality in the 15 years that have passed since the adoption of the Convention.[1] On the contrary, the issue became one of the most controversial within the CBD. For different reasons, the different actors in the area feel uncomfortable with this state of affairs. While provider countries and institutions point to frequent cases of misappropriation and biopiracy, researchers and industry complain about the distrust and arbitrariness of access procedures. Diplomats within the institutions of the CBD are engaged in lengthy and highly controversial negotiations. In order to better understand this difficulty, it appears to be necessary to clarify the purpose and the backgrounds of the Convention's ABS system. Furthermore, it will be submitted that different rationales apply to transactions between the private sector and states, and that they reflect different concepts of justice. Such analysis, it is hoped, may contribute to the ongoing negotiations on an international regime for ABS.

The **ABS** system as an environmental instrumentality?

As the ABS system is part of an environmental agreement, it is quite often considered to represent a mechanism that is directly aimed at fulfilling an environmental purpose. Today, the ABS system is still seen in this way by a number of observers, diplomats and academics. It is maintained that benefits to be shared may help to implement conservation measures and that the expectation of the potential later use of material as GRs may prevent decision makers from turning rainforest into farmland.

Understanding ABS in these narrow confines would have a number of implications when assessing its performance and possible means to improve it. Mainly, it would suggest the application of a clear-cut test of effectiveness and efficiency and, if necessary, looking for alternative mechanisms. Basically, effectivity would imply the degree to which an environmental measure achieves its goals and efficiency. In the case of ABS, the environmental objective would clearly be the conservation of biological diversity. If the ABS system were to be judged upon this criterion, the result would be disappointing.[2] There is no indication that, in the 15 years since the adoption of the Convention, the ABS system has had any significant impact on conservation, be that benefits being used to undertake certain conservation measures or halting of the human degradation of biodiversity.

Furthermore, the costs of the system would appear highly inappropriate. These costs include the cost of national implementation legislation, international negotiations, private transactions and costs resulting from activities related to GRs, which had not been undertaken because of the difficulties with the legal situation. Thus, from this perspective, it would be advisable to urgently seek alternative solutions. In this regard, it would be highly recommended to explore the potential of other economic incentives as addressed by Article 11 of the Convention. The Article envisages that each contracting party shall, as far as possible and as appropriate, adopt economically and socially sound measures that act as incentives for the conservation and sustainable use of components of biological diversity.[3]

However, there are doubts as to whether the ABS system can be analysed from such a perspective. A closer look at the Convention indicates that there are hardly any links between the conservation objective and related provisions on one side and the rules on ABS on the other. Clearly, ABS is related to the other main objective of the convention, which is the sustainable use of the components of biological diversity as is made clear by the wording of Article 1 of the Convention, which particularly emphasizes in this context 'the fair and equitable sharing of the benefits arising out of the utilization of GRs, including by appropriate access to GRs and by appropriate transfer of relevant technologies'. Indeed, the Convention endeavours to couple environmental protection and development, as is called for by the principle of sustainable development. However, the different objectives of the Convention and their related mechanisms are not strictly interlinked. Thus, the need to make the ABS system work is not put into question by pointing to its possibly modest contribution to conservation nor by reference to other and more suitable mechanisms to achieve that end.

The CBD's acknowledgement of sovereign rights to GRs: Part of a neglected story of appropriation of biotechnology and its inputs and products

Beyond the confines of a strict functionalist and environmental perspective, it becomes apparent that the ABS provisions of the CBD and most prominently its reference to GRs and related sovereign rights are closely related to other international and most controversial developments that concern entitlements to biotechnology, which have quite different conceptual underpinnings and implications.

The CBD's acknowledgement of sovereign rights to GRs

Article 15 para 1 CBD reads: 'recognizing the sovereign rights of States over their natural resources, the authority to determine access to GRs rests with the national governments and is subject to national legislation'. The provision refers to the long-standing sovereign right of states over their natural resources, which is based on the international law principle of territorial sovereignty and has been further developed by UN bodies, and reiterated once more in Article 3 CBD and – in view of biological resources – in the preamble paragraph 4.[4] Article 15.1 clarifies that GRs in the way that they have been defined by the Convention – namely, biological material with functional units of heredity that have an actual or potential value[5] – are subject to such sovereign entitlement. It is the very essence of state sovereignty over GRs that states can freely dispose upon such resources for their own uses and to provide them to third parties upon terms and conditions they may deem appropriate. The latter is referred to by the term 'access'. As Article 15 paras 1, 4 and 5 may indicate, 'access' has a legal connotation. It may be understood to refer to the entirety of entitlements, rights and legal authorization necessary for all the different activities involved in the search for, collection of, exportation and use of GRs as ruled upon by the respective state on the basis of its sovereign rights. As this indicates, access requires a complex legal structure within the resource state, which provides for rights and entitlements, procedures and remedies, and takes into account all relevant rights and legal interests involved. Even more, that structure has to lay ground for the negotiation, conclusion and execution of agreements on benefit-sharing, taking into account any possible legal interest in this regard. In sum, this legal structure may be referred to as an 'access mechanism'.

The CBD contains some provisions as to how such access shall be demanded for and granted. It refers to a determination of Governments (Article 15 para 1), envisages that such access be on mutually agreed terms (para 4), and requires prior informed consent of the relevant Contracting Party unless determined otherwise (para 5).

From common heritage to sovereign rights: The FAO-undertaking background

It is often neglected that the concept of GRs has not been newly invented by the drafters of the CBD, but indeed was introduced shortly before within the Commission on Plant Genetic Resources for Food and Agriculture of the Food and Agricultural Organization of the United Nations (FAO) as a result of a long and controversial process, which is still ongoing in a number of international forums.

As the early examples of tea and rubber may indicate, access to GRs has played a critical role in international relations in the past. More recently the issue has been raised in the Food and Agricultural Organization of the United Nations (FAO). In 1991, the FAO Conference resolved that 'nations have sovereign rights over their plant genetic resources'.[6]

This decision marks the end of a long development, which started with the adoption of a legally non-binding Undertaking on Plant Genetic Resources for Food and Agriculture (PGRFA) in 1983.[7] Most significantly, that undertaking in its initial version did proclaim that PGRFA are the 'common heritage' of mankind.[8] The common heritage principle, developed through UN negotiations concerning the uses of the deep seabed and its resources, contains little more than an idea of free access and an air of distributional justice.[9] The same holds true for the Undertaking: it had little to say in view of the fact that such plant GRs are not a wealth to be distributed, but require investments into the conservation of centres of origin and landraces as well as into improvement of breeds, which is mainly done by private breeders.

Today, it looks quite strange that the Undertaking proclaimed such a principle to be applicable to virtually any germ plasm with relevance for food and agriculture, including wild species, landraces and highly developed commercial varieties.[10] Indeed, the undertaking was soon modified. In 1989 the FAO Conference made it clear that plant breeders' rights under the UPOV (The Union for the Protection of New Varieties of Plants) Convention should not be affected by the Undertaking.[11] Also, it was stated that 'the term "free access" does not mean "free of charge"'.[12] In turn, developing countries successfully asked for recognition of the rights of farmers.[13]

Two years later, the FAO conference again modified the system of the Undertaking. At the FAO conference it was decided that 'breeders' lines and farmers' breeding material should only be available at the discretion of their developers during the period of development',[14] and thereby acknowledged the proprietary character of such lines. However, in turn, the Conference decided that 'nations have sovereign rights over their plant genetic resources'.[15] Thus, within just a few years, the former 'common heritage' has been divided up into various proprietary claims.[16]

It cannot be overlooked that the different claims made in this case clearly represent the conflicting interests involved. The recognition of plant breeders' rights and the proprietary character of breeding lines were of comfort to the breeding industry – which in those days was mainly situated in the North. The so-called 'farmers' rights' and the concept of a sovereign right on GRs can be roughly considered a counterclaim of the South. In sum, the example amply shows that plant GRs, which can be considered a public good, have become the subject of claims of different stakeholders. This is likely to cause conflicts in demand, intensive negotiation and result in inefficiencies. Meanwhile, a Treaty on PGRFA has been concluded to enable facilitated access to resources while accommodating the various different entitlements.[17]

As this may indicate, the term 'genetic resources' has been framed within those discussions in FAO. Also, it has to be remembered that the FAO decision was taken shortly before the negotiations on the CBD, which was adopted only ten months later. It is somehow odd that this coincidence has been barely referenced in the CBD process and related academic writings. A potential cause for that neglect may result from the fact that agricultural and environmental issues are often dealt with quite separately both nationally and internationally.

Strengthening intellectual property rights: The TRIPS agreement and beyond

It should be added that the development of different entitlements relevant to the use of GRs went on in other international regimes. Most importantly, three years after the conclusion of the CBD, the World Trade Organization (WTO) was established and its vast body of law entered into force. The latter includes the WTO agreement on trade-related aspects of intellectual property rights (IPRs) – the well-known TRIPS agreement. That agreement envisages a strong protection of IPRs. Inter alia, it eliminated the discretion of states to provide for exceptions from patent protection, which hitherto existed under the Paris Convention and was frequently used by states to exclude the patenting of inventions relating to the agricultural, food and health sector. Furthermore, it stipulates a strong protection for plant breeders' rights and contains new standards on the enforcement of IPRs. Also, it should be mentioned that, in a number of countries, including the USA, Japan and the Member States of the EU, patents were granted for genes and gene sequences, which have been isolated from biological material.

Proprietary rights, effectiveness and justice

Taken together, after abandoning the somewhat ill-fated concept of a common heritage, an amazing number of new entitlements in biotechnology and its applications have been introduced over the last few years. Taken to the extreme, a simple plant and its utilization can be the subject of sovereign rights to GRs, farmers' rights, rights of local or indigenous communities according to Article 8(j) of the CBD, breeders' rights and breeders' claims regarding their breeding lines and patents, including those concerning genes or gene sequences.[18]

Effectiveness – The sovereign claim to GRs in particular

These developments and the resulting array of rights can be judged in different ways. One way would be to question the effectiveness and efficiency of such entitlements. In this regard, some doubts arise in the case of farmers' rights and TK but also in view of the sovereign entitlement in GRs.

Farmers' rights and the rights of local and indigenous communities regarding their traditional knowledge represent somewhat a new type of entitlement in intellectual achievements. Unlike classic IPRs, they do not aim at providing incentives for generating innovations, but rather reward achievements of the past. Furthermore, they are both in their nascent state of development and in need of clarification in view of the proper assignment of the power of disposal and the beneficiaries, as well as in view of appropriate enforcement mechanisms. Also, in contrast to IPRs, they are not enforceable throughout the world.

Sovereign claims to GRs are based on territorial sovereignty. An additional justification in view of environmental effect is doubtful, as has been mentioned before. Also difficulties arise both in view of the proper definition of subject matter and enforcement. As regards subject matter, it is questionable whether the entitlement really captures what constitutes the value of GRs. According to the CBD, the entitlement relates to genetic material, whereas the economic value very probably relates to the informational content as explored and developed by subsequent steps of research and development. The resulting legal uncertainties can be best illustrated by the various attempts and discussions on extending the scope of the sovereign entitlement to derivatives and biological resources, and by the difficult and ongoing debate regarding the notion of GRs at all.[19]

However, aside from the fact that the entitlement to GRs hardly captures the relevant stages of utilization, its range of application is severely limited due to the fact that such entitlement is only enforceable within the confines of the jurisdiction of the resource state. In the likely case of utilization taking place in a third country, the entitlement scarcely offers any help.

Justice in a globalizing world

In a homogeneous and advanced economic and social system, the introduction of an entitlement to GRs would certainly meet with much scepticism on the grounds mentioned above. Also, more effective and efficient means would be available to acknowledge and honour the contributions of farmers and the achievements of local and indigenous communities. In a situation where genetic material and those valuable contributions and achievements are frequently translated into innovations and commercial success in other parts of the globe without consent, and even later may return back embodied in products and protected by IPRs, the question for justice arises. In the absence of more elaborate means of acknowledgement and compensation, affected groups, communities and states can hardly be blamed for putting forward proprietary claims for the sake of justice, whose merits in view of economic and legal effectiveness are doubtful.

The CBD as a blueprint for transactions in the utilization of GRs

The acknowledgement of a sovereign title in GRs set the stage for a number of provisions within the CBD, which are usually referred to as 'access and benefit sharing' (ABS). These provisions are informed by and spell out the objectives and principles in Article 1, which provides for the 'fair and equitable sharing of the benefits arising out of the utilization of GRs, including by appropriate access to GRs and by appropriate transfer of relevant technologies, taking into account all rights over those resources and to technologies, and by appropriate funding'.

Benefit sharing

The 'fair and equitable sharing of benefits' is addressed by Article 15.7 and other provisions.[20] From a more detailed perspective, it becomes clear that those provisions are not confined to a sharing of results, products, commercialization and other positive outcomes of the processing and other use of GRs.[21] Much to the contrary, those provisions also concern a participation in scientific research[22] and a transfer of technology,[23] and thus also address a participation or sharing of activities, methods and means that are undertaken or employed to achieve such results in the end. This distinction has important implications. It amply shows that the CBD does not confine itself to calling for a sharing of outcomes, but aims at enabling resource states to strengthen their national capabilities and capacities to engage themselves in the sustainable use of GRs. However, it is important to note that, unlike benefits *strictu sensu*, the participation in research and the sharing of technologies does not depend on any positive outcome of research, development or any other process being undertaken with the resources at hand.

Importantly para 2 adds to that in pointing out that parties should 'endeavour to create conditions to facilitate access to GRs for environmentally sound uses by other contracting parties and not to impose restrictions that run counter to the objectives of the convention'. It thereby reflects the statement in Article 1 that 'appropriate access to genetic resources' is an inherent element of such benefit sharing.

Intergovernmental cooperation and its concept of justice

From a legal perspective, the relevant provisions have a common structure. In substance, they envisage certain ends and objectives, namely, utilization of GRs in accordance with fairness and equity and – sometimes – with due regard to the state of development of the country at hand.[24] Also, obligations of states are defined, with a different degree of precision, mainly by explaining the kind of state measures required.[25] The provisions contain little guidance as to how states shall accomplish their obligations. Only in one case does the wording indicate that activities and obligations sought shall be effected by the private sector.[26] The obligations might also be honoured by bilateral and multilateral means of state cooperation. Furthermore, reference is made to the financial mechanism of the Convention.[27]

In sum, these provisions reflect what in more general terms is called the international law of cooperation. Such a law of cooperation, it is held, deviates from what is understood to be the classic international law structure of pure reciprocity and simple exchange relations – an understanding of justice, which is in line with the idea of an *iustitia commutativa* in legal thought, based on *Aristotle's Nicomachean Ethics* (Lutz-Bachmann, 2000, p2).[28] When defining common ends and agreeing to contribute jointly to their achievement, it is said a different type of relationship emerges among states, which includes states contributing in a different way to the common end, taking into account their different capabilities and resources. This kind of relationship reflects a different measure of justice, which includes some distributive elements and therefore comes close to what is known as *iustitia distributiva*.

This idea of cooperation of states and the specific measure of justice governing their relationships in these areas is reflected by the principle of common but differentiated responsibilities, as embodied by most modern environmental regimes, and particularly so by Article 20 of the CBD, and is implemented inter alia by the financial mechanism.

Very likely, this measure of justice has to be taken into account when applying or interpreting the CBD's provisions on benefit-sharing at a governmental and intergovernmental level, including the standards of fairness and equity. Clearly, this *iustitia distributiva* kind of logic may govern national or multilateral measures in view of technology transfer and will guide decision-making concerning the financial mechanism.

Utilization of GRs by private actors

The utilization of GRs, in most cases, is a matter for universities, research institutions and industry, including also small and medium-size enterprises and entities. Their relationship with providers of GRs has a different format and a distinct logic of justice. Utilization of GRs may, for instance, be driven by the specific rationale of a long-term academic cooperation, where academic institutions on both sides are undertaking a joint research project. Such cooperation often includes elements that reflect a distributional type of justice. For instance, projects might include training and capacity building, the provision of equipment and services, and the sharing of data, research results and joint publications.

Iustitia commutativa

Utilization of GRs may be driven by commercial interest on both sides, where a business entity seeks access to GRs in the course of its research and development activities. These latter types of transactions certainly are the ones that the drafters of the CBD had in mind in expecting that such commercial uses will primarily bring about the benefits that are needed to make the system work. However, these kinds of transactions have a number of noteworthy particularities. First of all, due to their commercial nature, both sides are likely to be governed by economic considerations and the resulting logic of bargaining and agreement will basically be one of market transactions, where parties give something in order to get what they want. Thus, the ratio of such relationships will be much more of an *iustitia com-mutativa*-type relationship as explained above. Thus, there may be a mismatch between the kind of law of cooperation approach reflected by the Convention's provisions and the business realities that govern the conduct of a business entity seeking access and particular transactions. Thus it will be difficult to expect commercial users to engage in a type of relationship that contains distributional elements.

Asymmetries between a commercial user and a state agency provider
These commercial transactions are also special because business skills and experience are required, while – typically – the partner providing GRs will be a government agency. Commercial ABS agreements contain many features and elements common to licensing agreements. In order to successfully negotiate, draft and execute such agreements, legal, technical and business skills, as well as a good knowledge of the local market and the scientific and technological merits and the commercial potential of a given access project, are required. Unfortunately, most of this information is kept confidential. Thus, information asymmetries are likely to occur, which affect the relative bargaining power, as well as the content of a potential agreement. In this situation, government officials in charge of negotiating ABS agreements act under considerable uncertainty and may be hesitant to conclude agreements. This may even be the case where some learning has taken place; as such officials also have to justify their decision within governmental agencies. The resulting reluctance on the side of provider institutions may in turn cause frustrations on the side of businesses, which frequently have to invest considerably in undertaking access procedures and related negotiations.[29]

Making ABS work

As has been shown above, the conceptual basis for ABS as stipulated by the CBD is complex. It cannot be overlooked that the drafters of the Convention developed quite an ambitious concept and that the provisions of the Convention barely address the manifold practical implications and difficulties involved. It took quite a while for the CBD institutions to explore the intricacies of ABS and to develop a common approach to address the shortcomings.

A case for provider states' measures alone?

In the early days of the convention, it was understood that an acceptable state of affairs could be achieved if only resource states would clarify the procedures and rules for access and mutually agreed terms by domestic legislation. Indeed, establishing competent authorities, adopting procedures and welcoming applications for access appeared to be an easy and reward-ing task in view of the benefits to be captured. Expectations were high in those days. However, it turned out that most of the biodiversity-rich countries experienced important difficulties in setting up efficient mechanisms for ABS. Apparently, there have been competing claims between institutions and entities as to the assignment of the authority to grant access and to receive potential benefits. Another point of difficulty has been the proper integration of procedures to grant access to traditional knowledge. Furthermore, there has been a critical lack of capacities, especially in view of drafting agreements and working out the kind and amount of benefits to be asked for. The resulting legal uncertainties, administrative deficiencies and delays result in a considerable frustration among those who are willing to adhere to the rules and to respect rights to GRs.

Addressing biopiracy

On the other hand, a number of cases became known where GRs had been used without proper prior informed consent and mutually agreed terms. This led to a long and controversial debate about biopiracy and resulted in mistrust and a lack of confidence on the side of the authorities that are com-petent to grant access. Indeed, as has been shown above, the sovereign entitlement to GRs suffers from an important shortcoming as, so far, it is only effective within the confines of the jurisdiction of the resource state. A number of proposals have been made to address this situation. They include the voluntary or mandatory requirement of the disclosure of the ori-gin of genetic material used for an invention in patent applications and certificates of origin.

Clarifying the options: The Bonn Guidelines

In this situation, the Conference of the Parties adopted the so-called Bonn Guidelines for Access and Benefit Sharing, which were meant to clarify aspects of the transactions and mechanisms. However, the guidelines failed to meet the growing need for more comprehensive guidance.

Heading for an international regime

This is why a more comprehensive and authoritative instrument was asked for by the 2003 Johannesburg summit.[30] The CBD Conference of Parties reacted by starting a process of discussion and negotiation by means of the open-ended working group on ABS.[31]

The working group has made progress in a number of ways. First of all, it has been acknowledged that the legal entitlement of resource states to their GRs is insufficient to control the utilization of material and to create the degree of confidence that is necessary to engage in and to maintain beneficial transactions in the use of GRs. In this regard, the misappropriation of GRs has been addressed, and means and ways are discussed of how states can support the legitimate interest in controlling the utilization of material by certain measures.

On the other hand, it is claimed that potential users of GRs need predictability and a fair and speedy procedure of granting access and agreeing on mutual terms. Some standards on mechanisms and procedures to grant access have been put forward in this regard, including the so-called international access standards.[32]

However, aside from these two elements – measures against misappropriation and international access standards – more guidance appears to be necessary in view of the content and the format of individual transactions. While acknowledging the differences between the profiles, the potentials and the needs of different sectors, some standard contractual terms appear to be helpful in order to reduce transaction costs and promote legal certainty. They might also be beneficial in view of the potentially asymmetric distribution of bargaining power and information.

Conclusion

A closer look at the CBD ABS system and its conceptual underpinnings reveals a complex structure of divergent, interrelated but also contradicting expectations, claims and rationales. While a pure environmental approach fails to adequately address these complexities, it turns out that justice plays an important role in view of both the acknowledgement of sovereign entitlements in GRs and the CBD's provisions on benefit-sharing and individual transactions concerning the utilization of resources. However, it is necessary to distinguish strictly between different concepts of justice and their implications. It has to be noted that ABS-related intergovernmental obligations under the CBD reflect a concept of distributional justice, whereas commercial transactions involving the utilization of GRs are driven by some sort of justice in exchange transactions, which is sometimes referred to as *iustitia commutativa*. The negotiations on an International Regime may significantly improve the situation, as they envisage allowing for a meaningful enforcement of rights in GRs while at the same time promoting certainty and security by way of defining international access standards. Standardized contract terms may help to reduce transaction costs and promote confidence. However, the kind of cooperation of states as envisaged by the CBD appears to necessitate some additional elements which respond to the rationale of distributional justice. Such additional elements might include capacity-building, the promotion of technology transfer, the joint uses of GRs and a transfer of research and development to the states of origin to reflect the endeavour for cooperation as stipulated by the Convention. If the negotiations succeed, they may bring about what was the very objective of the creation of a sovereign right over GRs: doing justice by establishing fair and equitable ways of joint utilization of GRs.

Notes

1 See for instance, Gehl-Sampath (2005), p2ff.
2 See, for an economic discussion, Simpson et al (1996) and Simpson (1997).
3 See generally the Millennium Ecosystem Assessment (2005).
4 See Wolfrum et al (2001) p8.
5 See the definitions contained in Article 2.
6 Resolution 3/91, of the 26th Session of the FAO Conference, adopted on 25 November 1991, ftp://ftp.fao.org/ag/cgrfa/Res/C3-91E.pdf.
7 ftp://ftp.fao.org/ag/cgrfa/iu/iutextE.pdf, see Himmighofen (2000).
8 Article 1 of the Undertaking reads: 'The objective of this Undertaking is to ensure that plant genetic resources of economic and/or social interest, particularly for agriculture, will be explored, preserved, evaluated and made available for plant breeding and scientific purposes. This Undertaking is based on the universally accepted principle that plant genetic resources are a heritage of mankind and consequently should be available without restriction'.
9 See Schrijver (1988).
10 Article 2.1 of the Undertaking reads: 'In this Undertaking: (a) "plant genetic resources" means the reproductive or vegetative propagating material of the following categories of plants: (i) cultivated varieties (cultivars) in current use and newly developed varieties; (ii) obsolete cultivars; (iii) primitive cultivars (land races); (iv) wild and weed species, near relatives of cultivated varieties; (v) special genetic stocks (including elite and current breeders' lines and mutants)'.
11 Resolution 4/89 of the 25th Session of the FAO Conference, adopted on 29 November 1989, ftp://ftp.fao.org/ag/cgrfa/Res/C4-89E.pdf.
12 FAO Res. 4/89 supra, Article 5(a).
13 Resolution 5/89 of the FAO Conference, ftp://ftp.fao.org/ag/cgrfa/Res/C5-89E.pdf.
14 Resolution 3/91, of the 26th Session of the FAO Conference, adopted on 25 November 1991, ftp://ext-ftp://ftp.fao.org/ag/cgrfa/Res/C3-91E.pdf.

15 Resolution 3/91, of the 26th Session of the FAO Conference, adopted on 25 November 1991, ftp://ext-ftp://ftp.fao.org/ag/cgrfa/Res/C3-91E.pdf.
16 See Correa (1994); Stoll (2004).
17 ftp://ftp.fao.org/ag/cgrfa/it/ITPGRe.pdf.
18 See Stoll (2008), p118.
19 See recently the Report of the Meeting of the Group of Legal and Technical Experts on Concepts, Terms, Working Definitions and Sectoral Approaches, UNEP/CBD/ABSWG/7/2.
20 Article 15 paras 6, 7; Article 16; Article 19 paras 1, 2; Article 21. See Stoll (1997).
21 This is specifically referred to by Article 15 para 7 – 'results of research and development and the benefits arising from the commercial and other utilization of genetic resources' and Article 19 para 2: 'results and benefits arising from biotechnologies based upon genetic resources'.
22 Article 15 para 6 and Article 19 para 1.
23 Article 16 paras 1–3.
24 See Article 15 paras 6, 7; Article 16 paras 3, 4; Article 19 paras 1, 2.
25 See, for instance, the different wording of Article 16 para 1: 'undertakes to provide'; paras 3, 4: 'legislative, administrative or policy measures'.
26 This is expressly provided for in Article 16 para 4.
27 Article 15 para 7.
28 These concepts of justice are referred to by De Jonge and Korthals (2006) at p145ff in regard to crop GRs and by Schroeder and Lasén-Díaz (2006) at p137.
29 As regards the market imperfections and information asymmetries see Gehl-Sampath (2005), at p63ff.
30 Paragraph 44(o) of the Plan of Implementation adopted by the Summit requests to 'negotiate within the framework of the CBD, bearing in mind the Bonn Guidelines, an international regime to promote and safeguard the fair and equitable sharing of benefits arising out of the utilization of genetic resources'. Furthermore, para 44(n) of the Plan envisages to promote 'the wide implementation of and continued work by the Parties to the Convention on the Bonn Guidelines on Access to Genetic Resources and Fair and Equitable Sharing of Benefits arising out of their Utilization, as an input to assist the Parties when developing and drafting legislative, administrative or policy measures on ABS as well as contract and other arrangements under mutually agreed terms for ABS', UN Document UNEP/CBD/COP/7/5, 25 March 2003.
31 Re-Established by COP 6 Decision VI/24.
32 The issue of 'international access standards' has been raised by the EU in a position paper on 'Concrete Options for the Further Negotiation of Substantive Items on the Agenda of the Fifth and Sixth Meetings of the Ad Hoc Open-Ended Working Group on Access and Benefit-Sharing, Submission by the European Community and its Member States of 28 November 2007 in Response to CBD Notification 2007-132', reported in: UNEP/CBD/WG-ABS/6/INF/3 at p28ff. The relevant parts read: 'In response to the demand for potentially binding international commitments to support compliance with ABS requirements through clearly specified measures, the EU has identified the need for developing international standards on national access law and practice as part of the ABS negotiations. The EU believes that it is difficult to consider additional and more specific international commitments to support compliance with ABS requirements if there is uncertainty about and a broad variety of what

exactly is to be enforced in countries with users under their jurisdiction. Following from this argumentation the international ABS regime needs to include international standards on national access law and practice and an international mechanism/ process for assessing whether or not national access frameworks meet international standards'. Later on the issue has been incorporated in the list of 'Main Components' of the International Regime by the Ninth Meeting of the Conference of the Parties in Decision IX/12, Annex 1, III. B. 2.2, see UNEP/CBD/COP/9/29 at p115.

References

Chen, J. (2006) 'There's no such thing as biopiracy ... and it's a good thing too', *McGeorge Law Review*, vol 37, pp1–35; available at http://ssrn.com/abstract=781824, accessed 29 May 2009

Correa, C. M. (1994), 'Sovereign and property rights over plant genetic resources', Background Study Paper No 2, Commission on Plant Genetic Resources, Rome, FAO

De Jonge, B. and Korthals, M. (2006) 'Vicissitudes of benefit sharing of crop genetic resources', *Developing World Bioethics*, vol 6, no 3, pp144–157

Gehl-Sampath, P. (2005) *Regulating Bioprospecting: Institutions for Drug Research, Access and Benefit Sharing*, Tokyo, New York, Paris, UNU Press

Himmighofen, W. (2000) 'Work going on in the FAO', in Wolfrum, R. and Stoll, P.-T. (eds) *European Workshop on Genetic Resources Issues and Related Aspects*, Berichte des Umweltbundesamtes, vol 5/00, Berlin, Erich Schmidt Verlag, pp173ff

Lutz-Bachmann, M. (2000) 'The discovery of a normative theory of justice in medieval philosophy: On the reception and further development of Aristotle's theory of justice by St. Thomas Aquinas', *Medieval Philosophy and Theology*, vol 9, pp1–14

Millennium Ecosystem Assessment (2005) *Ecosystems and Human Well-being: Synthesis Report*, Washington DC, Island Press

Schrijver, N. J. (1988) 'Permanent sovereignty over natural resources versus the common heritage of mankind: Complementary or contradictory principles of international economic law?', in de Waart, P. and Denters, E. (eds) *International Law and Development*, Dordrecht, Kluwer, pp87ff

Schroeder, D. and Lasén-Díaz, C. (2006) 'Sharing the benefits of genetic resources: From biodiversity to human genetics', *Developing World Bioethics*, vol 6, no 3, pp135–143

Simpson, R. D. (1997) 'Biodiversity prospecting: Shopping the wilds is not the key to conservation', *Resources*, vol 126, pp12–15

Simpson, R. D., Sedjo, R. A. and Reid, J. W. (1996) 'Valuing biodiversity for use in pharmaceutical research', *Journal of Political Economy*, vol 104, pp163–185

Stoll, P.-T. (1997) Access to genetic resources: Establishing an international order for the exchange and sharing of genetic resources, research, technology and benefits under the Convention on Biological Diversity', Temas de derecho industrial y de la competencia (Buenos Aires), vol 2, pp317–344

Stoll, P.-T. (2004) 'The FAO "Seed Treaty" – New international rules for the conservation and sustainable use of plant genetic resources for food and agriculture', *Journal of International Biotechnology Law*, vol 1, pp239–243

Stoll, P.-T. (2008) 'Global public goods – the governance dimension', in Rittberger, V. et al (eds) *Changing Patterns of Authority in the Global Political Economy*, vol I, New Institutions and the Provision of Global Public Goods, New York

Wolfrum, R., Klepper, G., Stoll, P.-T. and Stephanie, F. (2001) 'Implementing the Convention on Biological Diversity: Analysis of the links to intellectual property and the international system for the protection of intellectual property', in Hahn, A. von (ed) *Implementing the Convention on Biological Diversity Analysis of the Links to Intellectual Property and the International System for the Protection of Intellectual Property*, Bonn, German Federal Agency for Nature Conservation, BfN-Skripten 47

Towards Regional Common Pools of GRs – Improving the Effectiveness and Justice of ABS[1]

Gerd Winter

Introduction

The CBD acknowledges sovereign rights of states over their natural resources including GRs. User states are obliged to share the benefits derived from the utilization of GRs with states providing the GRs. Thus, provider states and user states are expected to create bilateral exchange relationships. Legal practice on international and national levels has proved that this individualistic approach lacks efficiency: while the scope and content of sovereign rights of provider states over their GRs are far-reaching, due to the territoriality principle they are hampered to control the downstream process of value creation. Enforcing their legal powers effectively would also cause high transaction costs. User states, on the other hand, could be asked to make leeway. While they are less hindered by the territoriality principle because research and development (R&D) activities related to GRs are largely under their jurisdiction, they also face substantial transaction costs if they use what powers they have in order to control the upstream process. However, even if the control by provider and user states is improved, questions of distributional justice arise. Many GRs have a geographical range shared by regions of states, suggesting that benefits should be shared among all states in which the GR is endemic instead of the first provider state taking all of the share. Therefore, for reasons of efficiency and justice, I propose regional common pools for GR management. They are meant not to replace but to complement bilateral-ism. This chapter outlines structures and functions of common pools and suggests national legislation supporting them. The focus is on GRs as such and leaves common pools of TK associated with GRs for further reflection.

Sovereign rights over GRs

The CBD attributes GRs to the individual realm of states hosting them, thereby rejecting earlier concepts of common heritage. On this subject Article 15.1 CBD states:

> *Recognizing the sovereign rights of States over their natural resources, the authority to determine access to genetic resources rests with the national governments and is subject to national legislation.*

In the following paragraphs, I will discuss what this means in terms of content and scope of sovereign rights.

The content of sovereign rights

'Determining access to genetic resources' is not just a means of ensuring administrative oversight of access, but rather constitutes a part of the sovereign rights of states. GRs are thus made the property of a state. This implies that the state has the right to (i) reserve the utilization of GRs for itself, (ii) exclude others from utilization, and (iii) make utilization dependent on conditions (or require the signing of a contract) obliging users to report about R&D steps and to share material and immaterial benefits drawn from the GR or derivatives.

It is true, though, that limits are established by the Convention itself, in particular by Article 15.2, which says:

> *Each Contracting Party shall endeavour to create conditions to facilitate access to genetic resources for environmentally sound uses by other Contracting Parties and not to impose restrictions that run counter to the objectives of this Convention.*

However, the wording 'shall endeavour to create conditions to facilitate' is very vague.[2] Nobody could object on this ground if the provider state links access to conditions prescribing meticulous reporting and benefit sharing. Some strengthening of the user country position may be derived from Article 15.4, which states that 'access, where granted, shall be on mutually agreed terms'. However, the clause 'where granted' acknowledges that the provider state has the power to decide about whether to grant access at all. This makes bilaterally agreed terms dependent on a unilateral decision of the provider country.

How has state legal practice implemented the sovereign rights of provider states? Those states that have adopted access and benefit-sharing

(ABS) legislation have usually established a regulatory framework. This framework requires that the access seeker must obtain a permit and agree on a contract on the transfer of the GR, the allowed uses, knowledge transfer and the sharing of benefits (material transfer and benefit-sharing agreements – MTAs/BSAs). States have hardly ever constituted private property rights in GRs. For instance, they might have framed GRs as intellectual property rights (IPR) sui generis; that is, absolute rights of utilization, which like patents, breeders' rights or trademarks are to be registered and can be exploited and traded.[3] But regardless of regulatory or property law, the instruments laid down by national laws have raised high expectations of remuneration for the transfer of GRs. However, these expectations have resulted in delusion.

Due to the territoriality principle, the control of access, transfer, utilization, knowledge transfer and benefit sharing is confined to the territory of the regulating state. The state is entitled to oversee access to and transfer of its GRs, while in relation to utilization and benefit sharing it can only impose conditions to the permit but has no powers to enforce such conditions in user states. It can only ask the user state to provide assistance. Instead, as stated previously, most provider states have opted for the conclusion of MTAs/BSAs. But in order to enforce the contractual obligation, the provider state must address the user state courts as a forum. If the parties have agreed on provider state courts as being competent, the execution of the judgement in the user state would still have to involve the user state courts.[4] In fact, hardly any case has as yet been reported where provider states have searched assistance of provider state administrations or have filed complaints at user state courts in order to pursue their authorizations or contracts.

The scope of sovereign rights

High expectations and in effect delusions are also characteristic of the scope of the sovereign rights of provider states. The scope of access determination is, by Article 15.1 CBD, delimited by the term 'genetic resources'.

Article 1 CBD defines GR as 'genetic material of actual or potential value' and genetic material as 'any material of plant, animal, microbial or other origin containing functional units of heredity'. In a nutshell, therefore, the sovereign right of access determination extends to the functional units of heredity contained in natural resources of a given state and being of value.

Functional units of heredity

'Units of heredity' could be organisms, cells, chromosomes, genes and deoxyribonucleic acid (DNA) fragments (Ten Kate and Laird, 1999, p18,

Box). I submit that all of these levels should be included because the heredi-tary function may already be attached to an extracted DNA fragment, or it may result from the combination of DNA fragments within a gene, or of genes in a cell or from cooperating cells within an organism.

'Functional' does not imply that the unit of heredity must be able to reproduce itself. It suffices that the unit is subjected to technological manipulation, such as genetic engineering. This means, of course, that the term 'functional unit' widens with the development of genetic technology (Ten Kate and Laird, 1999, p18).

It has been suggested that, in addition to the material genetic substra-tum, the intangible scientific information about the genetic function should also be included in the definition of GRs (Tvedt and Young, 2007, p62ff). This would have two consequences: (i) the access to and transfer of infor-mation created within a provider country could be made subject to the regulation of the provider state; and (ii) the provider state could extend its reporting and benefit-sharing claims to benefits drawn from the informa-tion. However, it appears that 'unit of heredity' is meant to be the material genetic substratum contained in the genes. Incidentally, as will be explained later, this does not exclude the obligation of user states to ensure the shar-ing also of those benefits that are drawn from intangible information.

Functional units of heredity are often not the immediate basis of benefi-cial utilization. Rather, DNA fragments or genes may be extracted and transferred into another organism which then provides benefits, or they may be synthesized as artefacts and as such trigger the benefits. Hybrid microorganisms, plants or animals derived from interbreeding of organisms may rather be used to gain benefits than the original organisms. Should such so-called derivatives be covered by the sovereign rights of states? The CBD mentions derivatives only in the definition of biotechnology, not in that of GRs. Indeed, as they are different from original units of heredity, they cannot be counted as property of the state. Therefore, the CBD does not grant states the right to exclusive use of derivatives. Instead, states may use their sovereign rights on GRs and grant access on the condition that the beneficiary agrees to share also those benefits that arise from derivatives.

Actual or potential value
Genetic material becomes a GR if it has (actual or potential) value. Considering that Article 2 CBD characterizes biological resources as being of value to humanity, it can be deduced that value for humanity is also meant in relation to GRs. Thus, value is broader than just commercial prof-itability: it covers exchange value as well as use value. The use value even extends to pure scientific interest, thus including access for scientific pur-poses in the access regime established by Article 15 CBD.

It must be the genetic material that creates value. This link helps to exclude from the ABS regime the use of biological resources for bulk purposes such as consumption, heat generation or construction work. Only the value resulting from the utilization of the genetic characteristics constitutes GRs. It is true, though, that the line between bulk use and the use of the genetic code is far from clear. The biochemical compounds are a particularly controversial case in point. Although they are results of functional units of heredity but not such units by themselves, some states claim that they are to be considered as GRs.

It has been suggested that the access seeker must have the intention to utilize the genetic material and thus realize its value in order to make genetic material a GR. This would exclude genetic material from the access regime for which the intention is different, such as consumption. However, the text of the CBD does not speak of such intentions. It clearly includes potential value, that is, uses not yet realized or intended. This means that a provider state may regulate access to biological resources that are presently used as bulk material, but have the potential to be used as genetic material too.

Once again, while the promise of individualization of resources is far-reaching, it is illusionary at the same time. Even with the exclusion of intangible information and of derivatives, the remaining scope of GRs is still very large. Given the fast development of biotechnology, there is hardly any biological material whose genetic code could not be used. Moreover, with the decline of biodiversity, the scientific interest in preserving genetic information increases, which implies that virtually any biological material becomes of actual or potential value. Therefore, provider states can regulate access to any biological material.

However, this would not much help provider states to get a share in the benefits. If resource states establish access regulations for every single specimen of biological material, transaction costs for the state and private actors would be enormous. Disrespect for the law would ensue. Moreover, access control would be ineffective because, as stated earlier, provider states would not have the power to enforce the obligations attached to the access permit. For instance, mutually agreed reporting duties would imply that the recipient of material must inform the provider state about any biotechnological treatment, sales of the material or benefit drawn from it – an entirely futile expectation given the possible multiplication of downstream users. On the whole, traceability from sources forwards to subsequent users appears to be an impossible task.[5]

Benefit sharing

Given the unrealistic expectations connected with the determination of access on the provider side, it has been suggested that the focus on access should be replaced by a focus on benefit sharing (Tvedt and Young, 2007, p62ff). This brings the obligations of user countries into play. Rather than provider states trying to pursue their interests in user states, the user states themselves are called to take their international duties seriously. In principle, user states are less impeded than provider states to ensure benefit sharing because most of the knowledge and added value is created within their jurisdiction.[6] Their duties follow from Article 15.7 CBD:

> *Each Contracting Party shall take legislative, administrative or policy measures, as appropriate, and in accordance with Articles 16 and 19 and, where necessary, through the financial mechanism established by Articles 20 and 21 with the aim of sharing in a fair and equitable way the results of research and development and the benefits arising from the commercial and other utilization of genetic resources with the Contracting Party providing such resources. Such sharing shall be upon mutually agreed terms.*

This article obliges states hosting the utilization of GRs to share R&D results and commercial or other benefits in an equitable way. As a consequence, the user state must introduce legislation concretizing this obligation. A major advantage of this approach also is that those benefits that arise from genetic material obtained without the consent of a country of origin can be controlled, as well as benefits from uses not intended at the time of exportation from the resource country. Mutually agreed terms, as Article 15.7 requires, can still be concluded at the utilization stage. The kind of needed legislation will largely be regulatory. Users of GRs must be obliged to keep provider states informed about new knowledge, technology and benefits obtained, and to share benefits. Details concerning trade secrets and intellectual property protection could be left to MTAs/BSAs or specified by regulation. Administrative oversight must be established. For this to be effective, importers and users of GRs must be submitted to notification and information duties. Administrative agencies must be enabled to track benefits back to provider states. In addition, court procedure law and the international private law of the user states must ensure that MTAs/BSAs are enforceable within their jurisdiction.

However, other than Article 15.7 suggests, user states have largely remained passive. They have almost exclusively relied on provider states' legislation and the MTAs/BSAs triggered by a permit of access requirements.

The instruments discussed thus far are either hardly effective or cause massive transaction costs. Contract claims at user state courts are possible but costly. The execution of provider state courts is subject to a double check of user state courts in terms of ordre public. It is true that some primary administrative law tools have been practised, such as guidelines of public research funding organizations and disclosure requirements, in procedures granting IPRs.[7] However, research guidelines do not capture private sector research and are also difficult to enforce once a research funding has been granted. Disclosure requirements would only be useful if the origin of the genetic resource had a material impact on the granting of an IPR (such as a patent or breeder's right), which has hardly ever been effected by any user state.[8] Patent applicants could even argue that disclosure is an unproportional intrusion into rights of free enterprise and profession if it does not serve a purpose.[9]

Certificates of origin and of compliance have been proposed as another means of user state control. However, they pose many problems of construction and function. The consequences of obtaining or lacking them are unclear; they can only be issued for a specific moment in time: a moment later, the GR may have changed shape due to technological treatment; the certificate cannot be physically attached to the genetic material once this has been turned into intangible information; unique identifiers of the genetic code and its origin would have to be developed with a worldwide scope – a task needing immense effort and cooperation to accomplish.

Thus there are many technical difficulties in tracing the utilization of GRs back to a provider country. Often genetic material changes hands before it is utilized in profitable ways. The chain of utilization can be very long reaching from the plant and the extracted gene to organisms modified by the original or synthesized genes. In the meantime, a strain of genes may have been replaced by newer ones that were obtained from another state or from another parent organism. These and other factors can easily blur the country of origin.

All this proves that even the focus on user states is largely illusionary. User state measures if called to take the lead easily reach limits of efficiency.

Common pools: More efficient, but also more just

As a way out of illusionary expectations, it is submitted that common pools should be established by states constituting biogeographical regions. Such common pools would not question the basic decision of the CBD that GRs are the property of host states. They would not replace bilateralism, but provide an opportunity for provider and user states to opt for more efficiency of

ABS regimes. The concept offers provider states a chance to make use of their property in a way that ensures a realistic return. User states can regard the concept as attractive because it simplifies their international duty to trace provider countries and organize the sharing of benefits.

In addition to providing efficiency, common regional pools would also enhance distributional justice. GRs are, by their very nature, not bound to the territories of states. As living organisms they migrate between states or live in ecosystems occurring in several states. The simple fact that the organism containing the GR was taken from the land or marine area of a state is in CBD terms ground enough to provide the state with the full right to control access and claim a share from the benefits: Article 15.1 CBD attaches determinative rights to the eventuality of access, Article 15.3 introduces the notion of resources provided by a state, and Article 15.7 states that the user state must share benefits with the provider state.

From the perspective of distributional justice this is hardly justifiable. The simple and often adventitious event of access in one provider state is no good reason for that state to entirely control the utilization of and benefit from the genetic material if the same genetic material also occurs in other states. Therefore, even if a state of origin operates a perfect system of monitoring and claiming benefit sharing, there remains the distributional question of whether that is just. These same doubts were, for instance, uttered in relation to Costa Rica, with the most successful ABS regime of a provider state, and its managing agent the National Biodiversity Institute (INBio):

> *Some time in the future, a pandemic genetic resource provided by INBio will become a blockbuster biotechnology. Citing the CBD, other countries in the region will challenge the legitimacy of the patent, inasmuch as they will not have received any 'fair and equitable' share of the benefit arising from the pandemic genetic resource. It is no small irony that the success of INBio lies in its failure to have a commercial hit (Vogel, 2007, p130).*

Additionally, there is a side effect of the radical individualization of property rights in GRs. The privileging of the provider state will lead to forum shopping; that is, access seekers approaching that state with the least demanding reporting and benefit-sharing duties (Brand and Görg, 2001). A regulatory competition would lower standards and jeopardize the very goal of the CBD: to ensure technology transfer and the sharing of commercial benefits. It may also jeopardize the goal of conservation and sustainable use because weak access legislation could attract more bioprospectors and thus increase bioprospecting pressure.

Looking once again at the text of the CBD, one can find certain hints that it is open for a regional concept. As noted before, Article 15.1 starts with the assumption 'Recognizing the sovereign rights of States over their natural resources'. The word 'their' is commonly understood to establish property of resource states as contrasting the concept of global common heritage. In addition to this 'negative' significance, 'their' could be understood to have a 'positive' meaning aiming at neighbouring states: the GR of states must be 'theirs' in the sense that it must have a genuine and exclusive link to the territory of the single state in question. Where this is not the case, the GR is either common good (as GRs found in the oceans or in Antarctica) or common to a region. With this reading, property shared by several states within a region can be regarded as a concept recognized by the CBD.

A difference should, however, be noted between GRs as such and TK associated with GRs. While the mere presence of an organism in one state is no good reason to recognize property in its entire genetic potential, this may be different for TK. Such knowledge has been created by individuals and communities; investigation of organisms and experience drawn from them; and creativity, time and labour spent on breeding and other activities to improve the resource. Thus the value of the GR is highly enhanced by human intelligence. In line with the basic ideas underlying intellectual property regimes, this fact would justify the application of a scheme of stricter individualization in relation to TK, allowing for a 'first takes all' approach. As a corollary, the voluntary pooling of TK might be considered, reflecting the fact that TK often spreads over several communities. However, such pools will primarily be a matter of internal national legislation. To the extent TK reaches over national borders, states may consider including TK into the pools of GRs as they are proposed here. This, however, requires more in-depth study this chapter cannot provide.

Looking for blueprints

A number of concepts have been proposed and sometimes put into practice, all of which aim at communal solutions. Three that appear to be particularly significant are discussed in the following sections.

Science commons

One approach is based on the science commons project. This project is destined to create a worldwide exchange of scientific data (Wilbanks and Boyle, 2006, pp9–12). The various sectors include one on biological material. A standard material transfer agreement (SMTA) shall be developed for all those who exchange material. Each concluded contract shall be registered and made accessible to the whole community so that every participant in the system knows who possesses what material. The material is not collected in a common bank but rather shipped bilaterally between contract partners. In addition, information on research results on the characteristics and effects of material shall be collected and made accessible. This will, however, require a meta-language and a huge effort of data collection that needs to be financed. Also, copyrights of publishers will have to be dealt with. The system will use semantic web language; that is, instead of referring to documents as the internet traditionally does, it will refer directly to genetic material and associated knowledge.

The project is attractive because it ensures the exchange of material and scientific knowledge at low costs. Everybody including researchers from resource-rich developing countries has access to the system. There are, however, drawbacks which make the system less suitable for the common pool here envisaged. Although the system could be designed to enable the tracking of individual genetic material back to the provider state (Buck, 2007, pp88–91), this would be difficult and costly to implement. Unique identifiers for genetic material would have to be developed and included in the MTA, although the knowledge necessary for this is not yet available at the stage of accession. Another difficulty is that unique identifiers, as applied to genetically modified organisms (GMOs) that have been registered under European Community (EC) law,[10] are still widely lacking in relation to genetic material as such. Furthermore, any further transfer and technological treatment of GRs would have to be registered, which is hardly enforceable.

International Plant Exchange Network

The International Plant Exchange Network (IPEN) is a network of botanical gardens facilitating exchange of plant GRs in line with the requirements of Article 15 CBD (Gröger, 2007, pp121–123). The IPEN website is provided by Botanic Conservation International at Kew, England. Ninety-one botanical gardens are members, all of which are European. The exchange is regulated by a Code of Conduct and every individual plant is documented. The 'maximum documentation' includes information about collection, source, taxonomy, type of material, permits related to the acquisition and any terms of the country of origin. This maximum documentation is kept by the first garden, which introduces the plant material into IPEN. This garden also tags an individual IPEN number to the plant material. The number, referred to as 'minimum documentation', follows the plant through descendants and transfers. The transfer to non-members requires the signing of an SMTA binding the recipient to the same terms as contained in the Code of Conduct. The exchange is confined to the use of the GR for scientific and conservation purposes. In the case of intended commercial use, the requesting institution must obtain prior consent of the original provider state. As a measure of trust building, IPEN extends this requirement also to that plant material which was accessed prior to the enactment of the CBD.

IPEN is an exemplary case of a system that ensures the backtracking of plant material to sources. It is successful in facilitating exchange; however, since it is destined for exchange only for conservation and scientific purposes, it has intentionally excluded any management of reporting on commercial utilization and benefit sharing. Any intention to make commercial use of a GR is referred back to the provider country. It has been considered whether the system could be opened to use management, but it is feared that provider countries would then refrain from providing material. In conclusion, the system is a model for a common pool for conservation and scientific purposes, but not for the sharing of commercial benefits.

International Treaty on Plant Genetic Resources

The International Treaty on Plant Genetic Resources (ITPGR) is the basis for a multilateral system of ABS for plant GRs comprising 35 food crops and 29 forage genera.[11] The system establishes a common pool of GRs agreed upon by the Contracting Parties 'in the exercise of their sovereign rights'.[12] The system aims at including all brands of the listed food crops and forage genera that are under the management and control or jurisdiction of the Contracting Parties.[13] It also includes GRs held in the ex situ collections of the International Research Centres (IARCs) of the Consultative Group on International Agricultural Research (CGIAR).[14] In addition, other states shall be encouraged to include their Annex I GRs in the system.[15]

The Contracting Parties and the IARCs are obliged to provide access to their GRs according to terms laid down by an SMTA. Access is generally free of charge. No tracking of individual accessions in provider states is foreseen. In exchange for the free access, the recipient is not allowed to claim or establish IPRs on the GR in the form received from the multilateral system. However, the recipient is free to seek intellectual property protection for newly developed brands suitable for such protection.

The treaty establishes far-reaching duties to share benefits, including the exchange of information, access to and transfer of technology, capacity building and the sharing of monetary and other benefits of commercialization.

With regard to information exchange, the Contracting Parties are obliged to make available to each other all relevant information including characterization, evaluation and utilization of Annex I GRs, respecting restrictions from intellectual property protection. The information shall be made available through the Global Information System on Plant Genetic Resources for Food and Agriculture, which includes more GRs than those listed in Article 17.

With respect to commercial benefits, the MTA states that recipients must pay an equitable share to the trust account of the system:

> *(A) recipient who commercializes a product that is a plant genetic resource for food and agriculture and that incorporates material accessed from the Multilateral System, shall pay to the mechanism referred to in Article 19.3f, an equitable share of the benefits arising from the commercialization of that product.*[16]

Hence the money does not flow bilaterally but is channelled into a common fund. Direct and indirect payments are to be made from the fund to farmers, especially in developing countries and countries with economies in transition.[17]

In conclusion, the multilateral system set up by ITPGR creates a global common pool of certain GRs destined to share the genetic material, knowledge and monetary benefits. The GRs are disconnected from the states of origin; that is, the material and knowledge is freely exchanged and the monetary benefits are shared among participants with no regard for the states of origin. Disregarding doubts concerning the implementation of monetary transfers,[18] the system appears to be highly appropriate for crops and forage that are truly global in relation to their origin and use: they originate from global human efforts of breeding, and they are utilized and consumed as a fundamental means of subsistence by almost everybody. However, to go further and extend the approach to all other GRs will hardly meet the source states' interests. The disregard for the origin of the GR and the sharing of benefits with all and not only source countries runs counter to the basic approach of the CBD, which is to privilege source countries in relation to benefit sharing.

Towards regional common pools of GR endemic to a region

The three concepts presented all suffer from specific drawbacks, which do not recommend them as a general model. The science commonly disconnects information flow about provider states and does not engage in the sharing of benefits other than the sharing of knowledge. IPEN does generate and store information about provider states, but is – like the science commons – not engaged in benefit sharing other than knowledge. ITPGR does not track GRs back to provider states. Although it arranges benefit sharing it does not do this by channelling shared benefits primarily to provider states.

This experience suggests that a regional approach establishing regional common genetic pools (RCPs) might serve the various interests best. Although much more thought is needed to make RCPs practicable, some suggestions shall be made in the following concerning a possible legal basis, the shape of RCPs and auxiliary national legislation.

Features of RCPs

To give RCPs shape, the following characteristics are suggested:

- Participants in the regional agreements setting up RCPs should be provider and user states, as well as international organizations related to the use and protection of GRs.
- Based on the international agreements, RCPs should be established as corporations with legal personality under national law. This would enhance their ability to act. They would be partners of the MTAs/BSAs and able to pursue such contracts in user states. They could be endowed with trusteeship for the GRs managed by the pool and as such claim tort liability in cases of misappropriation. They could also be given powers to take binding decisions under national administrative laws.
- RCPs should build up data banks (and a meta data bank linking it to other useful banks) on their GRs.
- Participating states should notify the RCPs of any GRs they wish to be managed by the common pool; these are primarily GRs endemic to several states. But a provider state may also notify GRs specific to it in order to benefit from the system's management capacity.
- The data banks should contain common names of organisms, a description of their genetic code, any scientific knowledge about their potential and actual uses, and any technology related to the utilization of the GRs. RCPs may develop a system of unique genetic identifiers.
- Participant states must ensure that any scientific and technological information is provided to the RCPs by scientists and industry under their jurisdiction.
- The RCPs are entitled to enhance their information basis by literature research and links with existing data banks.

- Names of organisms, the genetic code, scientific knowledge and technology will be freely available for scientists and industry under the jurisdiction of the participant states.
- RCPs will be in charge and empowered to conclude MTAs/BSAs with users.
- RCPs will manage the sharing of commercial benefits by:
  - establishing principles on calculating equitable shares of benefits (a) between the beneficiary and the states of origin, and (b) among states of origin;
  - deciding on an individual basis about the shares of benefits to be paid;
  - operating a trust fund collecting the money;
  - deciding on the individual shares for the countries of origin;
  - transferring the money to them.
- Participant states must ensure that the necessary information on commercial benefits is provided to RCPs.
- RCPs should be integrated on a global level.

National legislation

National legislation would still have to be elaborated on bilateral relations concerning ABS; although specific provisions would be necessary to support the RCPs. If a state decided that all of its GRs should be managed by RCPs, it could confine its legislation to this supportive function.

As noted above, user states, if called to take their obligations under the CBD seriously, would have difficulty in tracking GRs to provider states. With common pools this task will become easier because it will be sufficient to identify the involved GR and the RCP managing it. However, both regulatory and private law will have to be used in order to provide support. Users of GRs under the jurisdiction of RCPs will be obliged to keep RCPs informed about new knowledge, technology and accruing commercial benefits, and to share commercial benefits. The extent to which IPRs and trade secrets are respected has yet to be specified, but such duties would be supervised by administrative agencies. User states would have to establish a duty of users to conclude MTAs/BSAs with the pertinent RCP. In relation to commercial benefits, special efforts will have to be made because this is of particular concern to provider states and important as a means of creating trust. Disclosure requirements could help in this respect; they could be tied to the marketing of GR-based products. Alternatively, the supervisory potential of the taxation system could be used by requiring users to reveal the origin in their income tax declaration. The user state must also enforce RCP decisions ensuring that the beneficiary pays the due amount into the trust fund.

The task of source state legislation would also be eased in a common pool system. National legislation would not need to require access authorization and MTAs/BSAs in each individual case of access. A notification of access would suffice except in cases of possible environmental harm. General legal rules would submit any person acceding to GRs to a set of specific obligations, including duties to keep RCPs informed on new knowledge, technology and accruing commercial benefits, to conclude MTAs/BSAs with the RCP and to share benefits according to rules set up by RCPs.

Legal basis

As stated earlier, RCPs should be based on an international agreement between regional provider states and interested user states. The regional fisheries commissions of the southern hemisphere might be studied as a model for such pools. They bring Southern and Northern fishing states together in order to manage a regional resource (Applebaum and Donohue, 1999). Alternatively, the basis for RCPs could be the already existing regional international organizations (IOs) or regional branches of universal IOs such as the five regional Commissions of the UN Economic and Social Council (ECOSOC). Yet another basis might be international sectoral organizations – or more precisely their regional substructures – such as the Food and Agriculture Organization (FAO; for agriculture and fisheries) and the World Health Organization (WHO; for healthcare and cosmetics). A fourth option would be to found RCPs on (yet to be created) regional substructures of the Biosafety Clearing House (BCH) or UN Conference on Trade and Development (UNCTAD). The next CBD Conference of Parties might be asked to take a resolution endorsing the introduction of RCPs. This would greatly help to disseminate the idea.

Notes

1 This chapter has greatly benefited from a workshop held by the Japanese Bioindustry Association in Tokio on 30 September 30 and 1 October 2008. I gratefully acknowledge most valuable comments by Matthias Buck, Evanson Chege Kamau, Hiroshi Isozaki, John Kleba and Seizo Sumida.
2 See further on this clause Kamau and Winter, in this book.
3 See, on ways to do this, Gerstetter, in this book.
4 See Isozaki and Godt, in this book.
5 There is one case frequently cited as perfecting a fully controlled upstream concept: Costa Rica. See Gomez (2007), p85, and Cabrera, in this book. However, this may be a singular case, which eventually raises questions of distributional justice. See below.

6 Problems may, however, arise if R&D activities span over several states.
7 See, for example, the research guidelines of the DFG, http://www.dfg.de/forschungs-foerderung/formulare/download/1_021e.pdf/ accessed 20 May 2009, and Kamau ('Disclosure requirement'), in this book.
8 On doctrinal constructs to this effect, see Godt (2007), pp603, 653.
9 Disclosure of origin may be suited to reveal that the invention was known before. This may sometimes be the case with regard to TK, but rarely ever with regard to GRs as such because research is often underdeveloped in provider countries.
10 Commission Regulation (EC) No 65/2004 of 14 January 2004 establishing a system for the development and assignment of unique identifiers for genetically modified organisms, OJ L10/2004, p5. The unique identifier is composed of letters for the applicant, letters and numbers indicating the transformation event, and a verification number. See Annex to Regulation 65/2004.
11 The food crops and forage genera are listed in Annex I to the treaty. In addition to the common pool, the treaty somehow reinforces and concretizes the bilateral ABS regime of the CBD.
12 Article 10.2 ITPGR.
13 Article 11.2.
14 Article 11.5.
15 Article 11.2.
16 Article 13.2(d)(ii).
17 Article 13.3.
18 It seems that no monetary benefits have yet been channelled through the system.

References

Applebaum, B. and Donohue, A. (1999) 'The role of regional fisheries management organizations', in Hey (ed) *Developments in International Fisheries Law*, The Hague, Kluwer Law International, pp217–249

Brand, U. and Görg, C. (2001) *Zugang und Vorteilsausgleich – das Zentrum des Konfliktfelds Biodiversität*, Bonn, Forum Umwelt & Entwicklung/Germanwatch

Buck, M. (2007) 'The science commons project approach to facilitate the exchange of biological research material – implications for an international system to track genetic resources, associated user conditions and traditional knowledge', in Feit, U. and Wolff, F. (eds) *European Regional Meeting on an Internationally Recognized Certificate of Origin/ Source/LegalProvenance*, Federal Agency of Nature Conservation, pp88–94, www.bfn.de/0502_international.html/ accessed 12 March 2009

Godt, C. (2007) *Eigentum an Information. Patentschutz und allgemeine Eigentumstheorie am Beispiel genetischer Information*, Tübingen, Mohr Siebeck

Gomez, R. (2007) 'The link between biodiversity and sustainable development: Lessons from INBio's bioprospecting programme in Costa Rica', in Manis, C. (ed) *Biodiversity and the Law*, London, Earthscan, pp76–90

Gröger, A. (2007) 'Botanic gardens and the International Plant Exchange Network (IPEN) – A brief statement on an internationally recognized certificate', in Feit, U. and Wolff, F. (eds) *European Regional Meeting on an Internationally Recognized Certificate of Origin/Source/Legal Provenance,* Federal Agency of Nature

Conservation, pp49–59, www.bfn.de/0502_international.html/ accessed 12 March 2009

Ten Kate, K. and Laird, S. A. (1999) *The Commercial Use of Biodiversity. Access to Genetic Resources and Benefit Sharing*, Sterling, Earthscan

Tvedt, M. W. and Young, T. (2007) *Beyond Access: Exploring Implementation of the Fair and Equitable Sharing Commitment in the CBD*, IUCN, Gland, Switzerland

Vogel, J. H. (2007) 'From the "tragedy of the commons" to the "tragedy of the commonplace": Analysis and synthesis through the lens of economic theory', in Manis, C. (ed) *Biodiversity and the Law*, London, Earthscan, pp115–134

Wilbanks, J. and Boyle, J. (2006) 'Introduction to science commons', Science Commons, http://sciencecommons.org/wp-content/uploads/ScienceCommons_Concept_Paper.pdf/ accessed 29 March 2009

Chapter 3

The Diversity of Principles Underlying the Concept of Benefit Sharing[1]

Bram De Jonge and Niels Louwaars

Introduction

Benefit sharing is an international policy concept that originated in the 1970s with the aim of regulating the distribution of certain resources and the benefits derived from their use (De Jonge and Korthals, 2006). Benefit sharing in relation to GRs was first included in international law by the CBD in 1992. It features as one of three objectives of the Convention, alongside the conservation and the sustainable use of biological diversity. The CBD introduced the concept of national sovereignty over genetic resources as a means to regulate access to these resources, a prerequisite for creating a legal basis for benefit sharing (Andersen, 2007). The CBD has been ratified by almost all countries, but few countries have shared substantial benefits through CBD-based regulations.

So far, most studies on benefit sharing have focused on practical problems or opportunities, or evaluate operational benefit-sharing policies. We, however, intend to analyse why benefit sharing appears to be such a complex issue and why expectations are so rarely met in respect of plant genetic resources by analysing different approaches to the issue.

Methodology

The main data have been derived from 77 semi-structured interviews between March 2007 and July 2008 with experts and stakeholders in Kenya, Peru and the Netherlands, and some international organizations. The three countries were selected because they represent three major geopolitical cooperation organizations – the African Union, the Andean Community and the European Union – with their respective views and interests in genetic resources. Stakeholders included representatives from government and public organizations, the scientific community, industry and civil society. The international organizations included the UN Food and Agriculture Organization (FAO), the Global Crop Diversity Trust and the Consultative Group on International Agricultural Research (CGIAR). We focused on the concept of benefit sharing that underlies the views of the interviewees. The data collection was supported by meetings and workshops (CBD and its Ad Hoc Open-Ended Working Group on Access and Benefit-Sharing, plus two international workshops on access and benefit sharing (ABS) in Germany and India) and a survey of the literature, used primarily to contextualize the verbal reports and for referencing purposes.

Basic motivations, mechanisms and outcomes

The study resulted in the identification of six fundamentally different approaches to the issue of benefit sharing in the field of plant genetic resources. These represent six distinct strains of argumentation or reasoning in which the concept of benefit sharing is embedded, based on the following perceptions or motivations:

1 The South–North imbalance in resource allocation and exploitation
2 The need to conserve biodiversity
3 Biopiracy and the imbalance in intellectual property rights
4 A shared interest in food security
5 An imbalance between intellectual property (IP) protection and the public interest
6 Protecting the cultural identity of traditional communities

In order to detail the arguments informing the different approaches, the six sections that follow describe the basic motivations, established mechanisms and intended outcomes of each of the different interpretations of benefit sharing. This forms the basis for a reflection of the major differences and a discussion of consequences.

The South–North imbalance in resource allocation and exploitation

One major justification for benefit sharing can be described in terms of the transfer of plant genetic resources from the biodiversity-rich South to the North, where economic benefits are obtained through research and commercialization of seeds, medicines or chemical products. On the economic importance of plant genetic material and the global division of benefits through collection practices, Kloppenburg (2004, p169) concludes: 'It is no exaggeration to say that the plant genetic recourses received as free goods from the Third World have been worth untold *billions* of dollars to the advanced capitalist nations'. He is therefore of the opinion that 'It is highly ironic that the Third World resource that the developed nations have, arguably, extracted for the longest time, derived the greatest benefits from and still depend upon the most is one for which no compensation is paid' (Kloppenburg, 2004, p153).

Mechanism: National sovereignty over plant genetic resources
In the 1980s, global resistance against the free use of germ plasm originating from developing countries arose following the expanding biotechnology industry and the merger of traditional seed companies into major industrial conglomerates. This stirred an appreciation in developing countries of the value of their plant genetic resources. As a result, the CBD abandoned the assumption of plant genetic resources as a common heritage, declaring instead that states have sovereign rights over their own biological resources (UN, 1992, Preamble). The convention explicitly calls for the fair and equitable sharing of the benefits arising out of the utilization of genetic resources (UN, 1992, Article 1). Benefit sharing is an independent basic objective of the CBD, which can be explained in terms of a purely political economy argument: developing nations should be able to reap the benefits of their biological resources, as they can with other natural resources such as oil and minerals.

Intended outcome: Equity in international economic relations
The facts that (1) genetic resources are a natural resource of countries of the South which cannot be appropriated and traded by the country as part of natural wealth in the same way as other natural resources such as oil or minerals can, and (2) the benefits from these genetic resources are largely accrued in the gene-poor industrialized countries of the North, are important motivations for benefit sharing. Benefit sharing is thus supposed to encourage equity in international economic relations, being regarded as a compensation mechanism.

Table 3.1 *Summary of approach 1*

Basic motivation	Established mechanism	Intended outcome
The South–North imbalance in resource allocation and exploitation	National sovereignty over plant genetic resources	Equity in international economic relations

The need to conserve biodiversity

A second and related rationale underlying the concept of benefit sharing from genetic resources is the perception that investments have to be made to conserve biodiversity. In international agreements, the sharing of benefits derived from the utilization of genetic resources has always been connected to the conservation of these resources. The underlying assumptions are that: (1) genetic resources have a global importance; (2) economic and environmental developments create pressures that work against the conservation of biological diversity (including deforestation, climate change, the modernization of agriculture and the globalization of plant and animal breeding); (3) countries where these pressures are most severe have least financial opportunities to counter them; and (4) benefit sharing on the use of genetic resources can provide a sustainable source of funds, knowledge and technology to conserve biological diversity.

Contracting countries to the CBD have the obligation to conserve the biological diversity in their territory and they have the opportunity to share benefits. There is, however, no explicit link between the two. The FAO International Treaty on Plant Genetic resources for Food and Agriculture (ITPGR) is more specific in regard to the link between benefit sharing and conservation:

> *The Contracting Parties agree that benefits arising from the use of plant genetic resources for food and agriculture that are shared under the Multilateral System should flow primarily, directly and indirectly, to farmers in all countries, especially in developing countries and countries with economies in transition, who conserve and sustainably utilize plant genetic resources for food and agriculture (FAO, 2001, Article 13.3).*

Mechanism: Benefits to support conservation efforts
The basic idea of the CBD is that it promotes the conservation and sustainable use of biodiversity by, on the one hand, creating incentives (i.e. the promise of benefit sharing) for developing countries to protect their potentially valuable plant genetic resources and, on the other hand, assisting them in gaining access to the means for conservation by promoting the flow of technology, information and financial resources (i.e. the content of benefit sharing).

Critics, however, claim that benefits derived from systematic bioprospecting contracts may actually make it *less* necessary to conserve the resource. The chances of finding new genetic material after an ecosystem has been systematically screened are smaller than before the bioprospecting mission. Similarly, when all genetic diversity within a crop has been sampled and stored in a gene bank, less emphasis may be put on on-farm management of diversity. However, forward-looking governments will continue to conserve biodiversity and promote its continued evolution in situ for future generations to sample and research with new technologies and for new purposes.

Whether the funding strategy of the ITPGR will be able to generate enough funds to sustainably conserve crop genetic resources remains to be seen (Visser et al, 2005). The Global Crop Diversity Trust, which can be considered a supporting component of the treaty, is collecting significant amounts for the *ex situ* component of the conservation strategy. The objective of this trust is to be able to support the most relevant collections in order to keep them eternally available. The treaty is very specific that non-monetary benefits also may significantly contribute to the goals of conservation and the sustainable use of crop genetic diversity.

Intended outcome: Conservation and the sustainable use of plant genetic resources
Both the CBD and ITPGR have clear objectives that aim to support conservation and the sustainable use of plant genetic resources. Benefit sharing may provide the incentives and tools to conserve biodiversity. Large programmes such as the National Biodiversity Institute (INBio) in Costa Rica create a significant capacity for nature conservation and diversity-related research. Linking these developments to ecotourism seems to provide an effective, longer term financial capability to maintain the relevant forest reserves.

Biopiracy and the imbalance in intellectual property rights

A third interpretation or context in which discussions on benefit sharing take place concerns an asymmetry in allocations of intellectual property rights (IPRs) to and over plant genetic resources and related knowledge, and the subsequent acts or accusations of biopiracy. This asymmetry originated during the course of the 20th century. Industrialized countries started to expand their IP systems to include new plant varieties and genetic

Table 3.2 *Summary of approach 2*

Basic motivation	Established mechanism	Intended outcome
The need to conserve biodiversity	Benefits to support conservation efforts	Conservation and sustainable use of plant genetic resources

material starting with sui generis protection systems first in the USA (Plant Patent Act, 1930), followed by plant breeders' rights systems in a number of European countries (harmonized under the Convention on the Protection of New Varieties of Plants, 1961) as a means to encourage the development of new plant varieties and the international seed trade. From the 1980s onwards, it became possible in a growing number of countries to obtain patent protection on living organisms and components of heredity and on the methods and tools to manipulate these. The Agreement on Trade Related Aspects of Intellectual Property Rights (TRIPS) of the World Trade Organization (WTO, 1994) made the IPRs concepts of the industrialized countries a global obligation.

Patents protect inventions that satisfy the criteria of novelty, inventive step and industrial applicability. TK and plants that have been used and developed by local communities over centuries are thus not patentable, but newly derived or purified products developed in industry are. Such TK tends to be of a collective nature not easily attributed to an individual IP holder (Koopman, 2005). This asymmetry in allocations of IPRs was the basis of the concept of biopiracy, described as 'the appropriation of the knowledge and genetic resources of farming and indigenous communities by individuals or institutions who seek exclusive monopoly control (patents or intellectual property) over these resources and knowledge' (ETC Group, website). Examples and charges of biopiracy are currently central to many debates on benefit sharing.

This argument focuses on individuals and communities rather than on nations. Benefit sharing is based here on inalienable rights that communities have on their resources. These should be on a par with the IPRs that inventors in the scientific community have.

Mechanism: Countervailing rights systems and user measures

The rights of indigenous communities over their genetic resources are difficult to capture in legal terms. Debates within the CBD over suitable concepts for and interpretations of its Article 8(j) have been ongoing for many years. Problems may include aspects of democracy (Why would certain groups in the country have more rights than others?), demarcation (Does a person in the city still belong to an indigenous community and is that person allowed to share in benefits?) and representation (Who can negotiate on behalf of the community?).

In the meantime, evermore examples of (alleged) biopiracy appear. Communities that give access to certain resources hardly ever receive in return a share in IPRs on the products developed out of these resources (Hayden, 2007; Visser et al, 2005). Calls have been made for the establishment of indigenous and collective sui generis IPRs systems. One such system is Traditional Resources Rights (TRR), which aims to protect both the tangible and intangible qualities of such resources as germ plasm, knowledge and folklore, and even landscapes, through a bundle of rights taken from a variety of international agreements. IPRs are only one aspect of TRR, because 'Property for indigenous peoples frequently has intangible, spiritual manifestations and, although worthy of protection, can belong to no human being. Privatization or commoditization of their resources is not only foreign but incomprehensible or even unthinkable' (Posey and Dutfield, 1996, p95). This is opposite to approaches that aim at maximizing benefits through the use of strong IPRs (Herold, 2003).

At the international level, discussions related to IPRs for indigenous communities continue at the Intergovernmental Committee (IGC) on Genetic Resources, Traditional Knowledge and Folklore under the auspices of the World Intellectual Property Organization (WIPO). Meanwhile, the ITPGR has produced an agreed formulation of the concept of farmers' rights as rights arising from the enormous contribution that the local and indigenous communities and farmers of all regions of the world, particularly those in the centres of origin and crop diversity, have made and will continue to make for the conservation and development of plant genetic resources which constitute the basis of food and agriculture production throughout the world (FAO, 2001, Article 9.1). Benefit sharing by farmers is enshrined as a Farmers' Right: Article 9.3 seeks to balance different IPRs that may rest on seeds, providing a positive right instead of the relatively weak farmers' privilege provided in IPRs laws.

Another proposal to link benefit sharing with IPRs is disclosure of origin, source or legal provenance of the genetic resources and their associated knowledge by patent applicants (Barber et al, 2003; Tobin, 1997). This could be the basis for a subsequent requirement to provide evidence of prior informed consent and mutually agreed terms (including benefit sharing). This would shift the burden of proof from the weak shoulders of indigenous and farming communities to the stronger shoulders of industrial companies and research centres.

Intended outcome: Equity in legal rights over plant genetic resources
Different methods are described that try to counterbalance the perceived asymmetry in allocations of IPRs in order to stop biopiracy. These examples form a central part of many discussions on benefit sharing and are primarily concerned with a fight for recognition for the knowledge and resources that farmers and indigenous and local communities have managed, conserved and developed throughout centuries.

Table 3.3 *Summary of approach 3*

Basic motivation	Established mechanism	Intended outcome
Biopiracy and the imbalance in intellectual property rights	Countervailing rights systems and user measures	Equity in legal rights over plant genetic resources and related knowledge

Shared interest in food security

A fourth underlying objective of benefit sharing is related to the agricultural sector. The genetic resources for food and agriculture have been distributed around the world for millennia as a common heritage of mankind, as formally recognized by the International Undertaking on Plant Genetic Resources for Food and Agriculture, a non-binding agreement under the FAO (FAO, 1983). This idea was strengthened by the observation that all countries are interdependent with regard to their agricultural plant germ plasm (Flores-Palacios, 1997). While, for example, Latin America has given the world, amongst others, the potato, tomato, cacao and maize, it has received rice and soybean from east Asia, wheat from west Asia and coffee from Africa. Humans have selected and bred crops since the advent of agriculture. This has literally changed the food we eat. Because of population growth and the continuous threat of diseases, pests and environmental stresses, plant breeding is a never-ending challenge. The conservation and exchange of the building blocks of further crop improvement is thus considered essential for global food security.

It is in this context that the agricultural sector, including the seed industry (International Seed Federation, 2007), is in general critical of the CBD and its bilateral model of access and benefit sharing. Most nation states focus primarily on the protection of their plant genetic resources, creating barriers for exchange and increasing transaction costs. As a result, the number of new collection missions (Falcon et al, 2002) and the international transfer of plant genetic resources (Fowler et al, 2001) have declined dramatically since the ratification of the CBD. A decreasing exchange of genetic resources may seriously endanger food security in the long run.

Mechanism: Facilitated access and exchange of plant genetic resources
The specific characteristics of genetic resources for food and agriculture were recognized by the CBD (Stannard et al, 2004), but it was not until 2001 that new international rules were agreed upon to manage access and benefit sharing for plant genetic resources for food and agriculture. In that year, the ITPGR introduced a multilateral system of ABS that, while in harmony with the CBD, was better suited to the specific nature of the

agricultural sector. The multilateral system introduced a standard material transfer agreement for exchange of germ plasm of major crops and forages that is under the control of the signatory governments, thus avoiding the need for further negotiations and reducing transaction costs. It includes a multilateral benefit-sharing fund in which payment is liable when a commercial product is developed using resources from the multilateral system and the genetic resources of that product are not available under the same conditions; for example, if it is patented or bound by technical or other legal restrictions. Other benefit-sharing mechanisms are included, such as facilitating the exchange of information, access to and transfer of technology and capacity building (FAO, 2001, Article 13), in particular to help small farmers in developing countries. Furthermore, it is stated that the facilitated access to the plant genetic resources of the multilateral system constitutes itself a major benefit (FAO, 2001, Article 13). This reflects the idea of a common interest in food security as a basic rationale behind the system.

Intended outcome: Food security and sustainable agriculture
The objectives of the ITPGR are the conservation and sustainable use of plant genetic resources for food and agriculture, and the fair and equitable sharing of the benefits arising out of their use, in harmony with the CBD, for sustainable agriculture and food security (FAO, 2001, Article 1). The multilateral system of access and benefit sharing that the treaty introduces supports this aim by facilitating the free exchange of plant genetic resources for food and agriculture, and by stimulating the provision of the means for sustainable agriculture, especially to smallholder farmers in developing countries. These methods are not so much based on the rights of farmers as on a common concern for sustainable agriculture and food security.

An imbalance between intellectual property protection and the public interest

A fifth rationale underlying benefit sharing has to do with concerns about the rise of IPRs in the field of plant genetics and its effects on the public domain. The general worry is that current intellectual property legislation may block the equitable sharing of benefits of modern research and development within society.

Table 3.4 *Summary of approach 4*

Basic justification	Established mechanism	Intended outcome
A shared interest in food security	Facilitated access and exchange of PGRFA	Food security and sustainable agriculture

According to the WIPO, patents create incentives for innovation as inventors obtain recognition and commercial protection for their inventions. This then contributes to the continuing enhancement of the quality of human life (WIPO, 2005, p5). It is on this latter aspect that the present IP system is criticized; not all sections of society benefit from the research and development (R&D) thus promoted. Most research within the field of biotechnology has focused on commercial crops and large-scale production systems, and no serious investments have been made in the most important crops and smallholder systems in developing countries, creating a new divide between the industrialized and the developing countries (Fresco, 2003).

Another point of criticism of intellectual property protection is that since genetic material, knowledge and technologies can be protected, R&D in this field finds itself in an anticommons trap. The tragedy of the anticommons (Heller and Eisenberg, 1998) is a scenario in which too many entities have exclusive rights to a given resource, which makes the resource prone to under-use. Innovation can be blocked because it becomes too costly for innovators to obtain access to all the technology they need and concentration of the biotechnology-based industries is the result. This global concentration of power in the new life-science industry has created public concern that 'a small, authoritarian minority is now dictating what kinds of research are permissible and which technologies and products should be available in the marketplace' (Kloppenburg, 2004, p314). The fact that the products involved are basic needs for human life has only increased public unease on this issue.

Mechanism: Stimulating technology transfer and knowledge sharing
Recently, several initiatives have been developed that try to correct the intellectual property–public interest imbalance and focus on ways to share the benefits of modern R&D more equitably. Worth mentioning in this respect are:

- the open source movement in biotechnology, as represented by the Center for the Application of Molecular Biology to International Agriculture (CAMBIA), which emphasizes new collaboration and licensing tools to maximize the freedom to operate on biotechnologies, and thereby to empower both public and private sectors to develop health and agricultural products and processes of real relevance to all sectors of society (CAMBIA, website)
- the Public Intellectual Property Resource for Agriculture (PIPRA), which aims to improve agriculture by decreasing intellectual property barriers and increasing technology transfer (PIPRA, website)

- the employment of Humanitarian Use Licences, in which the rights holder allows the use of the technology for specific uses in development – for example, the license negotiated by Syngenta on Golden Rice that provides free access to the technology for resource-poor farmers; or the more far-reaching licensing arrangement agreed by the partners in the Generation Challenge Program that provides access to all technologies in the programme for research and use by the poor (Barry and Louwaars, 2005)
- public–private partnerships (PPPs), which are playing an increasingly important role in the fight against neglected diseases in developing countries especially – according to a Wellcome Trust report, pharmaceutical companies that had moved away from unprofitable research on neglected diseases are now returning to this area on a no-profit–no-loss basis (Moran et al, 2005), a success that warrants further research on the application of PPPs in the similarly neglected field of orphan crops.

These initiatives are not directly related to the exchange of plant genetic resources and therefore are not often referred to in the literature on benefit sharing. They are, however, aimed at sharing the benefits of modern R&D more equitably by stimulating technology transfer and knowledge sharing. This is exactly what the existing models of benefit sharing in the CBD and ITPGR aim to promote under the heading of non-monetary benefit sharing, which is largely regarded an important aspect of any benefit-sharing policy (Byström et al, 1999; Raymond and Fowler, 2001), but which implementation has proven rather difficult so far (Visser et al, 2005). One reason is that governments have to rely on various stakeholders in developing, financing and implementing such non-monetary benefit sharing. Linkages with the above mechanisms may therefore be very productive.

Intended outcome: Equity in distributing the benefits of research and development
The initiatives described above try to correct the imbalance between IP protection and the public interest by stimulating technology transfer and knowledge sharing. In so doing, they hope to re-establish an open and

Table 3.5 *Summary of approach 5*

Basic motivation	Established mechanism	Intended outcome
An imbalance between IP protection and the public interest	Stimulating technology transfer and knowledge sharing	Equity in distributing the benefits of research and development

stimulating environment for innovation and development, for the benefit of those in need. In reaction to the increasing enclosure and concentration of resources, alternative arrangements and partnerships are created to make available the necessary means for innovation and development in and for developing countries.

Protecting the cultural identity of traditional communities in a globalizing world

A sixth perspective on benefit sharing in the present debate is concerned with the cultural identities of traditional communities in today's globalizing world. This motivation is often linked to the arguments on the imbalance in intellectual property rights and the fight against biopiracy. The differences are, however, substantial. The major concern of approach 3 relates to the growing influence of IPRs used by the formal research sector and the means by which small farmers and traditional communities can protect themselves against biopiracy. In the opinion of many traditional communities, the concepts and regulations on access and benefit sharing are in themselves already a form of globalization, one that encroaches on their traditional lifestyles and cultures. Instead of reacting to these foreign pressures, and in that act adapting to them, perspectives and initiatives that focus on the cultural identity of traditional communities are prioritized. The starting point here is the world view of the traditional communities themselves and an articulation of what they think benefit sharing should be about.

This is reflected in, for example, different viewpoints on the concept of biopiracy. For many, ABS contracts are tools to stop biopiracy, to which end the CBD established its legal framework for bioprospecting. The indigenous non-governmental organization Association for Nature and Sustainable Development (ANDES) in Peru has a radically different view:

> *Contractual benefit sharing is like waking up in the middle of the night to find your house being robbed. On the way out the door, the thieves tell you not to worry because they promise to give you a share of whatever profit they make selling what used to belong to you. (Coalition Against Biopiracy, website).*

Another clarification of the problem at hand comes from Jack Beetson, an Aboriginal activist. He warns that traditional ways of life can be destroyed in the very effort of protecting them. Inviting indigenous communities to an international conference, to put on a suit and negotiate their interests in English, straight away asks them to abandon their traditional way of life. When talking about capacity building in this context, Beetson wonders whether this should be aimed at indigenous communities or whether the negotiators from governments, industry and other institutions should not instead build their own capacity – their capacity to go to the communities themselves, sit with them and discuss the issues in their language.[2]

Mechanism: Recognition for customary laws in ABS regimes
The concerns about the cultural identity of traditional communities have led to specific ideas about what benefit sharing should be about or how it should be incorporated in international and national legislation. Brendan Tobin, of the Association for the Defence of Natural Rights (ADN), argues that the customary law and traditional tenure rights that govern land and natural resources in many parts of the world are often undermined by culturally insensitive national legislation. This leads to the erosion of traditional authority and social structures within communities. He promotes a wider and more expansive view of the nature, role and values of TK and its relationship to traditional resource management systems in the ABS debate (Tobin, 2004), with the ultimate aim of ensuring the effective recognition, respect and enforcement of customary law in any international regime on ABS (Tobin, 2004, p1).

Argumedo (ANDES), who has been closely involved in setting up the Potato Park in the Peruvian Andes, speaks in this respect of reversing the ABS regime. Reversing means to put the interests and customary laws of the indigenous farmers as central by: (1) aiming to return to the local communities the plant varieties and associated knowledge once taken from them; (2) ensuring that the genetic resources and knowledge remain under their custody and do not become subject to IPRs in any form; and (3) recognizing the ability of the Andean farmers to conserve and develop the genetic resources for the benefit of their people and all mankind. This strategy is implemented in the Potato Park, where potato diversity is managed by six Quechua communities according to customary laws, including collective land tenure, community registers and resource management (Argumedo and Pimbert, 2005).

Intended outcome: Preserving and restoring traditional communities and their cultures
The ultimate goal of putting the cultural identity first is the restoration and preservation of the rights and traditions of the indigenous communities by reformulating ABS legislation according to their own world views and reconciling them with their own customary laws.

Table 3.6 *Summary of approach 6*

Basic motivation	Established mechanism	Intended outcome
Protecting the cultural identity of traditional communities	Recognition for customary laws in ABS regimes	Preserving and restoring traditional communities and their cultures

Reflection

Six interpretations and contexts in which the discussions on benefit sharing take place have been described. The different motivations are all valid, but this variety of approaches to what benefit sharing actually is and what it ought to accomplish implies a range of different implementation mechanisms of benefit sharing that in turn lead to widely different outcomes. If these outcomes were to point in roughly the same direction, it would be relatively easy to combine them in a common policy. We identify, however, significant friction between the different approaches in terms of their implementation and a complex situation in analysing stakeholder views and positions. This seems to explain the complexity of the current debates on benefit sharing and their general lack of productive outcomes.

Different approaches leading to a joint policy?

Expectations
The six different motivations and intended outcomes may be all valuable in their own right, but they do appear rather incompatible. The controversies about economic inequalities between North and South, for example, would need to be followed by very significant levels of benefit sharing before one could speak of equity (approach 1). This logically leads to the rejection of any system that would provide for benefits satisfying (only) conservation needs (approach 2). Differences in expectations regarding the magnitude of benefits at best blur the debate and may lead to an impasse.

Rights
Debates focusing on rights may be held at different levels: while the CBD primarily operates at the level of nation states, the issue of rights to and over genetic resources may give rise to debate at the sub-state level of communities that claim to have developed or be custodian to the genetic resources (under Article 8(j)), as well as at the private level (company, individual) based on IPRs (approaches 1, 3 and 5). When claims of right are made over

the same resources at different levels, tension or outright conflict ensues. Attempts to balance such rights may lead either to increasing total levels of rights (approach 3) or attempts to jointly reduce them (approach 5). Increasing the control level of genetic resource rights and rights to/over TK to bring them on a par with IPRs leads to hyperownership (Safrin, 2004) by those who have or can (afford to) buy control of the genetic resources. Such impacts are unlikely to be compatible with facilitated access to genetic resources for food security (approach 4) or with the protection of the public domain (approach 5), or the recognition of customary laws (approach 6). It is also far from clear how the scenario leading to hyperownership would stimulate conservation (approach 2).

Two meta-approaches

There is a basic division between the three approaches (1, 3 and 5) that are driven by the perception of imbalance and a motivation to increase equity (albeit in different ways, at different levels and to different ends), and the other three (2, 4 and 6), which concentrate on other aims, primarily nature conservation, food security and the preservation of traditional cultures. This division may to some extent be attributed to prevalence of legal specialists in the former and science-based policies in the latter. The intended outcomes of these two meta-approaches are fundamentally different, making coherent policies on the basis of a combination of these different ways of addressing the subject extremely difficult.

Different mechanisms

The mechanisms by which the various objectives are to be reached are fundamentally different and sometimes contradictory. The bilateral contract model that follows from the principle of national sovereignty may also be used at the level of community rights. This cannot be expected to lead to equity, unless the conditions enjoyed by the negotiating parties are themselves equitable; that is, when the suppliers and users have equivalent negotiating capabilities, information bases and financial resources with which to engage in conflict resolution. Such contract-based approaches are even more difficult, however, in the systems associated with purposes other than equity. Food security and conservation goals cannot be easily captured in contracts between two parties – the multilateral system is contract-based but in a standardized form. Similarly, incompatibilities can be observed between mechanisms that aim primarily at monetary benefit-sharing and others that explicitly value non-monetary benefits, notably approach 5, which concentrates on technology transfer, approach 6, which aims to protect the cultural identity of traditional communities, and approach 4, which identifies access to genetic resources as an important benefit in its own right.

Additional pressures: Another approach

New challenges are continually arising. Industry, for example, is following the international ABS negotiations with some concern, fearful of the negative consequences these may have on business. Companies do not oppose ABS measures in principle (American BioIndustry Alliance, 2008; Biotechnology Industry Organization, 2007), but they are worried about the lack of clarity and precision in the current regulations (IP-Watch, 2006) and about the possible introduction of inefficient regulations, such as disclosure measures. Industry argues that effective and competitive trade regulations, including strong IPRs, are needed in order to produce the benefits to be shared. In general terms, the argument is that if industry is flourishing everybody will gain, whether through ABS contracts or direct economic growth.

Industry thus prefers simple and liberal access regulations to secure the easy availability of resources for its businesses. This thus represents a new motivation in the current debates on ABS, one which reacts to the other motivations for benefit sharing described above and in obvious contradiction with some of them. With this reaction by business being based in part on the uncertainties resulting from the problematic reconciliation of other motivations and mechanisms, the variety of approaches are, somewhat ironically, causative of yet another approach: rather than leading to joint policy, the surfeit of approaches is working to complicate matters still further.

Stakeholder analysis

Linking each motivation to a particular stakeholder group would lead to a straightforward analysis of stakeholder interests and positions, which would facilitate the search for solutions to the contradictory mechanisms for ABS and intended outcomes. Unfortunately, however, such a one-to-one correspondence does not reflect reality. Stakeholders appear to pursue a mix of different aims and objectives in the debates on ABS:

- Indigenous communities seem to be mainly concerned with biopiracy issues and their rights over their genetic resources and associated knowledge (approach 3). Within this, however, some communities do not oppose the use of strong IPRs, while others fear that IPRs and ABS regulations threaten their cultural identity (approach 6). Also, most communities are highly interested in the conservation of biodiversity (approach 2) and issues of food security (approach 4).
- The FAO is primarily concerned about food security (approach 4), but it also has a stake in conservation (approach 2) and aims to stimulate

technology transfer (approach 5). In addition it has taken some measures to support traditional farmers and the preservation of their traditional cultures (approach 6).

- Most governments of developing countries focus first on their national sovereignty over plant genetic resources (approach 1). However, they differ widely in their attitudes towards the rights of indigenous communities (approaches 3 and 6), conservation issues (approach 2), IPRs (approach 5) and food security policies (approach 4).

- Industry is primarily interested in liberal ABS regulations that create enabling conditions for biotechnology. Those industries with a close dependence on genetic resources are also concerned with conservation (approach 2) and exert a certain level of social responsibility towards global food security (approach 4). Different attitudes exist towards the role of industry to overcome fears of intentional misappropriation (IP-Watch, 2006) (approach 3). The acceptable level of complexity of regulations related to this depends heavily on the size of the corporations involved, providing the smaller corporations with the largest challenges. The size of corporations may also have an effect on their attitudes towards IP protection policies (approach 5).

The fact that every stakeholder seems to have a mix of objectives and motivations with respect to benefit sharing is likely to further complicate any possibilities of reaching consensus in the international negotiations on this matter.

Ways forward?

Worldwide standstill

The range of different, not infrequently opposing, conceptions about what benefit sharing is and what it intended to achieve, together with the fact that the stakeholders involved pursue different combinations of motivations, has resulted in a slow advancement of the ABS negotiations, both at national and international levels. At the national level, governments have generally failed to arrive at implementing effective ABS regulations. ABS contracts signed are few in number and the efforts to reach such agreements often lengthy and costly. The negotiations for an International Regime on Access and Benefit-Sharing within the framework of the CBD to overcome the deadlock suffer from the same basic problem despite the promise to come to an agreement before the tenth meeting of the Conference of the Parties in 2010.

Searching for ways out

The search for common ground is essential for moving the international ABS negotiations forward. In this process, it is important that the different motivations and expectations of the stakeholders involved are clear. We have tried to show that there are several different understandings of what benefit sharing is and what it should contribute to. As long as these differences remain implicit, unstated, no joint outcome of negotiations is to be expected. The overview presented in this chapter can be a tool to map the different interpretations of benefit sharing and reflect upon the major contradictions involved.

It appears that an appropriate clarity and agreement (or balance) needs to be sought between the different motivations for benefit sharing and their intended outcomes if workable mechanisms are to be designed. Alternatively, different ABS mechanisms may need to be pursued for different types or uses of genetic resources in different contexts (such as the ITPGR). Other important initiatives are needed for balancing and connecting rights regimes to/over genetic resources and knowledge at the individual, community and national levels, with the aim to avoid the tragedy of the anticommons and a scenario of hyperownership.

A remarkable aspect of some initiatives (e.g. open source, potato park) is that they have been initiated by groups in society, without the help of the governments that negotiate the International Regime on Access and Benefit-Sharing. It is therefore imperative that governments take good notice of the benefit-sharing initiatives that are already being undertaken by different stakeholders in society and search for ways to support and facilitate them. The ABS debate includes a variety of perspectives emanating from different groups in society, and all these groups and their contributions will be needed if we are to arrive at an effective, workable system of access and benefit sharing in the future.

Notes

1   This article is the result of a research project of the Centre for Society and Genomics in The Netherlands, funded by the Netherlands Genomics Initiative; and the Netherlands Organisation for Scientific Research.
2   Presentation at the International Conference on Access and Benefit Sharing for Genetic Resources (New Delhi, India, 6–7 March 2008). See website: http://www.ris.org.in/icgr.htm/ accessed in 2008.

References

American BioIndustry Alliance (2008) *ABS Negotiation Principles*, http://www.abial-liance.com/index.html/ accessed in 2008

Andersen, R. (2007) *Governing Agrobiodiversity: International Regimes, Plant Genetics and Developing Countries*, PhD thesis, Department of Political Science at the Faculty of Social Sciences, Oslo, University of Oslo

Argumedo, A. and Pimbert, M. (2005) *Traditional Resource Rights and Indigenous People in the Andes*, London, International Institute for Environment and Development

Barber, C. F., Johnston, S. and Tobin, B. (2003) *User Measures: Options for Developing Measures in User Countries to Implement the Access and Benefit-Sharing Provisions of the Convention on Biological Diversity*, second edn, Tokyo, United Nations University Institute of Advanced Studies

Barry, G. and Louwaars, N. P. (2005) 'Humanitarian licences: Making proprietary technology work for the poor', in Louwaars, N. P. (ed) *Genetic Resource Policies and the Generation Challenge Programme*, Mexico D. F., Generation Challenge Program, pp23–34

Biotechnology Industry Organization (2007) *Options on the Substantive Agenda Items for the Sixth Meeting of the Working Group on Access and Benefit-Sharing*, http://www.bio.org/ip/international/20071130.pdf/ accessed in 2008

Byström, M., Einarsson, P. and Nycander, G. A. (1999) *Fair and Equitable: Sharing the Benefits from Use of Genetic Resources and Traditional Knowledge*, Uppsala, Tagstalund and Bjorkeryd, Swedish Scientific Council on Biological Diversity

CAMBIA (date unknown) 'Biological innovation for open society', http://www.bios.net/daisy/bios/g2/2442.html/ accessed in 2008

Coalition Against Biopiracy (date unknown) 'Biopiracy', http://www.captain-hookawards.org/biopiracy/ accessed in 2008

De Jonge, B. and Korthals, M. (2006) 'Vicissitudes of benefit sharing of crop genetic resources: Downstream and upstream', *Developing World Bioethics*, vol 6, no 3, pp144–157

ETC Group (date unknown) 'Biopiracy', http://www.etcgroup.org/en/issues/biopiracy.html/ accessed in 2008

Falcon, W. P. and Fowler, C. (2002) 'Carving up the commons – Emergence of a new international regime for germplasm development and transfer', *Food Policy*, vol 27, pp197–222

FAO (1983) 'International undertaking on plant genetic resources', Rome, FAO, ftp://ext-ftp.fao.org/ag/cgrfa/iu/iutextE.pdf/ accessed in 2008

FAO (2001) 'International Treaty on Plant Genetic Resources for Food and Agriculture', Rome, FAO, ftp://ext-ftp.fao.org/ag/cgrfa/it/ITPGRe.pdf/ accessed in 2008

Flores-Palacios, X. (1997) 'Contribution to the estimation of countries' interdependence in the area of plant genetic resources', Rome, FAO Commission on Plant Genetic Resources for Food and Agriculture, Background Study Paper No. 7

Fowler, C., Smale, M. and Gaiji, S. (2001) 'Unequal exchange? Recent transfers of agricultural resources', *Development Policy Review*, vol 19, no 2, pp181–204

Fresco, L. (2003) 'A new social contract on biotechnology', *Agriculture*, vol 21, http://www.fao.org/ag/magazine/0305sp1.htm/ accessed in 2008

Hayden, C. (2007) 'Taking as giving: bioscience, exchange, and the politics of benefit-sharing', *Social Studies of Science*, vol 37, no 5, pp729–758

Heller, M. A. and Eisenberg, R. S. (1998) 'Can patents deter innovation? The anticommons in biomedical research', *Science*, vol 280 (May), pp698–701

Herold, B. (2003) 'Fair and equitable benefit-sharing within the international treaty on plant genetic resources for food and agriculture: The view of the Berne Declaration', http://www.syngentafoundation.com/pdf/Contribution_Bernhard_Herold.pdf/ accessed in 2008

International Seed Federation (2007) 'Plant genetic resources for food and agriculture', Christchurch, International Seed Federation position paper, http://www.world-seed.org/Position_papers/PGRFA.htm/ accessed in 2008

IP-Watch (2006) 'Industry works to allay concerns on patenting of genetic resources', http://www.ip-watch.org/weblog/index.php?p=486&res=1280&print=0/ accessed in 2008

Kloppenburg, J. R. (2004) *First The Seed: The Political Economy of Plant Biotechnology*, Madison, University of Wisconsin Press

Koopman, J. (2005) 'Reconciliation of proprietary interests in genetic and knowledge resources: Hurry cautiously', *Journal of Ecological Economics*, vol 53, no 4, pp523–541

Moran, M. et al (2005) *The New Landscape of Neglected Disease Drug Development*, London, The Wellcome Trust, http://www.wellcome.ac.uk/assets/wtx026592.pdf/ accessed in 2008

PIPRA 'The public intellectual property resource for agriculture', http://www.pipra.org/en/about.en.html/ accessed in 2008

Posey, D. A. and Dutfield, G. (1996) *Beyond Intellectual Property Rights: Towards Traditional Resource Rights for Indigenous and Local Communities*, Ottawa, IDRC

Raymond, R. and Fowler, C. (2001) 'Sharing the Non-monetary benefits of agricultural biodiversity', *Issues in Genetic Resources*, no 5, Rome, IPGRI

Safrin, S. (2004) 'Hyperownership in a time of biotechnological promise: The international conflict to control the building blocks of life', *American Journal of International Law*, vol 98, pp641–685

Stannard, C. et al (2004) 'Agricultural biological diversity for food security: Shaping international initiatives to help agriculture and the environment', *Howard Law Journal*, vol 48, no 1, pp397–430

Tobin, B. (1997) 'Certificates of origin: A role for IPR regimes in securing prior informed consent', in Mugabe, J. et al (eds) *Access to Genetic Resources: Strategies for Sharing Benefits*, Nairobi, ACTS Press

Tobin, B. (2004) 'Customary law as the basis for prior informed consent of local and indigenous communities', Draft paper presented at the International Expert Workshop on ABS, Mexico, October 2004, http://www.ias.unu.edu/sub_page.aspx?catID=67&ddlID=69/ accessed in 2008

UN (1992) 'Convention on Biological Diversity', Rio de Janeiro, UN, http://www.biodiv.org/doc/legal/cbd-un-en.pdf/ accessed in 2008

Visser, B et al (2005) 'Options for non-monetary benefit-sharing: an inventory', Rome, FAO background study paper 30

WIPO (2005) 'Understanding industrial property', Geneva, WIPO publication no 895, http://www.wipo.int/freepublications/en/intproperty/895/wipo_pub_895.pdf/ accessed in 2008

WTO (1994) 'Agreement on Trade-Related Aspects of Intellectual Property Rights', Marrakesh, WTO, http://www.wto.org/english/tratop_e/trips_e/t_agm0_e.htm/ accessed in 2008

Chapter 4

ABS in Relation to Marine GRs

Alexander Proelss

Introduction

It was not long ago that the issue of marine GRs began to receive major attention within the realm of international environmental law and the law of the sea. Lyle Glowka was the first to analyse the legal regime of marine GRs beyond the limits of national jurisdiction in his pioneering article of 1996 (Glowka, 1996).[1] Meanwhile, three books (Friedland, 2007; Leary, 2007; Salamanca Aguado, 2003), two studies (Arico and Salpin, 2005; Korn et al, 2003) and a multitude of papers dealing with the subject matter were published. The academic discussion is accompanied by an intense debate led by the competent bodies of the United Nations (UN), whose outcome is still far from clear. Delegations particularly argue whether the preservation and management of marine GRs is appropriately covered by the existing legal rules, or whether the latter should be amended by an agreement explicitly addressing the organisms concerned.

By building on previous work published by the author (Proelss, 2007), the present contribution undertakes to take a closer look on the applicable legal rules and analyse the results which have so far been achieved in the course of current UN-sponsored debates. It questions the applicability and feasibility of existing access and benefit-sharing (ABS) systems with regard to the marine GRs within, as well as beyond, the limits of national jurisdiction. Furthermore, both relevant ABS state practice (with a view to areas under national jurisdiction) and interpretations submitted by legal scholars will be examined. To this aim, a brief survey on the factual background shall be given in order to clarify the situation to which the relevant legal rules apply or ought to apply.

Factual background

The issue of marine GRs is closely linked to the discovery of hydrothermal vent sites located on the ocean floor at depths of 1,800 to 3,700 m, mainly on the mid-oceanic ridges. These sites are characterized by the ejection of superheated water saturated with minerals from the underlying magma. The fluid, which is especially rich in polymetallic sulphides, exits up to 20-m-high columnar chimney structures often referred to as 'black smokers'. Polymetallic sulphides are the primary substance needed for a process called chemosynthesis. Organisms living at hydrothermal vent sites use energy from chemical oxidation instead of light (photosynthesis) to produce organic matter from carbon dioxide (CO_2) and mineral nutrients. The organic matter is then consumed by various organisms with the help of sulphide-oxidizing bacteria which live either in symbiosis with the vent fauna or in the surrounding environment. Relevant organisms, about 90 per cent of which are endemic, that is, exclusively native to these sites, include algae, sponges, fungi, molluscs, crabs, tunicates and giant tube worms. As the faunal biomass is estimated to be 500 to 1,000 times higher than that of the surrounding deep sea, hydrothermal vent sites have accurately been described as 'oases of the abyss' (Bernhard et al, 2000, pp77–80).

Both scientific and economic interests in accessing hydrothermal vent sites are based on the diverse metabolic, physiological and taxonomic structures of the vent fauna. Scientists as well as commercial entities are keen to investigate its biotechnological potential. In particular, due to their ability to survive under extreme temperatures and in a high hydrostatic pressure and toxic environment, marine organisms living at or nearby hydrothermal vent sites are expected to become useful in the development of therapeutic agents. Recent research indicates that the organisms concerned are likely to turn out to be important raw material for purposes of genetic engineering. This explains why they are commonly addressed as 'marine genetic resources'. It should be emphasized, though, that the vast majority of marine GRs with commercial value are located in shallow waters and reefs within the territorial seas of coastal states (McLaughlin, 2003, p308).

Notwithstanding the exact location of the relevant resources, according to a recent report, 67 patents worldwide were issued between 1999 and May 2003 for novel compounds for the pharmaceutical industry using marine natural products, the majority of which serve anticancer, antibacterial and anti-inflammatory purposes (Frenz et al, 2004, pp30–31).

As a matter of logic, and taking into account the extremely negative experiences of high seas fisheries, increasing research and commercial activities at hydrothermal vent sites involve the danger that any such activity is not conducted in a sustainable way and might thus result in a serious danger to the respective vent ecosystems (see Korn et al, 2003, pp19–25). Having said that, it should be noted that hydrothermal vents have the ability to 'grow' and 'recreate' quickly due to the continuous process of ejection of mineral-saturated fluid from the underlying magma chambers. Therefore, collapse of a black smoker, being an everyday phenomenon, does not have any long-term effects on the vent fauna.

Marine GRs under the Convention on Biological Diversity and the UN Convention on the Law of the Sea

One of the central questions to be examined in the following is whether the existing ABS regime under the 1992 Convention on Biological Diversity (CBD) is applicable to marine GRs at all and, in case of affirmation, whether any States Parties to the Convention have implemented the regime into national legislation with particular view to these resources. At first glance, the fact that the Conference of the Parties (COP) to the CBD, whose main duty is to keep under review the implementation of the Convention (Article 23(4) CBD), as well as its Subsidiary Body on Scientific, Technical and Technological Advice (SBSTTA), have repeatedly dealt with marine GRs within and beyond the limits of national jurisdiction suggest a positive answer to the questions presented.

Research on and management of marine GRs is, however, not only subject to the rules on the conservation of biological diversity and sustainable use of its components (see Article 1 CBD), but also addressed by the 1982 UN Convention on the Law of the Sea (UNCLOS). Article 22(2) CBD obliges ('shall') States Parties to the Convention to 'implement [it] with respect to the marine environment consistently with the rights and obligations of States under the law of the sea'. Thus, in the case of conflict between the two agreements, UNCLOS enjoys priority over the CBD. While it is true that the wording of Article 22(2) CBD suggests that this conclusion only applies if 'the rights and obligations' of states under the law of the sea are affected (Wolfrum and Matz, 2000, pp475–476; 2003, p125), Article 311(3) UNCLOS clarifies that between the parties to the Convention, the superiority of the law of the sea also extends to the basic principles and provisions contained therein from which derogation is incompatible with its object and purpose.[2] Additionally, Article 311(2) UNCLOS in rather general terms requires existing as well as future agreements (see Nordquist et al, 1989, p243) of the States Parties to be 'compatible with this Convention'. The scope of application of Article 22(2) CBD and that of Article 311(2) UNCLOS are, therefore, not identical (see Matz, 2005, pp191–192; Friedland, 2007, pp151–152). Consequently, it seems that repealing Article 22(2) CBD would not per se alter the relationship between the CBD and UNCLOS. The relevance of the superiority argument raised here becomes manifest in a recent statement made by one source, observing that '[t]he ABS debate ignores, to a great extent, the core provisions in the 1982 UNCLOS that could be useful in this debate' (Gorina-Ysern and Jones, 2006, p224).

Marine GRs within areas under national jurisdiction

As regards the applicability of the CBD to marine GRs, reference has to be made to Article 4 CBD. According to this provision, which governs the jurisdictional scope of the Convention, the CBD applies:

- *in the case of components of biological diversity, in areas within the limits of its national jurisdiction*
- *in the case of processes and activities, regardless of where their effects occur, carried out under its jurisdiction or control, within the area of its national jurisdiction or beyond the limits of national jurisdiction.*

Thus, the ABS regime of the CBD undoubtedly applies to those parts of the ocean which fall within the ambit of the coastal state's territorial sovereignty, that is, its internal waters and territorial sea (see Article 2(1) UNCLOS).

Internal waters and territorial sea
Under the relevant rules of the law of the sea, the coastal state has the competence to prescribe and enforce its national regulations in those maritime zones, subject to the right of third states of innocent passage through the territorial sea (see Article 17 UNCLOS). By its very nature, the coastal state's sovereignty extends to the exploitation and management of both living and non-living resources. States wishing to undertake scientific research with regard to marine GRs in a foreign state's territorial sea may only do so 'with the express consent of and under the conditions set forth by the coastal State' (Article 245 UNCLOS). This perfectly corresponds with the approach of the CBD according to which 'the authority to determine access to GRs rests with the national governments and is subject to national legislation' (Article 15(1) CBD). However, if one takes a closer look at the existing relevant national legislation in terms of Article 15(1), (2) and (7) CBD, there is not, as far as can be seen, too much room for optimism. If states have enacted any legal instruments on ABS and thereby implemented Article 15(1) CBD, these acts explicitly refer to marine GRs only in isolated cases.

Whether existing general ABS legislation (e.g. Australian Environment Protection and Biodiversity Conservation Act, www.cbd.int/doc/measures/

abs/msr-abs-au-en.pdf, as amended by regulations of 2000, www.cbd.int/ doc/measures/abs/msr-abs-au5-en.pdf, and 2005, www.cbd.int/doc/measures/abs/msr-abs-au7-en.pdf; Kenyan Environmental Management and Coordination Regulations of 2006, www.cbd.int/doc/measures/abs/msr-abs-ke2-en.pdf) includes marine GRs seems to be a matter of interpretation. As a general rule, one should expect that the relevant regulations are at least applicable within the internal waters and territorial seas of the states concerned. A third category of states has only adopted general strategies or policy documents without binding force. In this respect, the National Biodiversity Strategy and Action Plans of both Micronesia (www.cbd.int /doc/measures/abs/msr-abs-mi-en.pdf) and Niue (www.cbd.int/doc/measures/abs/msr-abs-nu-en.pdf) refer specifically to certain marine and coastal areas as areas of potential biological protection. In a similar way, the National Strategy of Biodiversity Conservation in Russia (http:// www.cbd.int/doc/world/ru/ru-nbsap-01-p3-en.pdf) mentions the need to create legislation regulating the 'water areas' to 'improve nature conservation in general and biodiversity in particular' (see p27 of the Russian National Strategy).

Exclusive economic zone and continental shelf

While it is not completely clear at first sight whether Article 15(1) CBD covers marine GRs located in the exclusive economic zone (EEZ) or on the continental shelf, since these zones are not part of the coastal state's territory, an interpretation of that provision in conformity with Article 4(a) CBD and Article 56(1)(a) UNCLOS, according to which the coastal state has 'sovereign rights for the purpose of exploring and exploiting, conserving and managing natural resources, whether living or non-living, of the waters superjacent to the seabed and of the seabed and its subsoil', suggests an answer to the affirmative. As only the coastal state is competent to exercise jurisdiction over the marine GRs located in the EEZ and on the continental shelf, these zones ought to be considered as areas within the limits of its national jurisdiction in terms of Article 4(a) CBD. It should be noted that Article 15(1) CBD also does not speak of 'sovereignty', but of 'sovereign rights of States over their natural resources'. Therefore, the coastal state is entitled to expand its national ABS legislation to the EEZ and the continental shelf (see Articles 1(I), 7(I), 14(I)(a) and 16 of the Brazilian Medida Provisória No. 2.186-16, www.cbd.int/doc/measures/abs/msr-abs-br-en.pdf).

It remains to be examined whether nationals of third states have the right to conduct marine scientific research or bioprospecting in a foreign EEZ. In this respect, one must take into account that under Article 246(3) UNCLOS, which only refers to non-profit-oriented research, the coastal state's jurisdiction to regulate and authorize respective activities by other states is limited insofar as the coastal state is, under normal circumstances, obliged to grant its consent (McLaughlin, 2003, p311). This rule clearly contrasts with the rules of the CBD, where the resource state can be said to have absolute discretion to grant or refuse access to its GRs. Due to Article 22(2) CBD, the discretionary powers of the resource state under the CBD must, therefore, be interpreted in a restrictive manner.

On the other hand, if the planned research 'is of direct significance for the exploration and exploitation of natural resources, whether living or non-living' and thus commercially oriented, the coastal state may at its discretion withhold its consent according to Article 246(5) UNCLOS. While admittedly the distinction between fundamental research and applied research is not easy to draw, UNCLOS contains some provisions which aim at preventing abuses of Article 246 UNCLOS by the research state. These rules include, inter alia, the duty to provide a full description on the nature and objectives of the planned research (Article 248 UNCLOS) and the obligation to make communications concerning marine scientific research, including clearance for the activities concerned through official channels (Article 250 UNCLOS; see Gorina-Ysern and Jones, 2006, pp240ff). Thus, the coastal state is authorized to evaluate the planned research, through a series of tests on its true nature. Having said that, Article 252 UNCLOS intends to support the research state by providing an implied-consent mechanism under which a state may proceed with a marine scientific research project six months after the date upon which the information required was provided to the coastal state. In any event, the clearance process as to marine scientific research should be integrated in the future ABS debate as an important tool relating to marine GRs.

Marine GRs located in areas beyond national jurisdiction

While, according to its Article 4, the CBD does not cover the deep seabed as such, it does apply to activities which are carried out on the deep seabed by the States Parties.[3] However, with a view to the ABS regime which is particularly relevant here, Article 4 CBD must be read in conjunction with Article 15(1) CBD. This provision emphasizes the 'sovereign rights of states over their natural resources' by recognizing that 'the authority to determine access to genetic resources rests with the national governments and is subject to national legislation'. In referring to the sovereign rights of states over their natural resources, Article 15(1) CBD clarifies that the ABS regime does not cover the areas beyond the limits of national jurisdiction (Glowka, 1999, p60; König, 2008, pp153ff).[4] This interpretation does not conflict with Article 4(b) CBD, since the general rule contained in that provision only applies provided that no special rule exists ('except as otherwise expressly provided in this Convention'). The result is that the ABS regime of the CBD is not applicable in respect of the marine GRs of the deep seabed beyond the outer edge of the continental shelf.

Applicability of the regime of the area
As indicated, the fact that the ABS regime of the CBD is not applicable vis-à-vis the marine GRs located beyond the limits of national jurisdiction does not mean that the management of these resources is not subject to any legal requirements at all. On the contrary, it should be noted that the Law of the Sea Convention, to which the CBD refers in its Article 22(2), has been concluded in order to 'settle ... *all* issues relating to the law of the sea'. It is especially noteworthy in the present context that Part XI UNCLOS, which deals with the seabed and subsoil beyond the limits of national jurisdiction ('Area'), establishes a resource exploitation regime that assigns all rights in the resources concerned to mankind as a whole, on whose behalf the International Seabed Authority (ISA) shall act (see Articles 137(2), 140 UNCLOS). In particular, according to Article 140(2) UNCLOS, the ISA shall provide for the equitable sharing of financial and other economic benefits derived from all activities in the area. However, the term 'resources' is defined as 'all solid, liquid or gaseous mineral resources' (Article 133(a) UNCLOS). Therefore, notwithstanding the fact that Article 136 UNCLOS declares the area and its resources as being the common heritage of mankind, the scope of the benefit-sharing regime contained in UNCLOS is limited to the exploitation of non-living resources. The same conclusion may be drawn from the use of the term 'activities in the Area' (see, e.g., Article 140(1)), which is defined in Article 1(1) No 3 UNCLOS as 'all activities of exploration for, and exploitation of, the resources of the Area', and which thus incorporates the meaning of 'resources' under Article 133(a) UNCLOS.[5]

Applicability of the regime of the high seas
It does not follow from the inapplicability of the regime of the area that the issue of marine GRs of the deep seabed was not addressed by UNCLOS. The Convention is based on the assumption that the regime of the high seas covers all activities carried out beyond the areas of national jurisdiction, irrespective of whether they are conducted in the water column or on the seabed, as long as the Convention itself does not contain any special rule to the contrary (Proelss, 2007, p653).[6] This interpretation, which emphasizes the liberal approach on which the law of the sea is generally based, is owed to the historical fact that prior to UNCLOS III, all resources of the areas beyond the limits of national jurisdiction, whether living or non-living, were

commonly regarded as being subject to the freedom of the high seas (Churchill and Lowe, 1999, p225). Further evidence may be found in Article 112 UNCLOS, according to which 'all States are entitled to lay submarine cables and pipelines on the bed of the high seas beyond the continental shelf'. Thus, since Part XI UNCLOS only applies to the exploitation of mineral resources, the regime of the high seas must necessarily govern the utilization of the marine GRs of the deep seabed.

The relevance of the high seas regime to the marine GRs becomes manifest in the provisions on the conservation and management of the living resources (Articles 116–120 UNCLOS). Even though some of their components seem to be intended to govern high seas fisheries only (maximum sustainable yield, etc.), there is no evidence in the *trauvaux préparatoires* that the regime of the high seas would not be applicable to other living resources such as hydrothermal vent organisms. It would not have been necessary to exclude sedentary species (which are usually not fish in the biological sense) from the regime of the EEZ (see Articles 56(3), 77(4) UNCLOS) if the term 'living resources', when used in UNCLOS, would not include other species than fish. The vast majority of legal writers support this wide interpretation of the term 'living resources' (see Churchill and Lowe, 1999, p239; Gorina-Ysern and Jones, 2006, pp258–259; McLaughlin, 2003, p309; Proelss, 2007, p653; Verhoosel, 1998, p97; see also Farrier and Tucker, 2001, p218; Oude Elferink, 2007, pp144–147). The consequence is that States Parties are obliged to cooperate in the conservation and management of marine GRs (Article 118 UNCLOS) and that marine scientific research on hydrothermal vent organisms is principally free (see Articles 256, 257 UNCLOS). Notwithstanding the fact that the provisions of Part VII UNCLOS, which embody the regime of the high seas, are, indeed, far from constituting a comprehensive and satisfactory set of legal rules, the 'legal lacuna' identified with a view to the conservation and sustainable use of GRs in maritime areas beyond the limits of national jurisdiction[7] does not exist due to the comprehensive approach on which the regime of the high seas is based (see also UN Document A/61/65, 2006, para 30). The same conclusion may be drawn from the applicability of the rules and principles relevant to the protection and preservation of the marine environment under Part XII UNCLOS (UNEP/CBD/SBSTTA/8/INF/3/Rev.1, 2003, paras 55ff).

Should Part XI UNCLOS be expanded to the marine GRs?

Against the background of what has been argued so far, it seems necessary to ask whether an equal treatment of the mineral resources of the area, on the one hand, and the marine GRs, on the other, is indicated. The majority of commentators support an expansion of the benefit-sharing regime contained in Part XI UNCLOS by referring to the need to prevent the development of a de facto monopoly on GRs by the technologically advanced states (Arico and Salpin, 2005, p32; König, 2008, p159ff; Matz, 2002, p295; Tanaka, 2008, p140). While one should not ignore that unregulated access is not an illegitimate motive per se as it may provide incentives for investment and exploitation of valuable resources (UNEP/CBD/ SBSTTA/8/INF/3/Rev.1, 2003, para 110), it is obvious that the developing states should benefit in one way or another from any future utilization of marine GRs. The question whether the common heritage principle is a suitable tool to achieve that aim is, however, generally not asked in legal literature. The SBSTTA addressed this issue in its 2003 study of the relationship between the CBD and UNCLOS with regard to the conservation and sustainable use of GRs on the deep seabed. It took the position that:

> *leaving deep seabed genetic resources unregulated and freely available to those that have the resources to collect and exploit them, as is currently the case, would provide an incentive for investment and exploitation of valuable resources. However, deep seabed genetic resources would be constantly under threat of over-exploitation, as is the case with the present regime for the living resources of the high seas, and there would be no guarantee that the benefits arising out of their exploitation will be shared on a fair and equitable basis amongst all states. Such a free-for-all would not only be [sic] contrary to the regulatory intent of the United Nations Convention on the Law of the Sea as a whole, it would also run counter to the specific regime of the Area, which was designed to carefully regulate and protect seabed resources designated as the common heritage of mankind. Therefore, it is suggested that a precautionary and equity-based approach could be adopted and a specific regime established (UNEP/CBD/ SBSTTA/8/INF/3/Rev.1, 2003, para 114 [footnote omitted]).*

The SBSTTA concluded that:

> *the inclusion of deep seabed genetic resources within the regime dealing with the Area and its resources would respond to the ideas of benefit-sharing and permanent management. However, in light of the differences between mineral and biological resources, one might consider using the regime as a model rather than copying the United Nations Convention on the Law of the Sea exactly (UNEP/CBD/ SBSTTA/8/INF/3/Rev.1, 2003, para 133 [note that the paragraph is falsely numbered and follows para 121 of the document]).*

Thus, while the subsidiary body argued in favour of making some kind of benefit-sharing regime applicable to the GRs of the deep seabed, it considered the situation as not being necessarily parallel to the exploitation of the mineral resources of the area (but see Tanaka, 2008, p140).

It is interesting to note that the issue of missing comparability has so far only been raised with regard to high seas fisheries. In this respect, some sources have argued that regulating marine GRs in accordance with the regime of the high seas (which is mandatory under the existing legal rules according to the author) is unsatisfying due to the fact that 'the acquisition and subsequent use of the Area's microbial GRs is not analogous to fishing, whether by technique, equipment, or nature' (Glowka, 1996, p168; see also Matz, 2002, pp290–291). That submission points to an important aspect: It is true that, compared with fish, marine GRs cannot be considered as a finite resource. Existing interests in the utilization of these microbial resources are not connected with the organisms themselves, but refer to the genetic information contained therein. If one accepts, however, that the marine GRs are not exploited but rather 'sampled for subsequent study in small discrete quantities of sediment or water' (Glowka, 1996, p169), advocating an expansion of the benefit-sharing regime under Part XI UNCLOS appears to be inconclusive, as that regime, similar to fisheries but unlike the marine GRs, deals with the exploitation of a finite resource. Therefore, on closer inspection, it seems that the issue at hand is not one of equitable sharing of benefits *stricto sensu*, but rather one of distribution of and access to information (which is not or at least not primarily governed by UNCLOS). Against this background, submissions militating in favour of expanding the scope of the common heritage principle to the marine GRs, while at the same time opposing the applicability of the high seas regime by reference to the missing comparability of fish and GRs, seem to be contradictory in nature.

Additionally, the benefit-sharing regime under Part XI UNCLOS has not yet had to pass any serious operationability test. If one tended to accept that contrary to what is indicated by Article 140 UNCLOS, benefit sharing does not constitute a mandatory element of the common heritage of mankind principle, the practical consequences of any expansion of that principle would remain completely unclear. It is interesting to recall that the CBD fiercely rejects the common heritage of mankind approach, but refers to the conservation of biological diversity as constituting the 'common concern of humankind' in its preamble. The overriding interest of the majority of states in the course of the negotiations of the CBD was to assert sovereignty over GRs and thereby to prevent the Convention from being used to compel developing nations to conserve, at their own expense, biodiversity for international benefit (see Gorina-Ysern and Jones, 2006, p264; Lesser, 1998, p4; McLaughlin, 2003, p299; Verhoosel, 1998, pp96–97). If one takes into account the comparably poor involvement of States Parties to UNCLOS in the work of the ISA (which comes to the fore, inter alia, in difficulties to secure the necessary participation relevant to the quorum according to Article 159(5) UNCLOS; see Leary, 2007, pp223–224), as well as the fact that the original approach of Part XI UNCLOS was changed radically by the provisions of the Implementation Agreement of 1994, the suggestion of expanding the deep seabed mining regime and/or the mandate of the Authority to the marine GRs should not be further pursued.

From a practical point of view, the outcome of previous discussions led by the competent bodies of the UN support the position advocated here. While it was stated in a report of the Executive Secretary of the CBD that the ISA's regulations on prospecting and exploration of polymetallic nodules in the area 'could be used as a model to develop regulations addressing the impacts of bioprospecting activities in the Area' (UNEP/CBD/SBSTTA/11/11, 2005, para 59), final agreement on the matter could not be achieved until this day. In particular, in the course of the Ninth Conference of the Parties (COP 9), the issue of marine GRs located beyond the limits of national jurisdiction was intensely and controversially debated within the marine biodiversity working group. While some delegations took the position that high seas marine GRs were part of the common heritage of mankind, others held that the organisms concerned fell under the regime of the high seas. With a view to a future regime, opinions ranged from keeping the status quo to establishing voluntary codes of conduct, or concluding a new implementation agreement. The EU pointed to the International Treaty on Plant Genetic Resources for Food and Agriculture (ITPGRFA) as a model for the marine GRs (see SEC(2006) 689_12, para 6.1). None of the suggestions was ultimately included in the relevant COP Decision IX/20. As a compromise between the conflicting views became impossible to achieve, that document rather focused on traditional approaches; for example, the development of scientific and technical criteria for the implementation of environmental impact assessments, and the adoption of scientific criteria for identifying ecologically and biologically significant marine areas in need of protection (including hydrothermal vent sites).

The fundamental disagreement of views on whether the marine GRs of the deep seabed are or ought to be subject to the regime of the area or to that of the high seas also became most manifest at the occasion of the first meeting of the Ad Hoc Open-ended Informal Working Group to Study Issues Related to the Conservation and Sustainable Use of Marine Biodiversity beyond Areas of National Jurisdiction, which was established by UN General Assembly Resolution 59/24. It should be noted, though, that consensus seemed to have existed that:

in general, ... a key priority should be to improve the level of imple-
mentation of existing *instruments, including the principles and tools*
available under those instruments to address the conservation and
sustainable use of marine biological diversity beyond areas of
national jurisdiction (such as the precautionary approach and the
ecosystem-based approach) (UN Doc A/61/65, 2006, para 50
[emphasis added]).

Conclusion

While the examination of the existing legal rules has shown that the utilization of marine GRs located in areas beyond the limits of national
jurisdiction is subject to the regime of the high seas, it appears to be mandatory to ask what legal rules *should* apply in future to any of the activities
concerned. Arguably, the poor state practice of the parties to the CBD with
regard to both state territory and maritime areas under national jurisdiction
strongly militates against the prospect of successfully expanding the ABS
regime of the Convention to the high seas. However, if the analysis of one
source is correct, according to which 'without this tool, it seems highly
unlikely that the CBD could have the potential to have any real effect on the
protection of biodiversity in developing countries' (Verhoosel, 1998, p97),
the quest for alternative mechanisms should essentially focus on the fields
of the law of the sea and intellectual property rights.

The high relevance of UNCLOS in respect of any future legal regime,
being formally embodied in the work of the open-ended informal consultative process on oceans and the law of the sea, has continuously been
emphasized in all relevant documents. While expanding the mandate of the
ISA to the marine GRs of the deep seabed by either amending Part XI
UNCLOS, or negotiating a further implementation agreement to the
Convention, does not seem to be a promising solution for the reasons stated
above,[8] the provisions of Part VII Section 2 of the Convention are 'of
greater substance than might appear at first sight' (Shearer, 1992, p257).
Indeed, attention should rather be directed at effectively implementing and
carefully amending the existing rules on protection of the living resources of
the high seas and of the marine environment than on negotiating a new
comprehensive treaty regime which might be difficult to achieve (see also
ISBA/8/A/5, 2002, para 53). In this respect, the issue of protection of
hydrothermal vent sites could be adequately dealt with within the frameworks of regional fisheries organizations. Expanding the mandates of these
organizations, if at all necessary, would not only correspond to the assignment of the organisms concerned to the regime of the high seas and their

treatment as 'living resources' in terms of Article 118 UNCLOS, but also draw a direct line to a recent practice of, for example, the North East Atlantic Fisheries Commission (NEAFC), which has since 2004 repeatedly decided to prohibit bottom trawling and fishing with static gear in certain particularly vulnerable high seas areas (for 2008, see Recommendation VII:2008, www.neafc.org/measures/current_measures/docs/07-rec_deep_sea.pdf, and Recommendation IX:2008, www.neafc.org/measures/current_measures/docs/09-rec_corals.pdf), and thereby established marine protected areas. One should also not ignore that the failure of those States Parties to UNCLOS and the relevant regional fisheries organizations that are engaged in high seas fisheries, or other activities such as bioprospecting to take measures necessary for the conservation of the living resources, constitutes a breach of treaty obligations (Pendelton, 2005, pp497–498). With a view to marine scientific research conducted in areas under national jurisdiction, it is shown above that, notwithstanding their relative vagueness, the rules contained in UNCLOS are better suited to achieve a fair balancing of conflicting interests involved in such research than those of the CBD. Having said that, the CBD and in particular the SBSTTA could become active in the development of standards for good scientific practice on relevant activities undertaken in areas beyond the limits of national jurisdiction. As emphasized by the Executive Secretary of the CBD, a prominent role should also be assigned to the scientific community itself (UNEP/CBD/SBSTTA/11/11, 2005, para 47ff).

Having regard to the need to achieve a fair and equitable sharing of information and research results deriving from the use of high seas marine GRs, it is submitted that given the general reluctance in the implementation of ABS rules on the international plane, the most promising way seems to be working towards establishing exceptions to intellectual property rights resulting from patents over health-related inventions from marine bioprospecting. Such an approach would not only reflect the emphasis put on the issue of intellectual property rights by the competent bodies of the UN (see UN Docs A/62/66, 2007, paras 219ff; A/62/66/Add.2, 2007, paras 232ff; A/62/169, 2007, para 49); rather it would also recognize the missing comparability identified above between the marine GRs on the one hand and fish stocks and mineral resources, respectively, on the other in terms of the results derived from their utilization. Additionally, through exceptions to intellectual property rights, developing states, which generally will not have the technology to become engaged in bioprospecting activities, could benefit from products such as pharmaceuticals, which have emanated from the biodiversity of areas beyond the limits of national jurisdiction, under fair terms. Against the background of the different underlying interests, the line of argument advocated here appears to be preferable to using the

ITPGRFA as a model, the more so as that instrument is not applicable to areas beyond the limits of national jurisdiction, nor does it cover chemical, pharmaceutical and/or other non-food/feed industrial uses (see Article 12.3(a)). Therefore, the approach on which the ITPGRFA is based may be applied to the marine GRs only subject to far-reaching adjustments of that instrument, which arguably might be as difficult to achieve as negotiating an implementation agreement to UNCLOS or the CBD.

The Agreement on Trade Related Aspects of Intellectual Property Rights (TRIPS), which provides minimum standards of intellectual property protection among the members of the World Trade Organization (WTO), permits States Parties to exclude plants and animals other than micro-organisms from patentability (see Article 27[3](b) TRIPS). While many of the marine GRs will fall within the scope of the term 'micro-organisms' for which the exemption from patentability does not apply,[9] amendments to the Agreement reached in the context of compulsory licensing for pharmaceutical products (Article 31 TRIPS in conjunction with the Protocol Amending the TRIPS Agreement of 6 December 2005) suggest that the rush to patent inventions arising from utilization of marine GRs might be effectively channelled for the benefit of developing countries. Corresponding adjustments would also be likely to have a strong impact on the CBD, since one of the main reasons for the developing states' reservations in implementing the ABS regime of the Convention lies in the still somewhat unclear relationship between the CBD and the international law on intellectual property rights codified in TRIPS. As one African head of state said, most of the developing countries 'find it difficult to accept the notion that biodiversity should flow freely to industrial countries while the flow of biological products from industrial countries is patented, were expensive and considered private property of the firms that produce them. This asymmetry reflects the inequality of opportunity and is unjust' (Kruger, 2001, p169). Thus, it seems that patentability will be a major issue in the quest for undoing the knot in ABS transactions in relation to marine GRs. Finally, as stated by the UN Secretary-General, measures taken within the framework of TRIPS could be complemented by tools such as open-source licensing (UN Doc A/62/66/Add.2, 2007, para 242) and experimental use exemptions (UN Doc A/62/66/Add.2, 2007, para 227).

Notes

1   When used in this chapter, the term 'areas beyond the limits of national jurisdiction' refers to the seabed, subsoil and the waters located seaward of the continental shelf and the exclusive economic zone (EEZ). Therefore, it encompasses the high seas and

the deep seabed ('Area') as defined by UNCLOS. Note that the EEZ and the continental shelf are, as regards their seaward limits, not necessarily identical due to the existence of the concept of the extended continental shelf under Article 76 UNCLOS.

2 Whether the CBD may be considered as an agreement under Article 311(3) UNCLOS depends on whether that provision's scope is limited to treaties which are intentionally concluded in order to modify or suspend certain UNCLOS provisions. While its wording militates in favour of a narrow interpretation, the object and purpose of Article 311 UNCLOS is 'to play the role similar to the one of article 103 of the United Nations Charter' (Vukas, 1998, p649), that conclusion speaking in favour of a wide scope of application (see also Wolfrum and Matz, 2003, p127).

3 It should be noted, though, that any such activity must not amount to an exercise of sovereign rights prohibited under Article 137(1) UNCLOS. The CBD does not contain any definition of what constitutes 'processes and activities'.

4 However, it has been argued that Article 16 CBD, dealing with transfer of technology, is not limited to GRs under national jurisdiction (Tanaka, 2008, p138).

5 Whether that is the end of the debate is not consistently answered. For a detailed analysis of whether Part XI UNCLOS in general and the common heritage principle in particular are applicable to the marine GRs, see Proelss, 2008; also Oude Elferink, 2007.

6 Parallel to the high seas, the regime of the EEZ in principle comprises the seabed and its subsoil (Article 56(1)(a) UNCLOS).

7 It is noteworthy that the 'legal gap' reasoning is mainly raised by those who wish to submit the management of marine GRs to an ABS regime. The political character of this line of argument is obvious, as the absence of ABS rules is in no way tantamount to the absence of *any* rule.

8 Even though an emerging interest of the ISA in biodiversity issues cannot be denied, it seems as if the authority confines its considerations of deep-sea biodiversity strictly to the terms of the existing mandate (see Leary, 2007).

9 It should be noted, though, that the obligation to patent micro-organisms is predominantly held to be subject to the general rule of patentability contained in Article 27(1) TRIPS ('patents shall be available for any inventions'). It has been concluded therefrom that Article 27(3)(b) TRIPS 'can be interpreted as applicable only to the genetically modified microorganisms, not to microorganisms in their natural form' (Millicay, 2007, p796).

References

Arico, S. and Salpin, C. (2005) *Bioprospecting of Genetic Resources in the Deep Seabed: Scientific, Legal and Policy Aspects*, New York, United Nations University – Institute of Advanced Studies

Bernhard, J. M., Buck, K. R., Farmer, M. A. and Bowser, S. S. (2000) 'The Santa Barbara basis is a symbiosis oasis', *Nature*, vol 403, pp77–80

Churchill, R. R. and Lowe, V. A. (1999) *The Law of the Sea*, Manchester, Manchester University Press

Farrier, D. and Tucker, L. (2001) 'Access to marine bioresources: Hitching the conservation cart to the bioprospecting horse', *Ocean Development and International Law*, vol 32, pp213–239

Frenz, J. L., Kohl, A. C. and Kerr, R. G. (2004) 'Marine natural products as therapeutic agents: Part II', *Expert Opinion on Therapeutic Patents*, vol 14, pp17–33

Friedland, J. C. (2007) *Der Schutz der biologischen Vielfalt der Tiefseehydrothermalquellen*, Baden-Baden, Nomos

Glowka, L. (1996) 'The Deepest of ironies: Genetic resources, marine scientific research, and the area', *Ocean Yearbook*, vol 12, pp154–178

Glowka, L. (1999) 'Genetic resources, marine scientific research and the international seabed area', *Review of European Community and International Environmental Law*, vol 8, pp56–66

Gorina-Ysern, M. and Jones, J. H. (2006) 'International law of the sea, access and benefit sharing agreements, and the use of biotechnology in the development, patenting and commercialization of marine natural products as therapeutic agents', *Ocean Yearbook*, vol 20, pp221–281

König, D. (2008) 'Genetic resources of the deep sea – How can they be preserved?', in Stoll, P.-T. et al (eds) *International Law Today: New Challenges and the Need for Reform?*, Berlin, Heidelberg, Springer, pp141–163

Korn, H., Friedrich, S. and Feit, U. (2003) *Deep Sea Genetic Resources in the Context of the Convention on Biological Diversity and the United Nations Convention on the Law of the Sea*, Bonn, Bundesamt für Naturschutz

Kruger, M. (2001) 'Harmonizing TRIPs and the CBD: A proposal from India', *Minnesota Journal of Global Trade*, vol 10, pp169–207

Leary, D. K. (2007) *International Law and the Genetic Resources of the Deep Sea*, Leiden, Brill

Lesser, W. (1998) *Sustainable Use of Genetic Resources under the Convention on Biological Diversity: Exploring Access and Benefit Sharing*, New York, CABI

Matz, N. (2005) *Wege zur Koordinierung völkerrechtlicher Verträge*, Berlin, Heidelberg, Springer

McLaughlin, R. J. (2003) 'Foreign access to shared marine genetic materials: Management options for a quasi-fugacious resource', *Ocean Development and International Law*, vol 34, pp 297–348

Millicay, F. (2007) 'A legal regime for the biodiversity in the area', in Nordquist, M. H., Long, R., Heidar, T. H. and Moore, J. N. (eds) *Law, Science & Ocean Management*, Leiden, Martinus Nijhoff, pp739–850

Nordquist, M. H., Rosenne, S. and Sohn, L. B. (1989) *United Nations Convention on the Law of the Sea 1982: A Commentary*, vol V, Dordrecht, Boston, London, Brill/Martinus Nijhoff

Oude Elferink, A. G. (2007) 'The regime of the area: Delineating the scope of application of the common heritage principle and freedom of the high seas', *International Journal of Marine and Coastal Law*, vol 22, pp143–176

Pendelton, G. D. (2005) 'State responsibility and the high seas marine environment: A legal theory for the protection of seamounts in the global commons', *Pacific Rim Law and Policy Journal*, vol 14, pp485–514

Proelss, A. (2007) 'Die Bewirtschaftung der genetischen Ressourcen des Tiefseebodens – Ein neues Seerechtsproblem?', *Natur und Recht*, vol 29, pp650–656

Proelss, A. (2008) 'Marine genetic resources under UNCLOS and the CBD', *German Yearbook of International Law*, vol 52, pp417–446

Salamanca Aguado, E. (2003) *La Zona Internacional de los Fondos Marinos. Patrimonio Común de la Humanidad*, Madrid, Dykinson

Shearer, I. A. (1992) 'High seas: Drift gillnets, highly migratory species, and marine mammals', in Miles, E. L. and Kuribayashi, T. (eds) *The Law of the Sea in the 1990s: A Further Framework for International Cooperation*, Hawaii, William S. Richardson School of Law, pp237–258

Tanaka, Y. (2008) 'Reflections on the conservation and sustainable use of genetic resources in the deep seabed beyond the limits of national jurisdiction', *Ocean Development and International Law*, vol 39, pp129–149

Verhoosel, G. (1998) 'Prospecting for marine and coastal biodiversity: International law in deep water', *International Journal of Marine and Coastal Law*, vol 13, pp91–104

Vukas, B. (1998) 'The Law of the Sea Convention and the Law of Treaties', in Götz, V., Selmer, P. and Wolfrum, R. (eds) *Liber amicorum Günther Jaenicke*, Berlin, Heidelberg, Springer, pp631–654

Wolfrum, R. and Matz, N. (2000) 'The interplay of the United Nations Convention on the Law of the Sea and the Convention on Biological Diversity', *Max Planck Yearbook of United Nations Law*, vol 4, pp445–480

Wolfrum, R. and Matz, N. (2003) *Conflicts in International Environmental Law*, Berlin, Heidelberg, Springer

PART TWO

TRADITIONAL KNOWLEDGE FROM NEW PERSPECTIVES

Chapter 5

Potential of TK for Conventional Therapy – Prospects and Limits[1]

Jack K. Githae

Introduction

Contrary to common Western assumptions, traditional ecological knowledge of indigenous people is scientific and holistic in that it is empirical, experimental and systematic. However, it differs in two respects from Western science. First, knowledge is highly localized. Its focus is the complex web of the relationships between humans, animals, plants, natural forces, spirits and landforms within a particular locality or territory. Therefore, indigenous people normally make better predictions about the consequence of physical changes or stresses within a particular ecosystem than scientists who base their forecasts on a narrow and non-holistic, generalized model and field observations of a relatively short duration.

Second, local knowledge has important social and legal dimensions. Every ecosystem is conceptualized as a web of *social* and *spiritual* relationships between the living and the dead members of the family, clan or tribe, and the other animate and inanimate ecological entities they coexist with. Hence the structure of the ecosystem is regarded as a negotiated order in which all the ecological entities are bound together by kinship and solidarity. It is common practice in many African communities to accord human personality to non-human entities such as wild animals, trees, rivers, mountains or rocks. This conception of the ecology is summarized in five legal corollaries:

1 Every dead and living human or non-human in the ecosystem bears a personal responsibility for appreciating, understanding and maintaining their relationships. Knowledge of the ecosystem is moral and legal knowledge.

2 Since knowledge confers heavy responsibilities, as well as the power to interfere with relationships between the dead and living, it must be

transmitted personally to an individual apprentice who has been properly prepared.

3 Knowledge is physically and instinctively transmitted between kin because it pertains to inherited responsibilities to their own ancestral territory.

4 Knowledge may sometimes be shared with visitors to the territory so that they can travel safely and subsist from local resources, but knowledge cannot be alienated permanently from the ecosystem to which it pertains.

5 Misuse of knowledge is tantamount to an act of war on other species, breaking their covenants and returning the land to a pre-moral and pre-legal vacuum. This is why indigenous people take a precautionary approach to the use of the ecosystems.

Consistent with these general principles, indigenous peoples possess their own locally specific systems of jurisprudence, which must be respected with regard to the classification of knowledge, proper procedures for acquiring and sharing knowledge, and the nature of the rights and responsibilities that are attached to possessing knowledge.

The complexity of the laws governing the distribution and utilization of indigenous and TK has important *political* implications because no individual, family or clan can possess sufficient knowledge to act alone. Decision making in sustainable distribution and utilization requires the sharing of knowledge.

Appropriate empowerment of indigenous people enhances maintenance and development of their own knowledge systems. Hence, indigenous people will undoubtedly share their medical and ecological sciences with other societies because generosity and reciprocity are the core values of indigenous cultures.

Definitions

Prospects

This term implies a forecast, a prediction and an anticipation of the various possibilities of traditional medicine acting as a resource for conventional medicine through the undertaking of an analysis of the historical and the current '*environment*' to TK and conventional medicine. This analysis should therefore revolve around the local and international environments of politics, economy, socio-culture, technology, ecosystem and religion, which all along have influenced various aspects of TK and conventional medicine.

Limitations

This implies the restrictions, limits, boundaries and the confines that may be encountered within the political, economic, socio-cultural, technological, ecosystem and religious environments, and which influence the possibilities of indigenous and TK acting as a resource for conventional medicine.

Culture

Culture is comprised of values, attitudes, norms, ideas, internalized habits and perceptions, as well as the concrete forms or expression they take in. It influences people's actions and interpretations of circumstances.

Traditional and indigenous knowledge

By way of definition, indigenous knowledge refers to the root, innate and natural knowledge characteristic to the original inhabitants of a particular ecosystem. We can also say that indigenous knowledge is community-, site- and role-specific epistemology governing the structures and development of the cognitive life, values and practices shared by a particular community (often demarcated by its language) and its members, in relation to a specific lifeworld.

African traditional medicine as an example of TK

This refers to the sum total of all knowledge and practices, whether explicable or not, used in diagnosis, prevention and elimination of physical, mental or societal imbalance, and relying exclusively on practical experience and observation handed down from generation to generation, whether verbally or in writing.

Herbal medicines

Processed and labelled medicinal products that contain as active ingredients aerial or underground parts of plants, or other plant material or combinations thereof, whether in crude state or as plant preparations, are known as 'herbal medicines'.

Conventional/allopathic medicine

Allopathic/conventional medicine operates from the basis of the scientific paradigm: it is based on observation and measurement. Diseases and illness are treated largely in isolation from the spiritual, psychosocial and mental determinations of illness. Conventional medicine involves aspects of Cartesianism, which separates the 'mind' from the 'body'.

The holistic healing principles

Holistic health care that is emerging in the 21st century encompasses the following indigenous principles:

- Use of safe, effective diagnostic and treatment options. These include education for lifestyle changes and self-care, complementary diagnostic and treatment approaches.
- Searching for the underlying causes of disease and prevention is preferable to treating symptoms.
- Illness is viewed as a manifestation of the dysfunction of the whole person, not as an isolated event.
- A major determining factor in the healing process encourages the patient to take responsibility for his or her health.
- Holistic physicians encourage their patients to evoke the healing power of love, hope, humour and enthusiasm, and to release the toxic consequences of hostility, shame, depression and prolonged fear, anger and grief.
- Optimal health is the conscious pursuit of the highest qualities of the physical, spiritual and social aspects of human experience rather than the absence of illness.

Community rights

Inherent rights of indigenous communities over biological resources, traditional medical knowledge and traditional methods they have discovered and developed are recognized without further legal or other assurances, and that these indigenous communities are the general owners, with primary and residuary title to:

- The formal or informal communal systems of innovation through which they produce, select, improve and breed a diversity of medicinal plants
- The medicinal plants' varieties, biological resources, traditional medicines, medical practices and devices and technologies produced through TK systems.

Property rights

Property may be divided into two forms, that is, real and intellectual property (IP). Real property comprises the tangible commodities capable of exclusive possession and clear delineation such as land or furniture. IP refers to 'all creations of the human mind or intellect'. It deals with information or services, which are intangible. It is property in the sense that it is owned by the inventor/discoverer and can only be used by other persons with the owner's authorization, by law and through legal procedures. The owner can be a person or a legal entity.

Duty bearers

Real and intellectual property, individual and community rights are linked to duties, accountability, obligations and responsibility. Duty bearers are the actors collectively responsible for the realization of these rights (Figure 5.1).

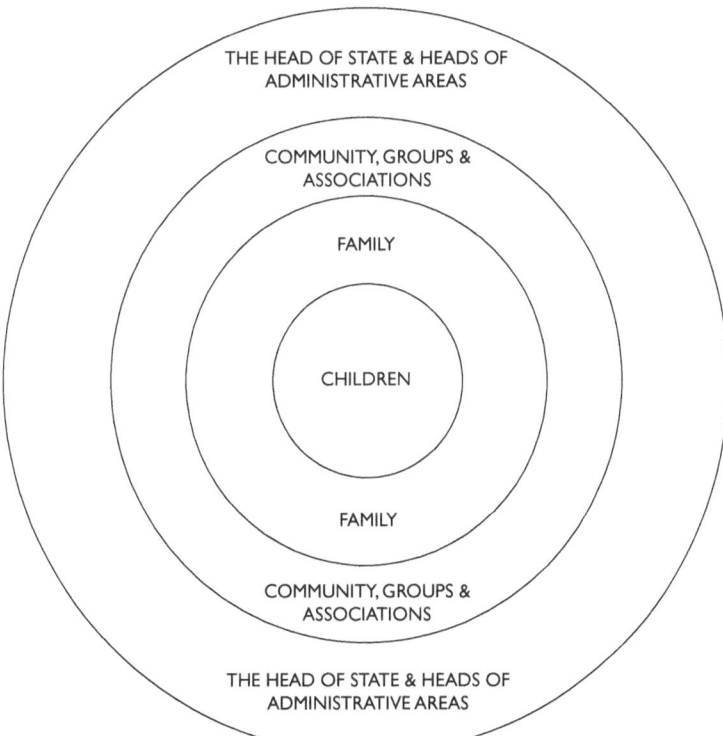

Figure 5.1 *Duty bearers (the custodians of real and intellectual property, individual and community rights)*

As Duncan Ndegwa illustrates in his book *The Sruggles of Mzee Jomo Kenyatta* (2006), it is important to note that ancestors are the spiritual custodians and are therefore duty bearers. Therefore, the family and the community must keep guard of the ancestral wisdom and secrets as they use it and sustainably pass it on for posterity.

Generation of TK

In order to delineate the prospects and limitations of TK acting as a resource for conventional medicine, there is a need to understand the political, economic, socio-cultural, technological, ecological and religious dynamics involved in the generation of TK, which in turn influences the pathways of the prospects and the limitations.

The analogy in Figure 5.2 illustrates that all knowledge and wisdom comes from God (Supernatural), then transcends to living things. Therefore, in a contemporary traditional African society, any philosophy, knowledge and wisdom in medicine that does not recognize God as the creator, giver of life and the ultimate physician, and the ancestors as the custodians of God's knowledge and wisdom, has limited impact, acceptability and applicability.

Brief on the branches of African traditional medical knowledge, skills and practices

African traditional medicine is an aspect of African indigenous knowledge that embraces a variety of effective culture diagnostic and treatment systems that have enhanced the health of African people since time immemorial. Africa is the cradle of mankind: as illustrated in Figure 5.2, the Africoid race is estimated to be more than 150,000 years old, while the Europoid and Mongoloid races are estimated to be 40,000 and 6,000 years old, respectively (Githae, 2005). That means Africa is the origin of modern medicine and human civilization. Man's survival depended on how effectively and how sustainably he maintained health delivery systems.

African traditional medicine is divided into the following branches:

Herbalists are traditional practitioners who utilize materials/and or extracts from plants, animals and minerals to manage disease.

Spiritualists are traditional medical practitioners who utilize their medium as a connecting vital force to the supernatural and are able to derive healing energy from the spiritual realm. They act as a link between the living, the dead and God.

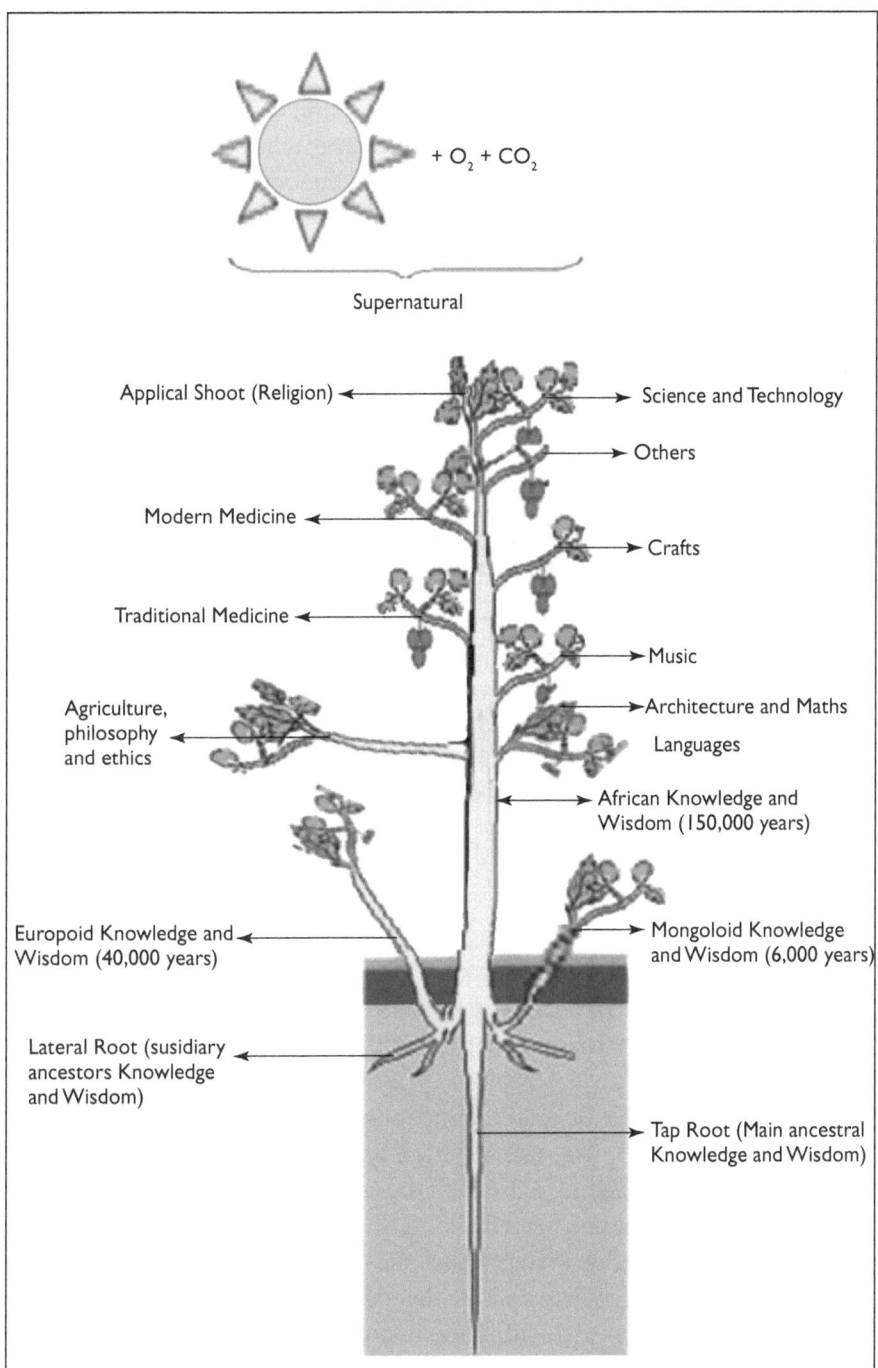

Figure 5.2 *Analogy of the tree of knowledge and wisdom (capturing African philosophy in as far as traditional perspectives, values and practices are concerned)*

Psychotherapists: To explain how psychotherapy works in African traditional medicine, I will use the example of the Zarma tribe from Niger. According to them, the human being, *boro*, is a unit comprised of three elements: the *ga*, or body; the *bya*, or double; and the *fundi*, or vital force, which infuses life into both the *ga* and the *bya*. The result of this conception is that there are two approaches to illness. The first is concerned with somatic ailments, those affecting the *ga*. The second is concerned with behavioural ailments, which are connected with the *bya*. The cure of this second kind of illness requires an array of ritual practices, whose psychotherapeutic value is undeniable.

Bone setters are traditional medical practitioners who can carry out the reduction of fractures through the use of manipulative movements with their fingers and the application of immobilization techniques that often involve the use of the bark of medicinal trees. In Kenya this means the tying of tree bark round the fractured bone with reeds derived from dried banana stems. As well as immobilizing, the bark provides medicinal drugs, which are absorbed through the skin to enhance pain relief and healing.

The spectrum of practice of *traditional surgeons* includes uvulotomy, circumcision, incision of abscess, treatment of pyomyositis, the dressing of wounds and ulcers, tooth extraction, cupping and hot iron burning.

Traditional birth attendants deal with antenatal examination, assisting delivery, management of pregnancy problems, management of neonatal diseases, family planning, fertility control and infertility management in women.

The prospects of establishing protocols for an indigenous-directed process of integration of TK and conventional medicine

Figure 5.3 illustrates the normal interactions that occur in nature and that control the production, use, transmission and transfer of knowledge. We shall take knowledge systems to be both the *contents* and the *processes* of that domain of lived experience we refer to as knowledge in any particular culture and time. Knowledge systems also influence the skills, values and the practice of a particular society and are diversified and dynamic. Knowledge and knowledge systems exist within the main realms of political, economic, socio-cultural, technological, ecological and religious/cosmic influences. Thus, under all circumstances, the people's knowledge and the ecosystem interact, as is shown by Figure 5.4.

In accordance with Figure 5.4, knowledge and knowledge systems interact with the indicated elements of the ecosystem. The interaction is two-way traffic, in that the knowledge and the knowledge systems can change the dynamics as well as the constituents of each of the ecosystem elements. Note that in Figure 5.3, all the elements of the ecosystem and those of the people interact in an intricate network, which also influences the outcome of the interaction of knowledge per se and the ecosystem. Thus, the network may promote or reduce overall the net interaction of knowledge and the ecosystem. In fact, this network of interactions is what is used to determine the ecosystem well-being index, human well-being index and ecosystem stress. This is also supported by Robert Prescott-Allen in *The Wellbeing of Nations* (2001). In a sustainable development model, the ecosystem well-being should not compromise human well-being and vice versa, and therefore people should strive to keep a balance between the two.

Thus, well-being is a combination of integrated sustainable development of the human being, as well as the ecosystem. This can be understood if only we could think of an egg as a model of well-being. The ecosystem surrounds and supports people much as the white of an egg surrounds and supports the yolk. Just as the egg can be good only if both the yolk and the white are good, so a society can be healthy and sustainable if both the people and the ecosystem are well balanced.

Because various indicators for each of the category of the factors to be assessed have different units of measurement, for example, land condition in hectares, water pollution in milligrams per litre, carbon dioxide emissions in metric tonnes of carbon, species diversity in percentages of threatened species, health in years of life expectancy and death rates, income in money, education in school enrolment rates, freedom in the observation of rights, it is necessary to find a common unit that does not distort the factors' qualities by the use of performance scores. The barometer of sustainability (Figure 5.5) was designed to measure human and ecosystem well-being without submerging one factor in the other. The barometer's unique features are:

- Two axes; one for human being well-being and the other for ecosystem well-being. This enables each set of indicators to be combined independently, keeping them separate to analyse ecosystem–human interactions.
- The axis with the lower score overrides the other axis. This prevents a high human well-being score from offsetting a low score for the ecosystem well-being (or vice versa), reflecting the view that people and the ecosystem are equally important and sustainable integrated development must improve and maintain the well-being of both.
- Each axis is divided into five bands. This allows users to define not just the end points of the scale but the intermediate points as well, for greater flexibility and control of the scale. Indeed the barometer of sustainability

The global ecosystem and its manipulation through enhancement of human survival, health, wellbeing and integrated sustainable economic development

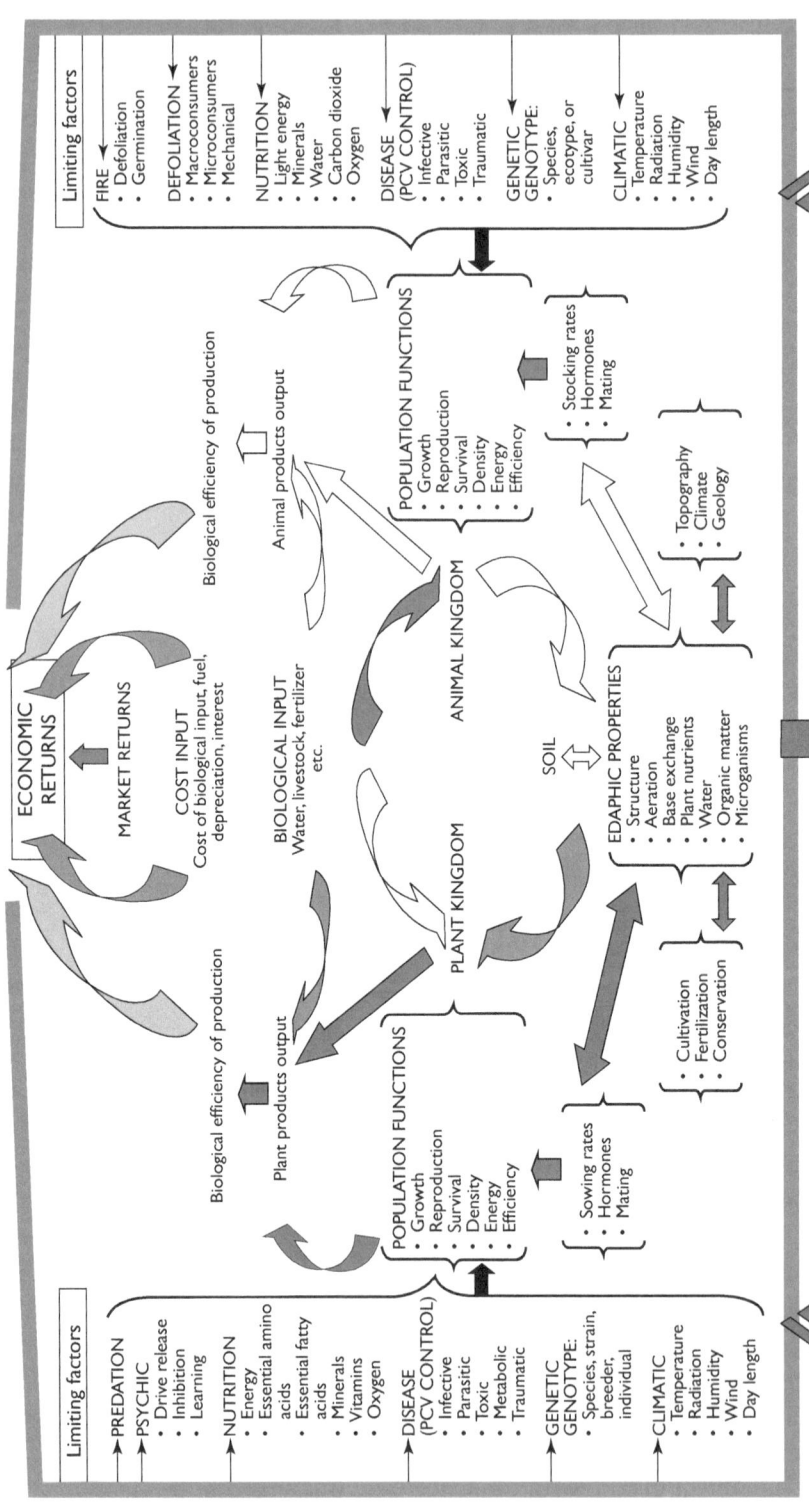

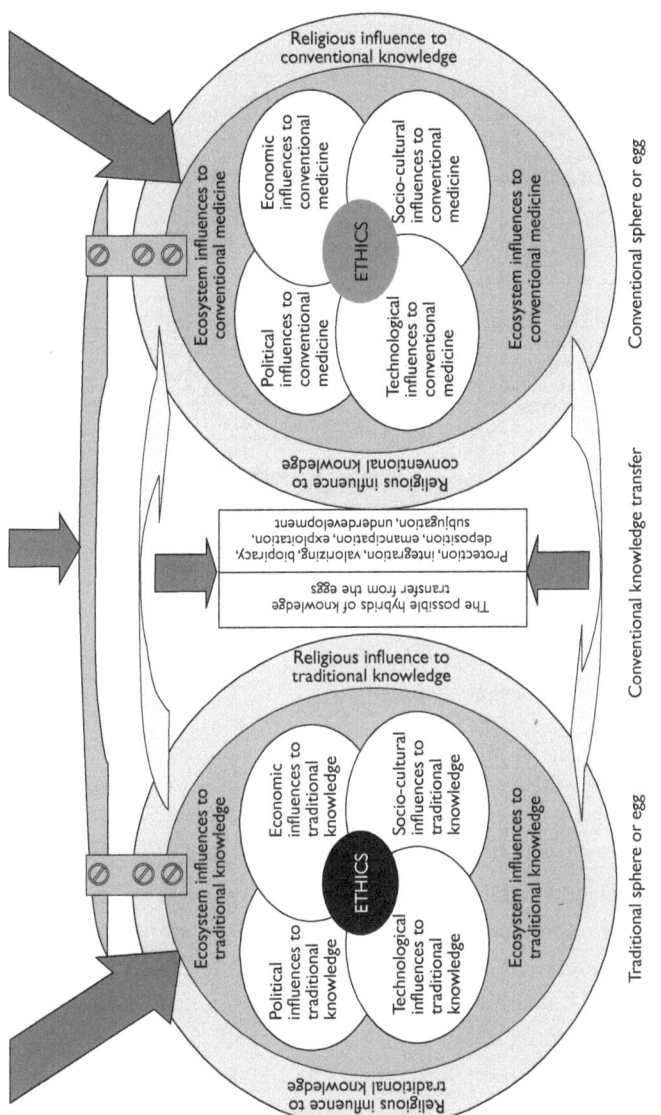

Figure 5.3 *An introduction to natural interactions that control the production, use, transmission and transfer of knowledge*

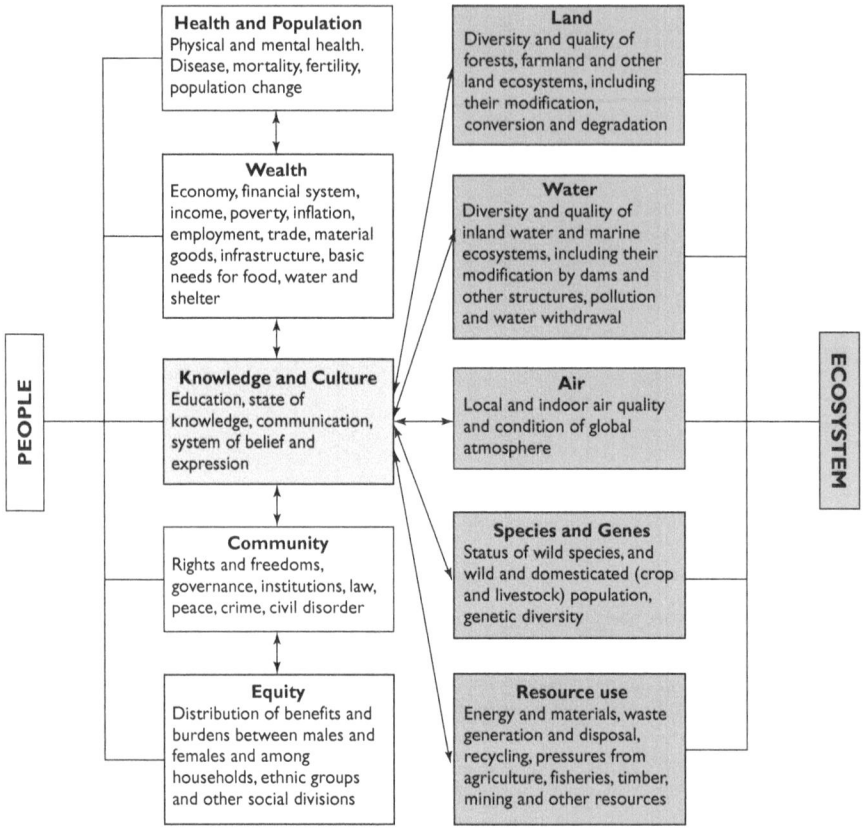

Figure 5.4 *The interaction of people, knowledge and the ecosystem in development*

is an indirect measure of the level of application of sound and integrated eco-ethics.

Thus, in normal circumstances, there will always be generation of knowledge either from the traditional side or the conventional side. Whatever the case, the knowledge so generated *and* transferred, as Figure 5.3 indicates, could result in either the *hybrid* or *deposition*, in which case learning does not occur because there is no voluntary will on the side of the learner. *Integration* therefore is an agreement to the knowledge use and propagation, *valorization* and/or *biopiracy*, in which case it is stolen without an agreement, by the owner, and therefore reimbursement, compensation, damages and amends could be sought by either side. Often it has been traditional and/or informal knowledge which has been *valorized* and *biopirated*. The hybrid could also be an *emancipation process*, leading to enhanced *development, health and well-being* or alternatively *subjugation* and

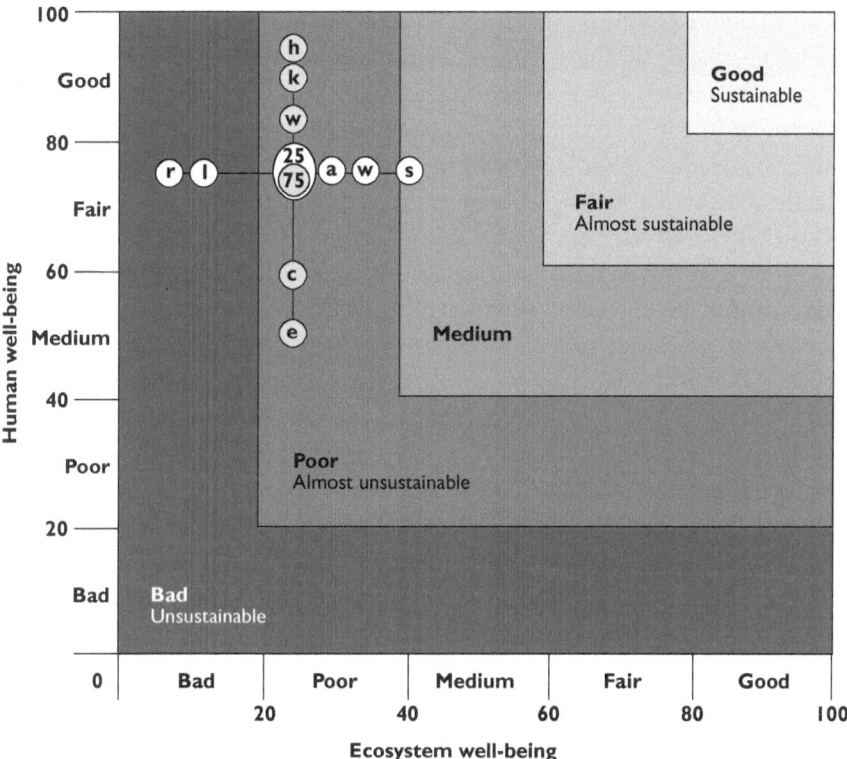

The well-being of a hypothetical country is shown on the barometer. The Human Well-being Index (HWI) is the yolk of the egg; the Ecosystem Well-being Index (EWI) is the white of the egg. The Well-being Index (WI) is the point where the HWI and the EWI intersect. Tinted circles in the vertical axis represent the scale of the human dimensions: c, community; e, equity; h, health; p, population; k, knowledge; and w, wealth. White circles on the horizontal axis represent the ecosystem dimensions: a, air, land, resource use; s, species and genes; and w, water.

Figure 5.5 *Barometer of sustainability*

exploitation processes, leading to loss of knowledge and thirst and consequently to *under-development*. Thus knowledge as a resource should be guided by the following legal ethics:

• Every individual human or non-human in the ecosystem bears a personal responsibility for understanding and maintaining their relationships. Knowledge of the ecosystem is moral and legal knowledge, and the adepts are not only expected to teach their insights to others, but also to mediate conflicts between humans and other species.
• Knowledge must be transmitted personally to an individual apprentice who has been properly prepared to accept the burdens and to use the power with humility.

- Knowledge is ordinarily transmitted between kin because it pertains to inherited responsibilities to their own ancestral territory. Since knowledge is localized it is not necessarily applicable to other ecosystems. Moreover, it could be dangerous for outsiders to obtain information that could be used to meddle with what is regarded as the internal affairs of the local human and non-human 'extended family'.
- Knowledge may sometimes be shared with visitors to the territory so that they can travel safely and subsist from local resources, but knowledge cannot be alienated permanently from the ecosystem to which it pertains. Knowledge can only be lent for a specified time and purpose.
- Misuse of knowledge can be catastrophic, not only for the individual who abuses it but for the people, the territory and (potentially) the world. Misuse of knowledge is tantamount to an act of war on other species, breaking their covenants and returning the land to a pre-moral and pre-legal vacuum. This is why indigenous people take precautions in the use of ecosystems. Any human activity that goes beyond the bounds of known relationships among species involves a risk of triggering retaliation and chaos.

Further, it is important to realize that according to the analogy of the spheres or eggs, as is illustrated by Figure 5.3, the *shell of the egg* according to indigenous knowledge and TK represents *religion and religiosity*, which has already been explained in the preceding sections about the generation of knowledge systems. Religion and religiosity can also determine the level of prospects and the limitations of knowledge transfer.

In Figure 5.3, the *egg white*, which represents the *ecosystem*, also influences the generation as well as the prospects and the limitations of the knowledge transfer, as illustrated by the rectangular drawing above and the two spheres of the eggs. The rectangular drawing shows the intrinsic *ecosystem interactions* determining the generation, transfer and the prospects, as well as the limitations. Therefore, it is a deeper explanation of the ecosystem's role indicated in the two spheres/eggs in as far as generation, transfer, limitations and prospects of the transfer are concerned. This figure shows that indigenous knowledge and TK have maintained harmony and balance in nature. The plant kingdom relied on the soil for its survival, but this was also balanced by the animal kingdom, which relied on the plants for its survival, though both kingdoms had to contribute to the edaphic properties of the soil, as well as the constancy of the atmosphere through the natural cycles of the environment such as the nitrogen cycle, carbon dioxide cycle and water cycle. Through this then, the population functions of both the plant kingdom and the animal kingdom co-exist at levels that the ecosystem can support. There are also natural limiting factors set up by nature to control population explosions in both the plant kingdom and the animal kingdom. Thus, these diagrams themselves portray some kind of harmony and balance between the left-hand side and the right-hand side, which must be taken into account in all endeavours of integrated sustainable indigenous knowledge transfer and sharing.

Conventional knowledge, as is well known, is *consumer oriented* and *market oriented*, and therefore profit and *economic returns oriented*. In such a scenario then, it would be expected that the environment would be over-exploited at the expense of the natural forces that maintain harmony and balance. Thus all the control forces that are indicated by *broad arrows in the rectangular diagram of Figure 5.3* will always be moving towards the side of economic returns, and thus lead to global climate change due to depletion and degradation of the environment. This in turn leads to the *environmental burden of human, animal and plants diseases and genetic degradation (human unwell-being)*, which is illustrated by the table in Figure 5.6, showing *indicative values for environmental attributable fractions, by specific environmental risk factor and disease risk*. On the other hand, TK is generated as a response to an ecological need in order to enhance survival. As such, with varied ecological zones in Africa, it is expected that the various types of knowledge generated are also varied depending on the needs to be met. This aspect of ecological variance may then limit the prospects of the wholesale transfer of certain knowledge systems to other, different ecological zones.

For the purpose of this chapter, we can briefly define the environment as *all the physical, chemical and biological factors external to the human host, and all related behaviours, but excluding those natural environments that cannot reasonably be modified.* This definition excludes behaviour not related to the environment, as well as behaviour related to the social and cultural environment, genetics and parts of the natural environment. This definition then aims to cover those parts of the environment that can be covered by environmental management.

There is also the yellow egg yolk, which in Figure 5.3 represents the people and their *political, economic, socio-cultural and technological* rights to knowledge – a people who also possess determined knowledge systems' generation and the prospects, as well as the limitations of the transfer.

On the other hand, in Figure 5.3 there is also the aspect of the nucleus of the egg yolk, the *ethics*, which represents the various codes, ascribed roles, values and taboos for example, which are determinants of knowledge generation and sustainable utilization and the prospects and limitations of transfer.

Traditionally as an oral culture, with many of the practices persisting, knowledge was held and kept by people. People were the repositories for

Disease or risk \ Risk factor	Water, Sanitation and Hygiene	Indoor air pollution	Outdoor air pollution	Noise	Other housing risks	Chemicals	Recreational environment	Water resources management	Land use and built environment	Other community risks	Radiation	Occupation	Climate
Lower respiratory		○	■		■								
Upper respiratory		☆	■		■								
Diarrhoeal diseases	○						■						■
Malaria								○	☆			■	■
Intestinal Nematode infections	○												
Trachoma	○												
Schistosomiasis	○												
Chagas Disease					○								
Lymphatic filariasis	○							☆					
Onchocerciasis								☆					
Leishmaniasis					☆								
Dengue					○								■
Japanese encephalitis								○					
STD's												☆	
HIV												☆	
Hepatitis B and C												■	
Tuberculosis					☆							☆	
Perinatal Conditions	■	■	■		■							■	
Congenital anomalies			■		■						■	■	
Malnutrition	○								■				
Cancer	■	■	■		■				■		■	☆	
Neuropsychiatric disorders				■	■	■						☆	■
Cataracts	■											■	
Deafness												☆	
Cardiovascular diseases	■	■	■	■		■	■		■			☆	
COPD		☆	■									☆	
Asthma		☆	■									☆	
Musculoskeletal diseases												☆	
Physical inactivity									☆				
Road Traffic Accidents									○			☆	
Falls					☆					☆		☆	
Drowning							○			☆		■	
Fires					☆							■	
Poisonings						■	○					■	
Other unintentional injuries					■		■		☆	■		■	■
Violence					☆	■			☆				
Suicide						■	☆					■	

Fraction attributable to the environment: ■ < 5%   ☆ 5–25%   ○ > 25%

Figure 5.6 *Table of indicative values for environmental attributable fractions, by specific environmental risk factor and disease risk*

the storage, transport and transmission of cultural content. With knowledge went the responsibility for its safe-keeping, continuity and proper use. The transmission of this knowledge within and outside the group was also subject to many different considerations and protocols of internal and external management of knowledge. Incidentally the practice of *ethical values* has been much grounded in indigenous and TK systems, in contrast to the conventional knowledge systems. This is why in Figure 5.3 the circle of ethics is represented in the traditional setup by a darker colour than the conventional side. The lack of the practice of ethics and/or its absence have subjected the traditional rural holders of knowledge to epistemological disenfranchisement by the combination of colonial, neo-colonial and apartheid practices buttressed by commercial greed and attitudes, ethos and practices of the scientific community. We need a group of academics and scientists who can act as catalysts and agents of change. They should in a participatory process initiate a process of the gradual transformation of conventional scientific ethos, which will enable ethics and practice to emerge, while developing a strong and committed system of protocols for developing and protecting indigenous knowledge systems, biodiversity and especially the intellectual property rights (IPRs) of the local communities. The ethical values should be based on the ten theses for biological diversity and ethical development sourced from Goulet (1993). Thus:

1 Ethical or authentic development requires biological diversity.
2 Ethical development also requires cultural diversity.
3 Ethical development requires plural modes of rationality for two reasons:
   –to destroy the monopoly of the legitimacy appropriated by unethical scientific and technological rationality.
   –to integrate technical, political and ethical rationalities in decision-making in a circular pattern of mutual interaction.
4 Ethical development requires plural modes of development. There is no single and necessary path to development predicated on energy-intensive, environmentally wasteful, culturally destructive and psychologically alienating models of progress.
5 Ethical development requires a non-reductionist approach to economics. As Schumacher (1973) insists in *Small Is Beautiful*, 'we must conduct economics as if people mattered'.
6 Ethical development requires pluralistic and non-reductionist approaches to technology. Technology is not an absolute value for its own sake that has a mandate to run roughshod over all consideration and must be demythologized.

7 Ethical development requires an approach to human beings that is not exclusively instrumental. Human beings are useful to other human beings and, to some degree, are used as aids in satisfying needs. But human beings have their ultimate worth independently of their instrumental value. Indeed, if one universal value exists in human life, it is that humans are precious for their own sake and on their own terms, independent of their utility to others.

8 The biosphere must be kept diverse both as an instrumental value to render ethical development possible and as a value per se. Like cultural diversity, biological diversity is a value for its own sake, although it's neither a transcendental nor an absolute value.

9 The question, 'Is it possible to have piety toward nature without accountability to nature's creator and to a supreme judge of human acts?', cannot be answered definitely and absolutely. One recalls, however, that all the great religions have preached the stewardship of the cosmos and responsibility for nature's integrity and survival, based on ultimate human accountability to nature's creator or providential conductor.

10 If ethical development is the only adequate support system for biological diversity, reciprocally, biological diversity is the only support system for ethical development.

It is worthwhile noting that the role of ethics in integrated sustainable development is grossly overlooked and/or underrated. Thus focusing on an indigenous-directed partnership approach to the on-going negotiations of the recognition, privileging, positioning, decolonizing, protection and involvement of indigenous/TK and practice, we should be aware of the wider and historical context of the discourse on these and associated matters, that is, colonialism, neo-colonialism, subjugation, dislocation and exploitation. This approach is essential in effecting encompassing transformation in a *world view and ethics of human kind* at this critical, challenging and pregnant state in history, where the presumed superiority of Western knowledge and knowledge systems have failed to solve all the inadequacies in political, economic, socio-cultural and technological development matters. Therefore, in the approach we need to be *objective* in order to get the objective facts of the past and the present, and reflect on these facts even though they are emotionally precipitating. These facts should then undergo an *open, transparent and participatory interpretive process* from which the *designed integrative processes* can effect a paradigm shift through action. This is not an accommodation but it amounts to a forward-looking liberation of substance, a *shared paradigm shift* in the lifestyles and values of the African and the Westerner. After all it is quite obvious from the preceding pages that modern medicine and indeed human civilization emanated from Africa, and by bringing this concept to the forefront we will ensure that there is an agreed understanding of where *we have come from, where we are now, where we went wrong, where we intend to go* and *how we intend to reach there.* This can better be guided by Figure 5.7, which illustrates possible and viable options of ethically guided integration of knowledge, knowledge systems as well as the elements of people, and the ecosystem explained in the previous sections of Figure 5.4. Figure 5.7 shows the possible stage-by-stage process of the possibilities for beginning and continuing with the ethical integration to ultimate human and ecological well-being depending on an analysis of the questions: Where have we come from? Where are we now? Where did we go wrong? Where do we intend to go and how do we intend to reach there?

One other critical guide for us in this thinking process is the question: could it be that in the process of their acquiring traditional medical knowledge from Africa (Egypt), the West did not take time to analyse the interactions of the various forces involved in the management of an illness leading to the inadequacies of the current practice of modern medicine? Otherwise how would you explain that 85 per cent of the population of Africa occasionally seek expensive and cumbersome African traditional medicine and not the conventional medical service even when the latter is free?

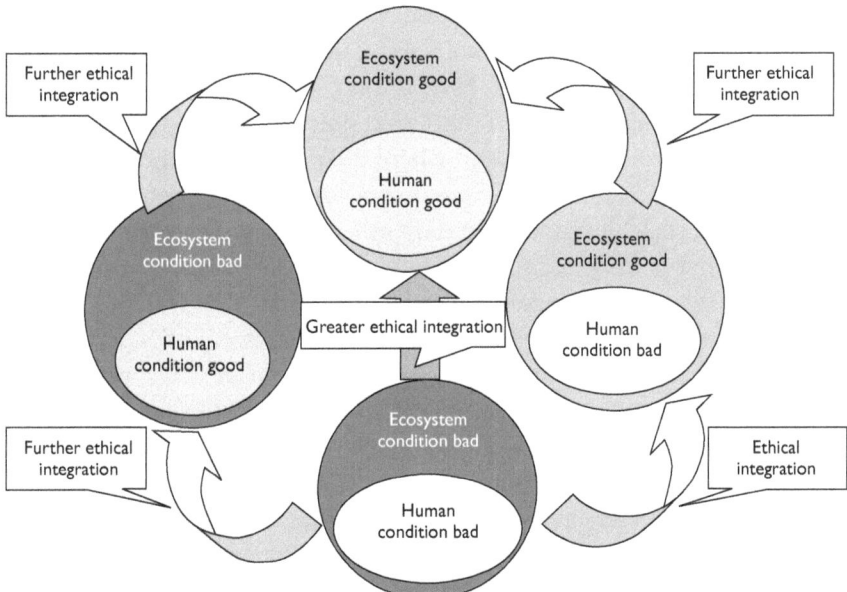

Figure 5.7 *Towards ethically integrated sustainable development (equal treatment of people and environment)*

Thus the process, in summary, should undergo the steps outlined in the subsequent subsections.

Examples of historical logical contribution of indigenous TK to conventional medicine: African and Egyptian traditional medical practice

- *The Papyrus Ebers*, a 110-page-long medical text, contains 887 sections in the Wreszinki version. This document from the early 18th dynasty (1567–1085 BC) represented by far the most sustained attempt in pre-Hellenic antiquity to understand the human body. It includes a treatise on cardiology and describes such practices as pulse taking.
- *Papyrus Edwin Smith*, another document from the early 18th dynasty, is a treatise on external pathology and bone surgery. It represents a systematic study of injuries, bruises and fractures affecting the entire body, from the head to the foot.
- *Mesopotamian medicine* was far below the level of ancient Egyptian Pharaonic medicine. And before the advent of Hippocrates, the father of modern medicine, it was ancient Egyptian medicine that influenced the medical schools of Asia minor. Hippocrates drew some of his expertise from the Egyptian tradition of medical science. This historical linkage is of decisive importance.
- The medical school of Alexandria was dominated by the names of Herophile (300 BC), best known as a specialist in anatomy, and Erasistratus, the physiologist. Now it was precisely in Egypt, where connections between the heart and the pulse beats had already been formulated for 2,500 years, and where the portable water clock (clepsydra) had also been in use for centuries, that Herophile counted pulse beats using the clepsydra. There are other medical practices mentioned in the Papyrus Ebers of Ancient Egypt, such as *medical incubation and aromatherapy*, whose ingredients came from the Land of Punt, which might be the present-day Puntland in Somalia, north of Kenya in East Africa. Other practices included *hypnotherapy, tooth filling*, with the use of several formulae based on resins and mineral substances, and *prescriptions for various illnesses*.
- The *Papyrus Ebers* also contain a store of ancient Egyptian knowledge, embracing: precise anatomical information, therapies based on vegetable, animal and mineral substances, invocations of *magical powers*, 'fetish' and religious forces, and the use of *incantatory therapies* based on the perceived powers of the spoken word; professional specialization in the treatment of illnesses; and the integration of metaphysical and

cosmic aspects in the treatment of illnesses. In a word this Pharaonic knowledge has contributed to and survived in the genuine medical and healing traditions of Africa and is alive today.

- The old territory of Buganda, now part of Uganda, a Kenyan neighbour in East Africa, was famous for the various specializations of its doctors. The Basawo Baganda were general medical practitioners, *musawo we musole* specialized in the treatment of snake bites, *mukozi we ddagala eganda* were the pharmacological experts who specialized in the concoction and prescription of medicines, and *musawo muyunzi* were the bone doctors, capable of managing all forms of fractures.
- In Kenya, the same kind of specialization of medical practices also exists; for example, there are brain surgeons who come from Kisii, an area in the western region of Kenya. These surgeons are capable of performing brain surgery to remove haematomas and other tumours with high precision and success.

In conclusion we can say that African medicine, a body of knowledge dating back to ancient Egypt, is an immense treasure of science, filled with ideas and practices rooted in the black African cosmobiological world view of the recent and remote past.

The study of medicine and the history of science would benefit from increased, widely propagated and integrated knowledge of that heritage. However, this can only happen if the respect for IP, community and individual rights are respected by individuals themselves, the world conventions, World Trade Organization (WTO) and other international bodies, including pharmaceutical companies.

Examples of limitations

It is sufficient to say that the following stand out as limitations to the process of knowledge transfer in most of the knowledge areas:

- That the CBD is widely considered an unenforceable protocol that relies on voluntary compliance among the parties. Although the CBD is a legal document, it lacks bite. There is no way to enforce it. Distrust is a problem. Some people think that bio-prospecting agreements are inherently unfair and out-priced for indigenous peoples from the outset. They assume that the cards are stacked against them.
- Regardless of the CBD and its principles, various environmental groups and researchers continue to view bio-prospecting as an initiative of

the North meant to globalize the control, management and biological diversity of resources which lie primarily in the Developing World.

- At present, 188 nations are parties to the CBD, of which 168 are signatories. A party nation is one that has signed, ratified, acceded to, accepted or approved the CBD. According to the CBD, ratification, accession, acceptance or approval all signify a country's willingness to be bound by the protocol. However, signing of the protocol does not, in itself, establish consent to be bound, hence the further act of ratification. The United States is the only signatory nation not bound by the CBD – the Senate has yet to ratify it after President Clinton signed it on 4 June 1993. Other than the United States, there are only six other sovereign entities that are not parties to the CBD. They are Andorra, Brunei Darussalam, the Holy See, Iraq, Somalia and Timor-Leste. Nations that did not sign the CBD during the time it was open for signature can only accede to it. Of the 188 parties, 20 fall under this category. The terms 'acceptance' and 'approval' are of more recent origin and mean the same as 'ratification'. The use of these terms stems from the diversity of the world's legal systems. This then gives a leeway to the nations who have not ratified the CBD to flout the regulations to this convention.
- A constructive and positive relationship has been barred by a tension between rage and reason, revolt and conciliation. There have also been and still are the not so honourable or respectable motives and actions that have been part of the experience and historical context. This has been the dominant reality throughout generations, even during the time the transfer of knowledge was occurring in the ancient Egyptian Pharaonic period. Anger, frustration, despair and justifiable mistrust have been arguably the main effects on Africa's indigenous people since colonization.
- That discrimination, disempowerment, oppression and marginalization that colonialism and neo-colonialism entail are sustained by racism, which is a living force that is embedded in Western culture and institutions. Some institutions such as the WTO and the transnational corporations have in them a culture of racism, which perpetuates subjugation and exploitation of African TK for their own selfish ends.
- Lack of systems of property rights in the indigenous communities.
- Corruption on the part of the duty bearers or their agents thereof who are given money by the bioprospecting companies so that biopiracy can be effected. This all boils down to laxity in establishing proper systems of checks and balances.
- Lack of awareness in the indigenous communities on their rights in as far as bioprospecting is concerned.

- Unscrupulous custodians of traditional medical knowledge in the practice itself.
- One of the central agreements of the WTO, the agreement on Trade-related Aspects of Intellectual Property Rights (TRIPS), obliges its member states to adopt patents, a sui generis system or a combination of both, for the protection of new plant varieties. The patenting of living organisms or their parts or components means legally granting private monopoly control rights over them and over their offspring.
- For Africa, the patents or other forms of IPRs on living organisms have profound implications for communal livelihoods that have sustained the continent for generations. The CBD recognizes the role and achievement of local and indigenous communities in the conservation of biological diversity and so recognizes the importance of biodiversity as an essential area within which to reaffirm and protect community rights. The TRIPS Agreement, however, is in direct conflict with the basic tenets of CBD, in that it formalizes the trend in which IPRs confer private, individual and exclusive ownership on life forms. There is a growing consensus that the current IPR regimes cannot protect indigenous technologies, innovations, practices and biodiversity. These systems encourage biopiracy and constitute a process of double theft. They steal creativity, innovations, technologies and practices of local communities by claiming collective innovations and practices to be their own, and then rob the community of the economic benefits derived from such products.
- There are limitations in terms of flouting religious beliefs of some communities because some traditional medical know-how operate under secret religious codes, and especially so in the secret societies as found among the Red Indians.
- The current talks are dwelling so much on the integration of the traditional practices and yet these practices have been passed through participatory educative processes. Unless the participatory educative processes become integrated, there may be limitations to the effectiveness and acceptance of the integrated practices.

Note

1   I would like to express my gratitude to the organizers and participants of the ABS workshop titled 'Undoing the Knot in Access and Benefit-Sharing Transactions. In Search of Amicable Solutions', which was held in Bremen, Germany, on 15–16 February 2008, where a paper examining these issues was first presented. I also thank my research assistants, Beatrice Kiambati, Thomas Kioko, Irene Ndungu and Lilian Wangui for their endless toil and unconditional support during the preparation of this publication.

References

Githae, J. (2005) 'Renaissance of indigenous knowledge and technology: The case of African traditional medicine', paper presented at the annual conference and workshop of the African Technology Policy Studies Network, Sun n Sand, Mombasa, Kenya, 28 November to 4 December 2005

Goulet, D. (1993) 'Biological diversity and ethical development', in Hamilton, L. S. (ed) *Ethics, Religion and Biodiversity: Relations Between Conservation and Cultural Values*, Cambridge, UK, The White Horse Press, pp17–19

Ndegwa, D. (2003) *Walking in Kenyatta Struggles*, Kenya Leadership Institute, Nairobi

Prescott-Allen, R. (2001) *The Wellbeing of Nations: A Country-by-Country Index of Quality of Life and the Environment*, Washington DC, Island Press

Schumacher, E. F. (1973) *Small is Beautiful: Economics as if People Mattered*, London, Blond & Briggs

Setting Protection of TK to Rights – Placing Human Rights and Customary Law at the Heart of TK Governance

Brendan Tobin

Introduction

Indigenous peoples insist that they be recognized as 'peoples' not 'people'. The 's' distinction is very important, because it symbolizes not just the basic human rights to which all individuals are entitled, but also land, territorial and collective rights, subsumed under the right to self-determination.

Darrell Posey

Despite more than 15 years of intense international debate, the rights of indigenous peoples over their traditional knowledge (TK) remain largely unprotected. Recent years have seen a welcome increase in national and international efforts to rectify this situation. This has included the adoption of a range of law, policy and administrative measures. It also includes a wide range of projects, programmes and processes designed to strengthen the role of TK in local, national and international development planning, environmental conservation and mitigation strategies and sustainable use of biological diversity.

Indigenous peoples have consistently argued that TK cannot be divorced from its cultural, spiritual, biological, environmental, territorial, legal and epistemological foundations. Effective protection of TK based upon this perspective can only be achieved within the framework of national and international recognition and protection of 'a basket of rights', including 'rights to land, traditional territories, sacred sites, biological and other resources, as well as indigenous peoples' rights to freely carry on their religious practices, organize themselves according to their own criteria and to apply their own customary laws' (Posey, 1996). Protection of TK will also be dependent upon realization of fundamental human rights to life, food and health, education, culture and self-determination.

Ongoing negotiations relating to TK protection at the Convention on Biological Diversity (CBD), World Intellectual Property Rights Organization (WIPO) and World Trade Organization (WTO) provide important opportunities to advance protection of TK. There are concerns, however, that these negotiations are tending to focus too much on regulation of trade issues and overlooking the fact that TK can only be effectively protected by addressing the wide range of threats it is exposed to. Indigenous peoples and local communities have cautioned against adoption of measures for protection that lean towards commoditization of TK, which they believe will exacerbate the pressures they already face in protection of TK. Instead they argue the international community should adopt an approach focusing on maintenance and nurturing of their knowledge and innovation systems, customary laws and traditional resource management practices through which TK has been developed, regulated and maintained for centuries.

This chapter focuses on the relationship between indigenous peoples' human rights, customary law and protection of TK. It concludes that realization by indigenous peoples of their civil and political as well as economic, social and cultural human rights is inextricably linked to protection of TK. It further concludes that effective respect and recognition for customary law is crucial for and dependent upon realization of indigenous peoples' rights to self-determination.

TK covers a very wide range of areas including knowledge relating to biological diversity, its conservation, management and uses, traditional cultural expressions such as song, dance and stories, and areas such as language, education, law and astronomy. Here the term 'TK' will be used primarily to refer to those aspects of indigenous peoples' knowledge which relates to biological diversity. Although focusing specifically on the rights of indigenous peoples, the current analysis also has relevance for protection of local communities' rights over their TK.

Indigenous peoples' human rights

International human rights law pertaining to indigenous peoples has evolved slowly. Early European attitudes towards indigenous peoples varied from recognition of their rights based on natural law to their treatment as uncivilized barbarians whose subjugation was a right if not an obligation of colonial powers. On the one hand, the Dominican Francisco de Vitorio (1480–1546) rejected the notion that first discovery provided a basis for claims over inhabited lands, taking the view that it 'did not destroy natural or human law which were the basis of ownership and dominium' (Thornberry, 2002). On the other hand, Juan Gines de Sepulveda, chaplain to the Spanish Emperor, took the view that indigenous peoples were barbarians lacking any science, written laws or even private property (Thornberry, 2002).

Over time indigenous peoples began to be treated as unfit to manage their own affairs, leading to development of a doctrine of guardianship which eventually found its way into international treaties. The Covenant of the League of Nations, for instance, holds that the development of peoples of colonies and territories, which gained independence after World War I and who 'are considered not yet able to stand by themselves', will best be realized where they come under the 'tutelage' of advanced nations.[1]

A similar paternalistic approach is apparent in the International Labour Organization (ILO) Convention 107, on Indigenous and Tribal Populations, adopted in 1957. Convention 107, the first international treaty specifically focusing on human rights of indigenous peoples, promotes their assimilation into the wider national population. While recognizing the right of indigenous and tribal populations to retain their own customs and institutions, the Convention limits this right where it is considered incompatible with the national legal system or the objectives of integration programmes (Article 7.2).

Assimilation of indigenous peoples came in time to be seen as inimical to realization of their human rights. This is reflected in ILO Convention 169 on Indigenous and Tribal Peoples in Independent Countries, adopted in 1989, which recognizes indigenous peoples' rights to their cultures, the importance of customary law and their rights to retain their institutional structures and distinctive customs, where not incompatible with fundamental human rights (Article 8). In essence recognizing significant rights to self-determination. It also recognizes their rights to their lands, territories and natural resources, as well as to participate in decision-making processes which affect them.[2]

Recognition of indigenous peoples' human rights received a major boost with the adoption in September 2007 of the United Nations Declaration on the Rights of Indigenous Peoples (UNDRIP).[3] The Declaration recognizes a broad range of rights for indigenous peoples over their lands, resources, culture and TK. This includes the right to maintain, protect and develop the past, present and future manifestations of their cultures, including their TK and any intellectual property over it. States are required to give due respect to the customs, traditions and land tenure systems of the indigenous peoples concerned when giving legal recognition and protection to indigenous peoples' rights. Where resources have been confiscated, used or damaged without the prior informed consent (PIC) of indigenous peoples, states are obliged to provide restitution or fair and equitable compensation.[4] The Declaration also obliges states 'to provide redress through effective mechanisms ... developed in conjunction with indigenous peoples with respect to their cultural, intellectual, religious and spiritual property taken without their free prior and informed consent or in violation of their laws, traditions and customs'.[5]

Convention 169, UNDRIP and other human rights instruments together establish a wide range of binding and 'soft-law' commitments, which will need to be taken fully into account by negotiators in the development of any international law affecting TK, including regulation of access to genetic resources and benefit sharing (ABS).

Protection of TK and human rights

Indigenous peoples have consistently argued that realization of their human rights is vital for the protection of TK. TK forms a significant part of the body of knowledge that humankind has developed over centuries. As such, it has intrinsic value as part of 'mankind's cultural patrimony'.[6] More importantly it is fundamental to cultural survival and the basis for subsistence and development strategies of a large proportion of the world's population. A wide variety of human rights, including rights to life, health, food, education, culture, land and resources, intellectual property, self-determination and participation in decision making are all directly linked to the maintenance, transmission and continuing development of TK.

Rights to life, health, food and education

TK is inextricably linked to protection of the right to life and rights to health care in many countries. WHO has estimated that up to 80 per cent of the populace of nonindustrial countries depends primarily on traditional medicinal practices for their health needs (Posey, 1999). Centralization of advanced medical facilities in major cities, cultural legitimacy and recognized efficacy, as well as economic affordability, are all factors that make traditional medicine the preferred and often the only source of health care for indigenous peoples. In some African countries, there is less than one doctor per 33,300 inhabitants, while there is one traditional healer for every 200 (http://www.oecd.org/dataoecd/56/39/40997324.pdf, accessed 2 June 2009).

TK plays a crucial role in securing local and global food security through development and conservation of local crop varieties (landraces), domesticated animals and wild biological diversity (Tobin and Taylor, 2009). The maintenance of traditional diet based on TK and conservation of these resources is of much importance for securing indigenous peoples' rights to health. Research has shown, for example, that indigenous peoples who adopt a Western diet, typically high in fat and refined carbohydrates, 'seem particularly vulnerable to suffer lifestyle diseases, like obesity, certain kinds of cancer, diabetes, high blood pressure and coronary heart disease' (Damman, 2005). Traditional diet also holds a central place in indigenous peoples' cultural and spiritual life, telling 'part of the story of who they are, their relationship to nature, and play[s] an important role for their cultural identity' (Damman, 2005).

The right to education including bi-lingual education is crucial for protection of TK. In order to respect, protect and fulfil the essential features of the right to education, states need to ensure its availability, accessibility, adaptability and acceptability. To fulfil the acceptability obligation, 'they must take positive measures to ensure that education is culturally appropriate for … indigenous people and of good quality'.[7] Ensuring appropriateness will require the development of curricula which incorporate TK into daily learning and teaching practices. Provision of opportunities for those traditionally responsible for transmission of indigenous knowledge, including women and elders, in formal education systems is likely to play a central role in protection of culture and the continuing utilization and development of TK.

Rights to culture, land and resources

TK of indigenous peoples finds its expression in their social and political structures, legal regimes, institutions, land and resource management practices, housing, traditional medicine and innovations systems, as well as in their song, dress, dance, music and stories. All of these form part and parcel of indigenous peoples' cultural diversity. The protection of culture in human rights law addresses scientific, literary and artistic pursuits of society, and the right of peoples to practice and continue shared traditions and activities.[8]

Until recently, the international community has focused primarily on protection of rights over cultural expressions, access to benefits of science and rights of authors. Treaties on culture have focused on the protection of valuable artifacts and buildings for example, leading one commentator to claim that 'the further the distance from the cultural losses of World War II, the more evident it becomes that the right to culture is not simply a right to culture but a right to *cultural property*'.[9] As with the protection of cultural sites, protection of authors' rights, primarily through use of intellectual property regimes, has tended to focus attention on material values rather than on the intrinsic spiritual, social, environmental and aesthetic values of culture. Commodification of cultural manifestations of TK, and in particular the use of intellectual property as a means to exert control over TK, has been strongly opposed by indigenous peoples. They argue that international law should focus more attention on development of measures to secure their cultural integrity, rather than on those designed to facilitate trade and meet cultural priorities set by big business, states and national elites. International instruments on indigenous peoples' human rights have responded to these concerns by adopting an alternative approach to culture, treating it as an activity rather than goods, whose protection is necessary to secure human dignity.[10] UNDRIP, for instance, recognizes the rights of indigenous peoples to practise and revitalize their cultural traditions, including the right to maintain, protect and develop the past, present and future manifestations of their cultures.[11]

Culture is based upon and interwoven with traditional practices, knowledge systems and sacred rites which cannot be easily maintained without protection of rights to lands, territories and resources. Speaking on the occasion of the adoption of UNDRIP, Professor Rodolfo Stavenhagen, Special Rapporteur on the situation of human rights and fundamental freedoms of indigenous peoples, highlighted this interrelationship between land and culture, saying: 'Indigenous peoples' ancestral lands and territories constitute the bases of their collective existence, of their cultures and of their spirituality. The Declaration affirms this close relationship, in the framework of their right, as peoples, to self-determination in the framework of the States in which they live'.[12]

The collective right of indigenous peoples to maintain and practise their culture does not always sit easily with the promotion of individual human rights, leading to sometimes diametrically opposed views of the impact of culture on human rights. On the one hand, protection of culture manifested in traditional institutions and legal systems are seen as crucial for realization of the right to self-determination. On the other, cultural relativism, customary law and traditional authority are seen as the screen and tools for continuing infringement of individual human rights, particularly the rights of women.

International law limits the application of rights to self-determination and to apply customary law where they would run counter to fundamental human rights. Similar qualifications are found in many national constitutions. Securing a balance between respect for customary law and rights to self-determination and individual human rights cannot easily be secured without the support of indigenous peoples' decision-making authorities. Furthermore, a general lack of effective mechanisms for enforcement of human rights suggests the need to promote realization of individual rights amongst indigenous peoples by 'persuasion rather than force ... cultural transformation rather than coercive change' (Engle Merry, 2006, p586).

Realization of international human rights law, including any law relating to culture and TK, will be enhanced where full participation of indigenous peoples in relevant international decision-making forums is secured. In this vein, international negotiations on ABS, TK and intellectual property will need to enhance indigenous participation at all levels of decision making, implementation and enforcement of relevant law and policy.

Intellectual property

Recent years have seen increased attention being given to the relationship between intellectual property and human rights. The 1948 Universal Declaration on Human Rights and 1966 International Covenant Economic Cultural and Social Rights recognize the rights of everyone to protection of the moral and material interests resulting from any scientific, literary or artistic production of which he [or she] is the author, and to share in scientific advancement and its benefits.[13]

Intellectual property rights have, however, the potential to conflict with human rights. The UN Sub-Commission on the Protection and Promotion of Human Rights, for instance, has stated that 'actual or potential conflicts exist between the implementation of the WTO Agreement on Trade Related Intellectual Property Rights (TRIPS) and the realization of economic, social and cultural rights'.[14] The uncertainties arising from these seemingly conflicting perspectives, in which intellectual property is seen variously as supporting and undermining human rights, 'highlight the need to develop a comprehensive and coherent "human rights framework" for intellectual property law and policy' (Helfer, 2007).

Intellectual property has been seen by indigenous peoples as a tool for biopiracy, which may be loosely defined as the unapproved and/or uncompensated use of genetic resources (GRs) and TK, for commercial and scientific purposes. These include grants of patents on products and inventions based on the neem tree, tumeric, basmati rice, maca, quinoa and ayahuasca. Such cases have led to general mistrust amongst indigenous peoples towards intellectual property systems, which they believe legitimize 'misappropriation of our peoples' knowledge and resources for commercial purposes'.[15] They argue that existing mechanisms are insufficient for protection of their intellectual and cultural property,[16] and that new legislative frameworks and sui generis systems to protect TK should be based on customary law and governance systems.[17] Indigenous peoples view 'all aspects of ... intellectual property (determination of access to natural resources, control of the knowledge or cultural heritage of peoples, control of the use of their resources and regulation of the terms of exploitation) ... [as] ... aspects of self-determination'.[18]

Rights to self-determination and participation

The human right to self-determination is set out in identical terms in two binding UN human rights instruments, the International Covenant on Civil and Political Rights (CCPR) and the International Covenant on Economic, Social and Cultural Rights (CESCR). Article 1 of each Covenant provides that:

> 1   *All peoples have the right of self-determination. By virtue of that right they freely determine their political status and freely pursue their economic, social and cultural development.*
>
> 2   *All peoples may for their own ends, freely dispose of their natural wealth and resources... In no case may a people be deprived of their means of subsistence.*
>
> 3   *The States Parties to the present Covenant ... shall promote the realization of the right of self-determination, and shall respect that right in conformity with the provisions of the Charter of the United Nations.*†

Numerous declarations by indigenous peoples begin with iteration of their right to self-determination as the basis for realization of their social, cultural and economic rights,[19] including the protection of rights over TK.[20] The Mataatua declaration, for instance, presents the right of self-determination as being inextricably linked to TK, arguing that 'in exercising that right [indigenous people] must be recognized as the exclusive owners of their cultural and intellectual property'.[21] Self-determination is seen as embracing the 'rights of indigenous peoples to live in [their] own territories, with respect for [their] distinct cultures, political institutions and customary legal systems, while allowing [them] the means to carry out their own sustainable self-development'.[22]

Indigenous peoples have stated their willingness to offer access to their TK and intellectual property for the benefit of all humanity provided

---

† See Article 1 of the International Covenant on Civil and Political Rights, United Nation Treaty Series, vol 999, p171, and Article 1 of the International Covenant on Economic, Social and Cultural Rights, vol 999, p3.

their 'fundamental rights to define and control this knowledge are protected by the international community'.[23] Realization of their rights to self-determination will therefore require empowerment of indigenous peoples to 'determine when, where and how [their TK] is used'.[24] Effective protection of indigenous peoples' rights will require attention not only to issues of ABS relating to TK, but also the need to strengthen and secure the knowledge systems within which it has traditionally been developed, maintained and nurtured.

TK and TK systems face a wide range of external and internal threats. These include not only biopiracy but often more pernicious threats arising from loss of language, land and resources, environmental degradation and the application of inappropriate development policies in areas such as health, education, agriculture and fisheries.[25] Responding to these threats will require a multi-sectoral approach, guided by commitment to securing indigenous peoples' human rights, including their 'rights to self-development', which depend upon recognition of their rights to self-determination.[26]

The link between self-determination, protection of TK and respect for indigenous peoples' customary laws has been clearly set down in Principles and Guidelines for the Protection of the Heritage of Indigenous Peoples, elaborated by the Special Rapporteur to the Sub-Commission, Mrs Erica-Irene Daes. These state that:

> 2 *To be effective, the protection of Indigenous Peoples' heritage should be based broadly on the principles of self-determination, which includes the right and the duty of Indigenous Peoples to develop their own cultures and knowledge systems, and forms of social organization.*

> 4 *International recognition and respect for the Indigenous Peoples' own customs, rules and practices for the transmission of their heritage to future generations is essential to these peoples' enjoyment of human rights and human dignity.*[27]

Similarly, Agenda 21, which devotes an entire chapter (26) to 'Recognising and Strengthening the Role of Indigenous People and their Communities', proposes that governments should adopt or strengthen appropriate policies and/or legal instruments that will protect indigenous intellectual and cultural property and the right to preserve customary administrative systems and practices.

Indigenous peoples have clearly identified an interrelationship between the struggle for self-determination and for protection of TK. On the one hand self-determination has been portrayed as 'a strong counter-force to ... intellectual property rights systems'[28] which threaten TK; while on the other hand protection of TK is seen as so central to indigenous peoples' struggle to secure their rights over land, resources and culture that it is viewed as being 'just as important as the struggle for self-determination'.[29] The extent to which customary law is effectively recognized, respected and incorporated into international and national TK regimes will serve as a clear indicator of the commitment to securing indigenous peoples' rights to self-determination and over their TK.

Effective exercise of rights to self-determination is dependent upon realization of rights to participation. UNDRIP recognizes indigenous peoples rights: to participate in decision making in matters which would affect their rights; to choose their own representatives; and obliges states to consult with them in good faith to obtain their free PIC before adopting or implementing legislative or administrative measures that may affect them.[30] In a similar fashion, ILO Convention 169 creates obligations on states to consult indigenous peoples prior to granting any rights for exploration or exploitation of resources on their lands.[31]

Both CBD and the WIPO Intergovernmental Committee on Intellectual Property, Genetic Resources, Traditional Knowledge and Folklore (IGC) have adopted measures to secure greater participation of indigenous peoples in their debates on TK protection. This has included: the establishment of funding mechanisms to secure indigenous participation as observers at international negotiations; requests that member states consider inclusion of indigenous representatives in national delegations; and the establishment of processes for securing greater involvement of indigenous peoples in negotiations, as has been done at the CBD working group on Article 8(j) (WG 8(j)). However, their role is still marginal and indigenous peoples have argued that although they 'are not states they are nations',[32] and should be given special recognition in forums such as the WIPO IGC. The extent to which indigenous peoples are incorporated into decision-making processes on TK will be indicative of commitments to realization of rights to self-determination,[33] and to respect and recognition for customary law and governance structures.[34]

Customary law and protection of TK

As seen earlier, international human rights law recognizes the right of indigenous peoples to their customs and institutions, and requires that customary law be taken into consideration in the development of national law, which may affect them. Many pluricultural and multiethnic states have already enshrined recognition of customary law in national constitutions and/or laws, leading to a diverse and rich body of legal experience in building bridges between customary law and other legal traditions.[35]

CBD and the WIPO IGC have both stressed the importance of customary law for TK protection. National practice is also showing a tendency to include a role for customary law in national TK law. This role has been primarily geared towards resolution of internal disputes between or among communities. However, CBD and many national and regional ABS regimes require users to obtain PIC of indigenous peoples as a condition for access to and use of their TK. This creates an opportunity for indigenous peoples to extend the remit of customary law by defining procedures for access and use of TK in accordance with their own internal legal regimes. A proposal to amend TRIPS in order to include requirements for disclosure of PIC in patent and other intellectual property grant procedures would, if adopted, further strengthen indigenous peoples' ability to control TK in accordance with their customary laws. This proposal is now supported by a majority of WTO member states.[36]

If customary law is going to play a central role in international TK protection, any ABS and TK regimes will need to create obligations for its recognition and enforcement in those countries where TK custodians reside and in countries where their TK is used. ABS and TK governance may as a result lead to customary laws finding their way into judicial, administrative and alternative dispute resolution proceedings, as well as into legislative decision making. These may involve not only strict commercial dealings and questions of misappropriation and compensation, but also questions of moral rights, cultural patrimony, spiritual integrity and environmental protection. Recognition of a vast multiplicity of customary law regimes will require flexibility, sensitivity, imagination and, above all, respect for its place amongst the sources of law that form part of a global intercultural and pluralistic legal order.[37]

In order to achieve culturally and spiritually appropriate and effective protection of TK, legal pluralism cannot be envisaged as mere acceptance of co-existence of legal regimes, with customary law applicable only to indigenous peoples within their territories and in relation to their own internal affairs. Rather it will require incorporation directly or indirectly of principles, measures and mechanisms drawn from customary law within national and international legal regimes for protection of TK. Achieving such an end makes it imperative that full and effective participation of indigenous peoples is secured from the outset in the development, implementation, monitoring and enforcement of relevant law and policy.

Efforts to develop measures for TK protection based upon customary law face the challenge of how to provide respect and recognition for largely oral-based legal systems, which frequently lack any central decision-making authority, and where divisions amongst representative organizations, traditional authorities and within and between communities may make the identification of those entitled to grant PIC and resolve disputes hard to identify. Those faced with the challenge will need to avoid the temptations of seeking a technical solution to what is primarily a cultural issue.

Recognition of a role for customary law in the management of TK is just one step, although a crucial one, for securing effective protection of TK. It will need to be complemented by measures for recognition and protection of indigenous peoples' rights to control their lands, resources, languages, education, health, cultures and religions. The effectiveness of any such measures will depend in no small part on the extent to which they incorporate compliance mechanisms to ensure their enforcement. Compliance will also depend upon the capacity of national and international authorities and institutions to adapt to and give due recognition to systems which require the application of customary law and legal principles. That capacity and the willingness to apply customary law and legal principles are likely to be affected by issues of legal certainty. Amongst the areas of law that will require examination and potential modification if customary law is to be respected, recognized and enforced outside the jurisdiction of indigenous peoples will be: contract, equity and tort law; legal principles of the public domain, fiduciary obligations and misappropriation of TK; rules governing examination of evidence and prior art; recognition of foreign judgements based upon interpretation of customary law; and enforcement of decisions (of a judicial nature) made by indigenous peoples' and local communities' traditional authorities.

In developing national and international law, attention should be given to national and regional experiences in the preparation, adoption and implementation of TK legislation. One particularly interesting example, which is discussed briefly below, is that of Peru.

The use of positive and customary law for TK protection in Peru

In August 2002, Peru became the first country to adopt a comprehensive regime for protection of TK related to biodiversity. Law 27811, which established a regime to protect the rights of indigenous peoples over their collective knowledge relating to biological diversity, is declaratory in nature, recognizing the existence of ancestral rights over TK. It requires PIC of indigenous peoples for access to and use of TK, and establishes a range of measures to secure fair and equitable benefit sharing and to prevent and mitigate acts of misappropriation. It also recognizes the existence of subsisting rights of indigenous peoples over TK in the public domain and their rights to apply customary law to resolve internal disputes. The law recognizes TK to be cultural patrimony, which implies that it must be managed for the good of present and future generations. However, the law goes on to establish that any single community may enter into an agreement for its use. The result is a tension between rights over collective cultural patrimony and the rights of an individual community to enter into a commercial agreement for the use of TK. The Peruvian TK law's provision that indigenous peoples may resort to customary law to resolve their internal conflicts provides a partial solution to this problem. However, a community lured by offers of commercial benefits associated with entering into a TK agreement may potentially decide to act alone, and, unless it decides to submit to a customary law-based decision-making forum, it cannot be obliged to do so.

The formation of the Peruvian law was informed by the fact that many indigenous peoples in Peru have traditionally lacked any central authority, and their current political organization includes numerous local and regional organizations and a number of national organizations, which are far from unified. Requiring approval of all custodians of TK as a condition for entering into any agreement for its use would, in all likelihood, lead to a virtual moratorium on bioprospecting involving TK in all but a few cases. In cases where indigenous peoples lack any central decision-making authority or accepted process to resolve intercommunity and inter-organizational disputes, customary law may not initially offer any solution. Customary law is not, however, a stagnant body of law, but is open to constant refinement and adaptation to meet changing cultural, economic, social and legal realities.

Bioprospecting, biopiracy and progressive national and international regulation of TK are amongst the current challenges to which indigenous peoples and customary law need to respond. One means adopted by indigenous peoples for addressing these challenges has been the progressive development of a range of legal and quasi-legal instruments, including research guidelines and sui generis contracts based upon customary law to regulate access to and use of TK. Such instruments have individually and collectively come to be known as community protocols. Indigenous peoples in Peru have utilized customary law and traditional decision-making forums in the development of a number of precedent-setting ABS-related agreements. These include an agreement for repatriation of native potato varieties between Andean communities and the International Potato Center and a bioprospecting agreement involving communities of the Aguaruna people of the northern Peruvian Amazon, negotiated within the framework of the International Biodiversity Group (ICBG) Program.[38]

Based upon their experience with the ICBG Program, Aguaruna communities participating in a series of workshops to review the Peruvian TK law came to the conclusion that indigenous peoples with shared TK should work together to develop community protocols defining procedures for issues such as PIC, benefit sharing and dispute resolution based on customary law. They proposed that a common community protocol on ABS and TK should be developed by the Aguaruna, Huambisa, Shuar and Achuar indigenous peoples, who all form part of the Jivaro ethnic group and share much of the same TK. The fact that their traditional territories are found on both sides of the border between Peru and Ecuador was seen as giving even more importance to the need for development of a common protocol.

Development of community protocols – in particular those in the form of guidelines or community by-laws setting down procedures on issues such as applications for access to TK and GRs, identifying authorities entitled to receive applications and to convey decisions on PIC – can send a strong message to national, regional and international authorities on how indigenous peoples wish to see issues of PIC, benefit sharing and dispute resolution addressed in TK legislation. In doing so they are likely to be influential in helping to define modalities for building functional interfaces between positive and customary law and their respective decision-making and enforcement authorities. At the regional level processes such as the ongoing negotiations to develop a sui generis TK regime within the four countries of the Andean Community of Nations (CAN), implementation of the African Model Law, and development of model TK regulation in the Pacific region, would all benefit from clearer definition by indigenous peoples of modalities for respecting and recognizing customary law.

Conclusions

If TK law and policy is to empower rather than disenfranchise indigenous peoples, it will need to build functional interfaces between international, national and indigenous peoples' legal regimes and their respective decision-making and enforcement mechanisms. Full participation of indigenous peoples in decision making, adoption and implementation of relevant law, policy, programmes and projects will be crucial in order to ensure the identification of common objectives and coherency, and appropriateness of legal and other measures for TK protection.

Rights to life, food, health, culture, traditional territories, resources, work and development are all relevant to and variously dependent upon protection of TK. Realization of these rights will require concerted and coordinated action across many sectors to address the wide range of external and internal forces that threaten TK. The international community and national authorities will, therefore, need to broaden their approach to TK protection to include not only control of third-party commercial use, but also adoption of measures and mechanisms, including necessary funding, to strengthen and invigorate TK systems.

Development of TK protection measures will need in particular to recognize, give deference to and support realization of the fundamental right of indigenous peoples to self-determination. While national authorities have the primary responsibility for taking measures to ensure that indigenous peoples' human rights are respected, protected and fulfilled, the international community can support their realization by establishing a global framework for TK protection that is based on a human rights approach, and gives due respect and recognition for indigenous peoples' customary laws and institutions.

Notes

1 Article 22, Covenant of the League of Nations.
2 A majority of countries which were parties to Convention 107 have now ratified Convention 169. However, the former Convention continues to apply to a number of mainly African and Asian countries. Eighteen countries have ratified Convention 169: Argentina, Bolivarian Republic of Venezuela, Bolivia, Brazil, Colombia, Costa Rica, Denmark, Dominica, Ecuador, Fiji, Guatemala, Honduras, Mexico, the Netherlands, Norway, Paraguay, Peru and Spain.
3 UN Declaration on the Rights of Indigenous Peoples (UN General Assembly Resolution A/RES/61/295 of 13 September 2007), Article 11(1).
4 Article 8, ILO Convention 169.
5 UN Declaration on the Rights of Indigenous Peoples, Article 28.
6 Declaration of Principles ratified by the IV General assembly of the World Council of Indigenous Peoples, in Posey, D. (ed) (1999) *Cultural and Spiritual Values of Biodiversity*, Nairobi, IT/UNEP, p556.
7 Committee on Economic, Social and Cultural Rights General Comment 13 (Article 13). Report on the 20th and 21st Sessions, E/2000/22, para 50.
8 Human Rights Education Associates (HREA), 2003. 'The right to culture', available at http://www.hrea.org/index.php?base_id=157, accessed 21 March 2009.
9 Barthel-Bouchier, D. (undated) 'A right to culture?' All Academic Research. Available at http://www.allacademic.com//meta/p_mla_apa_research_citation /0/9/6/8/9/pages96897/p96897-1.php, last accessed 29 December 2008.
10 Holder, C. 'Culture as an activity and human right: an important advance for indigenous peoples and international law', Bnet, available at http://findarticles.

com/p/articles/mi_hb3225/is_1_33/ai_n29426311/print?tag=artBody, accessed 29 December 2008.

11 UNDRIP Article 11.

12 Statement of Rodolfo Stavenhagen, following adoption of UNDRIP, Geneva, 14 September 2007.

13 Universal Declaration of Human Rights, Article 27, G.A. Resolution 217A(III), UN GAOR, Third Session, First len. Mtg., UN Doc A/810 (10 December 1948), International Covenant on Economic, Social and Cultural rights, Arts, 15(1)(b), (c), 16 December 1966, 933 UNT.S. 3,5.

14 UN Economic and Social Council [ECOSOC] Sub-Commission on Promotion and Protection of Human Rights, Intellectual Property Rights and Human Rights, Resolution 2000/7, UN Doc.E/CN.4/Sub.2/RES/2000/7 (17 August 2000). The UN Sub-Commission on the Promotion and Protection of Human Rights was abolished by General Assembly resolution 60/251 of 15 March 2006, which transferred its mandate, functions and responsibilities together with those of the Commission on Human Rights to the new Human Rights Council.

15 Statement from the International Consultation on Intellectual Property Rights and Biodiversity, organized by the Coordinating Body of Indigenous Peoples of the Amazon Basin (COICA), Santa Cruz de la Sierra, Bolivia 1994, in Posey, 1999, p571.

16 Mataatua Declaration on the Cultural and Intellectual Property Rights of Indigenous Peoples, June 1993, in Posey, 1999, p564.

17 Leticia Declaration and proposals for action from the international meeting of indigenous and other forest-dependent peoples on the management, conservation and sustainable development of all types of forests, Leticia, Colombia, 13 December 1996, in Posey, 1999, p581.

18 Statement from the International Consultation on Intellectual Property Rights and Biodiversity, organized by the Coordinating Body of Indigenous Peoples of the Amazon Basin (COICA) Santa Cruz de la Sierra, Bolivia 1994, in Posey, 1999, p571.

19 See, generally, Declarations of Indigenous Peoples Organizations, in Posey, 1999, p555. On economic rights; see recommendations from the conference 'Voices of the Earth: Indigenous Peoples, New Partners and the Right to Self-determination in Practice', Amsterdam, 10–11 November 1993, in Posey, 1999, p566.

20 Leticia Declaration and proposals for action from the international meeting of indigenous and other forest-dependent peoples on the management, conservation and sustainable development of all types of forests, Leticia, Colombia, 13 December 1996, in Posey, 1999, p581.

21 Mataatua Declaration on the Cultural and Intellectual Property Rights of Indigenous Peoples, June 1993, in Posey, 1999, p564.

22 Contribution submitted by the International Alliance of the Indigenous Tribal Peoples of the Tropical Forests to the Intergovernmental Panel on Forests (IPF/II), Geneva, 1996.

23 On willingness to share TK, see the Mataatua Declaration on the Cultural and Intellectual Property Rights of Indigenous Peoples, June 1993, in Posey, 1999, p564; on willingness to share intellectual property, see Julayinbul statement on Indigenous Intellectual Property Rights, Jingarrba, Australia, 27 November 1993.

24 The Charter of the Indigenous-Tribal Peoples of the Tropical Forests, Penang, Malaysia, 15 February 1992, in Posey, 1999, p556.

25 Tobin, B. and Swiderska, K. (2001) *Speaking in Tongues: Indigenous Participation in the Development of a Sui Generis Regime to Protect Traditional Knowledge in Peru,* London, UK, IIED.

26 Recommendations from the conference 'Voices of the Earth: Indigenous peoples, new partners and the right to self-determination in practice', Amsterdam, 10–11 November 1993, in Posey, 1999, p566.

27 See Annex to *Final report of the Special Rapporteur, Mrs. Erica-Irene Daes, in conformity with Subcommission resolution 1993/44 and decision 1994/105 of the Commission on Human Rights,* 21 June 1995. E/CN.4/Sub.2/1995/26.

28 UNDP consultation on the Protection and Conservation of Indigenous Knowledge, Sabah, East Malaysia, 24–27 February 1995, in Posey, 1999, p574.

29 UNDP consultation on the Protection and Conservation of Indigenous Knowledge, Sabah, East Malaysia, 24–27 February 1995, in Posey, 1999, p574.

30 United Nations Declaration on the rights of Indigenous Peoples, Articles 18 and 19.

31 ILO Convention 169, Article 15.

32 Intervention of Stuart Wuttke, acting director of environmental stewardship for the Assembly of First Nations, the national governing body representing First Nations communities in Canada, at 13th meeting of the WIPO IGC, reported by Mara, K., in 'Indigenous people seek recognition at WIPO meeting on their rights', Intellectual Property Watch, 23 October 2008, www.ip-watch.org/weblog/index.php?p=1286, accessed 21 March 2009.

33 Posey, D. (undated) 'National laws and international agreements affecting indigenous and local knowledge: conflict or conciliation?', http://lucy.ukc.ac.uk/ Rainforest/SML_files/Posey/posey_1.html#Section0, accessed 21 March 2009.

34 Leticia Declaration and proposals for action from the International Meeting of Indigenous and Other Forest-Dependent Peoples on the Management, Conservation and Sustainable Development of all Types of Forests, Leticia, Colombia, 13 December 1996, in Posey, 1999, p581.

35 Tobin, B. (2009) *The Role of Customary Law in ABS and TK Governance in Andean and Pacific Island Countries,* UNU, WIPO (forthcoming).

36 Kohr, M. (2008) *TRIPS: Majority of WTO members now support disclosure proposal,* TWN Info Service on WTO and Trade Issues (Mar08/18), 19 March 2008, Third World Network. Available at http://www.twnside.org.sg/title2/wto.info/ twninfo20080318.htm, accessed 21 March 2009.

37 For discussion of customary law and its place amongst legal traditions of the world, see Glenn, H. P. (2000) *Legal Traditions of the World,* Oxford, New York, Oxford University Press.

38 For detailed discussion of the application of customary law to TK protection in Peru, see Tobin B. and Taylor E. (2009) *Across the Great Divide: Customary Law and the Protection of Traditional Knowledge: A case study of complementarity and conflict between national sui generis legislation and customary governance of traditional knowledge in Peru,* Lima, SPDA, IDRC.

References

Damman, S. (2005) 'The right to food of indigenous peoples', in Eide, W. B. and Kracht, U. (eds) *Food and Human Rights Development Volume I: Legal and Institutional Dimensions and Selected Topics*, Antwerp, Oxford, Intersetia

Engle Merry, S. (2006) 'Human rights and gender violence', extract reprinted in Steiner, H. P. Alston and Goodman, R. (2008) *International Human Rights in Context*, third edition, Oxford, Oxford University Press

Helfer, L. R. (2007) 'Toward a human rights framework for intellectual property', *University of California Davis Law Review*, vol 40, pp977–1020

Posey, D. (1996) *Traditional Resource Rights: International Instruments for Protection and Compensation for Indigenous Peoples and Local Communities*, Gland, IUCN

Posey, D. (ed) (1999) *Cultural and Spiritual Values of Biodiversity*, Nairobi, IT, UNEP

Thornberry, P. (2002) *Indigenous Peoples and Human Rights*, Manchester, Melland Schill Studies in International Law, Juris Publishing/Manchester University Press

Tobin, B. and Taylor, E. (2009) *Across the Great Divide: Customary Law and the Protection of Traditional Knowledge: A case study of complementarity and conflict between national sui generis legislation and customary governance of traditional knowledge in Peru*, Lima, SPDA, IDRC

Chapter 7

A Socio-legal Inquiry into the Protection of Disseminated Traditional Knowledge – Learning from Brazilian Cases[1]

John B. Kleba

Introduction

The Convention on Biological Diversity[2] (CBD) is considered the most important binding instrument in international law for protection of *traditional knowledge associated to genetic resources*[3] (TK) (Hahn, 2004, p114). It recognizes the interdependence between the conservation of biodiversity and the protection of local communities and indigenous populations (Preamble). There is a double vulnerability to be protected, as both nature and the culture of local populations are threatened by the way modern economy is expanding. Hence, the CBD recognizes the innovative nature of the knowledge and practices of traditional communities, and recommends their involvement and approval in utilizing such knowledge. In addition, seeking to ground on ethical bases the interactions between traditional peoples and economic market agents (Khor, 2002), the Convention requires a fair and equitable sharing of benefits arising from the commercial uses of TK (Article 8(j)).

In Brazil, access to TK for research and for commercial purposes is regulated by Provisional Measure (MP) 2.186-16/2001.[4] Access to TK must be authorized by the Genetic Patrimony Management Council (CGEN),[5] based on a Prior Informed Consent Agreement (PIC)[6] and a Contract for Genetic Patrimony Use and Benefit Sharing (BS contract), to be signed by both providers and users.

Brazil is a country known for the magnitude of its biological and cultural diversity, where the hope to join biotechnological development with fair treatment for TK holders has faced disenchantment. Researchers and companies are discouraged to access TK because of the high transaction costs, the legal uncertainties and the risks of public blame, a trend against the interests of many traditional communities in favour of fair exchanges (Kleba, 2008). A gap between reality and legality has grown. The CGEN implemented the access and benefit-sharing (ABS) legislation slowly,[7] partly due to the proactive use of public consultations.[8] The first bio-prospecting project involving commercial use of TK was authorized only at the end of 2007.[9] From 2002 to 2008 the CGEN have authorized only two applications involving commercial use of TK and two applications were deferred.[10] Meanwhile a draft bill law, published by the federal government for public comments late in 2007,[11] proposed many changes in the ABS regulation,[12] among others, a new concept of *disseminated traditional knowledge* (DTK).

In this chapter, I will focus on the controversial questions related to forms of DTK – a debate that is still lacking in the ABS literature[13] – examining the social and legal implications of this concept in light of empirical cases. The first section of the chapter asks to whom does TK belong when its dissemination level is unclear. In the second section a dispute will be investigated as to whether a particular accessed knowledge is protected TK or knowledge in the public domain. The third section is about what defines a TK holder. In the final section I discuss the disseminated forms of TK, systematizing their empirical variations in a new typology and considering their socio–legal implications.

Shamans and ethnopharmacology – Demarcating who owns the rights over TK in the case of the Federal University of São Paulo and Krahô[14]

How should the boundary between legitimate TK holders and non-holders be demarcated in cases where the dissemination of knowledge is unclear? The following case concerning the legal risks involved in the exclusion of potential holders outlines this question.

In the year 1999 researchers from the Department of Psychobiology, Federal University of São Paulo (UNIFESP), São Paulo State, initiated a bioprospecting project about ritual uses of medicinal plants used by the Krahô Indians, with emphasis on the psychoactive effects of their use (Rodrigues, 2001). The Krahô comprise 2,000 people living in an indigenous reservation in the State of Tocantins, in the Cerrado region – the Brazilian tropical savannah, considered a biodiversity hotspot. The Krahôs speak the language Timbira, from the Jê language family.[15] Their records concerning the effects of traditional herbal medicine on the central nervous system are surprisingly rich: 'the seven shamans indicated 98 formulas, consisting of 45 plant species that had 25 uses apparently related to psychoactive properties' (Rodrigues and Carlini, 2006, p279). The whole project envisaged two phases. The first was an investigation into the traditional medicinal knowledge of the Krahô and the collection of plant samples. It was concluded in 2001 with the publication of a doctorate thesis by E. Rodrigues, which synthesized the principal findings (Rodrigues, 2001). The end of this phase provoked accusations in the media concerning the procedures of PIC and claims for compensation (Kleba, 2008, p18). The second phase was meant to develop medicaments, but was not implemented due to failed negotiations on BS (Kleba, 2008, p8ff).

A critical view of the case shows that ownership over TK was constructed in a manner in which political choices had more weight than the principle of impartiality. The scientist, Rodrigues, began the research in 1999, shortly before the first Brazilian ABS law was enacted. Thinking pragmatically, the scientist got informed consent from the Krahô association Vyty-Cati (or Wyty Katy) and organized information meetings with three Krahô villages,[16] since the seven interviewed shamans of these villages were sufficient for her research purposes (Rodrigues, 2001, p36). The accusation of illegal behaviour was raised by another Krahô association, the Kapéy, in an 'open letter of the Krahô people'.[17] The main point of the letter was that a PIC should have been given by all the Krahô villages, 17 at the time (today 18), and not only by the three consulted villages, because the concerned TK was owned by the whole Krahô. A negotiation process began with the participation of new mediators such as the Federal Public Prosecutor's Office (MPF).

But how should legitimate representation for a PIC in this case be defined? Both associations, Kapéy and Vyty-Cati, claimed to represent their people, and later three more associations were founded by the Krahô: each of them representing different groups of villages (Avila, 2006, pp149–150). At the beginning of Rodrigues' research, the Krahô did not have an established representation form. In the following years the indigenous pattern of decision making was organized through deliberation meetings with the participation of the modern associations, with skilled leaders side by side with traditional forms of *pahís* (chiefs), seniors, shamans and women (Avila, 2006, p153).

In a sociological analysis, it is remarkable that a process of *naturalization* of a political preference took place, when the dominant interpretation was turned into the only possible one. I will argue that the fundament of this choice is not of an ontological, but of a political, nature. The starting point was the *political act* from the Kapéy association, claiming the ownership rights of the concerned knowledge for the whole Krahô and by implication only for them. But is this claim empirically justified? Brazilian ABS law defines the TK ownership as of a collective nature: 'any traditional knowledge related to genetic heritage may be deemed to be held by the community even if only one member of this community holds this knowledge' (Article 9, Sole Par) [official translation]. But it does not define the procedure of how this community shall be demarcated. There are ethnic groups which recognize the autonomy of each of their villages to realize BS contracts.[18] Other Indian leaders, for example in the case of the medicinal use of the secretion of the *Phyllomedusa bicolor* frog,[19] prefer to attribute a particular TK to a broader set of geographically related ethnic groups in the Brazilian State of Acre and in Peru, despite the fact that its use is recognized as being more closely associated with a few specific groups (Lima, 2008, p173).

As for the collective character of the medicinal knowledge of the *wajacás* (Krahô shamans), it is partially family tradition, as the medicine men learn their professional skills amongst their close kinship (Rodrigues, 2001, p51ff). They are partly individual, as each *wajacá* has his own spiritual guide (*pahî*), is specialized in healing certain illnesses and searches individually for new phytotherapeutic agents (Rodrigues, 2001). Partly it is knowledge shared between the Krahô and the other four groups of the Timbira language family, the Apinajé, Canela, Krikati and Gavião, and with a few geographic neighbours such as the Xerente (Avila, 2002, p147; Rodrigues, 2001, p30). So the Vyty-Cati association claims to represent the whole Timbira people. Moreover, the identity 'Krahô' seems homogeneous only for white people. During rivalries of cultural identity between the Kenpocatêjê (today self-named Krahô) and the Mãkraré (Krahô with Mãkraré origin), that persisted for over two centuries, each of them sought to be the true Krahô (Avila, 2006, p144).

In a debate with the Krahô, Rodrigues reaffirms that ancestors of many *wajacás* 'belong to other ethnic Timbira and non-Timbira groups',[20] justifying her negotiations with the Vyty-Cati association. The reply that in this case the legitimate owners of the concerned traditional medicinal knowledge 'must include all Brazilian ethnic groups', or indeed 'all American ethnic groups',[21] is, however, a rhetorical statement, since recognizing some degree of DTK does not imply in turn that it is boundless. Not political statements or sophistry, but only empirical research can define how disseminated a particular knowledge is, as the ethnomedicine research has done.[22]

The claim of the Kapéy association that the knowledge of seven *wajacás* of three villages should belong to all the Krahô villages, and only them, involved the inclusion of different ones in a homogeneous unity (the Krahôs as equal owners) and the exclusion of other Timbira groups, which were not consulted. This boundary setting was legitimated through the expressed indigenous will of the Krahô alone and became functional to implement the law, as it hid the difficult question of the real distribution of the traditional medicinal knowledge. Such regulatory outcome implies, however, a legal risk. Suppose that the recipe X acquired from the *wajacá* Z becomes a requested global drug and that the concerned BS contract benefits exclusively the Krahô. In this case, an Apinajé Amerindian could attest evidence in the courts that he shares the same intergenerational knowledge, demanding participation in a new BS contract.

Considering those problems, two concepts of the Brazilian draft bill represent advances. First, detailed BS contracts should be realized only when a product is close to the market, freeing the bureaucratic burden of prospective research on GRs. A second idea is to establish a regional fund of benefit sharing, providing that benefits are going to a broader knowledge-sharing collective. An additional improvement is the requirement that mandatory anthropological expert opinions,[23] already produced during applications for access to TK in Brazil, shall include the question of the empirical dissemination of TK.

The fragrance of *Protium pallidum* – Disputes between free public and protected common rights

This section discusses the criteria used to decide whether a bioprospecting project is accessing legal protected TK or knowledge in the public domain. Although the engagement of Natura Company in ABS issues is known (CBD, 2008), this chapter examines for the first time, through an extensive analysis of administrative and juridical files,[24] the legal disputes of the company with two potential TK holders – the riverside dwellers of the Iratapuru river and the herb vendors of Belém.

Natura and the riverside community of Iratapuru

Natura Company (Natura Cosméticos SA) is a Brazilian cosmetics company headquartered in Cajamar, São Paulo, that has assumed in recent years the leadership of the cosmetic sector in Brazil.[25] At the CGEN it has been negotiating nine BS contracts.[26] Of the many products Natura developed using Brazilian biodiversity as part of its Ekos line, the essences of three Brazilian plants, breu branco (*Protium pallidum*), priprioca (*Cyperus articulatus L.*) and cumaru (tonka beans, *Dipteryx odorata*), became the centre of a controversy. I will focus solely on breu branco (which actually means 'white resin' or tar), as each one of the plants is connected to a different story.

In early 2001, at the request of Natura Company, a company called IFF Essências e Fragrâncias Ltda, headquartered in Barueri, São Paulo, prospected for breu branco, engaging a riverside and extracting community in the State of Amapá.[27] The community lives at the Sustainable Development Reserve[28] of the Iratapuru River. The riverside dwellers have worked with Natura since 1999 to source brazil nuts and transform them into oil, and today they are delivering copaiba, brazil nuts and breu branco to the company (CBD, 2008, pp80–82).

In the contracts with Natura, the community of Rio Iratapuru had chosen to be represented by its Cooperative COMARU (Rio Iratapuru Producers and Extracting Cooperative). In June 2004, independent of any requirement of the CGEN, the company signed a PIC declaration and a BS contract with COMARU, and in December, after certain roadblocks had been overcome,[29] it signed an add-on to the BS contract with the Amapá State Environmental Secretariat, the SEMA.[30] Actually the company was only legally required to sign the contract with this agency, which is the custodian for the Reserve, as the company had not applied for access to TK, but to genetic patrimony alone.[31] The BS with the Iratapuru community included lump-sum payments, 0.5 per cent of the net revenue from the sale of products containing breu branco resin,[32] as well as support for certified forestry according to the Forest Stewardship Council. The BS contract with the State of Amapá does not demand additional benefits.

In July 2004 the CGEN formally registered Natura's application to regularize[33] its access to breu branco, and in March 2005 issued it with the *Authorization of access to samples of a component of the genetic patrimony for the purposes of bioprospecting and technological development* (Decision CGEN no 94/2005). The Executive Secretariat of the CGEN asserted that 'this authorization does not apply to access to the associated traditional knowledge used by the Company, because the CGEN has no regulations on benefit-sharing in such cases'.[34] The cases considered here as being in the grey legal area were those of *disseminated TK*. The issue was deferred.

Traditional perfumery – Natura and the herb vendors of the Ver-o-Peso market

In July 2003, Natura employees produced a documentary based on the interviews with herb and perfume vendors working in stands at the traditional Ver-o-Peso ('True Weight') market in the capitol of Pará State, Belém, Amazon region.[35] The Ver-o-Peso is a reputable popular market, with historical and cultural roots.[36] The women who sell these herbs are known as '*erveiras*'. The Natura employees filmed the *erveiras* and asked them about the use of the herbs, how they are prepared and who supplies them. At the time, it signed a two-year contract with six of the *erveiras* for copyrights of their images and voices, paid in advance.[37] One decisive point of the future dispute is the fact that it is not clear whether the company did non-official consultations with the *erveiras* earlier, in 2001.[38] This is because by the time of the interviews with the *erveiras* in July 2003, Natura had already launched two products using breu branco: *Perfume do Brasil* and *Água de Banho*.[39] When the *erveiras* realized that Natura was selling products based on some of the herbs they had been interviewed about, they felt victimized and in April 2005 they consulted the State office of the Brazilian Bar Association (OAB/PA). The claim was that there had been use of TK without the statutory obligations of PIC and a BS contract.[40] A complex process of dissent began through minutes of meetings, reports by commissions, official notices, legal opinions, technical notes and administrative processes. The Pará State Bar Association and the MPF sustained the *erveiras'* position. The company took the opposite stand.

The CGEN, as a competent authority, decided that access to TK had indeed taken place.[41] Finally, in a settlement at the MPF, in June 2006, an agreement with Natura was formalized.[42] The herb and perfume traders were recognized as providers of accessed TK involving breu branco, pripri-oca and cumaru, as were respectively three communities that had supplied each of the three plants.[43] In October 2006 the company signed a declaration of PIC and a BS contract with the Ver-as-Ervas ('True Herbs') association, which represents the *erveiras*.[44] The signed documents provided for up-front payments, as well as a percentage of net profits, once the fragrance ingredients were highlighted on the label (0.15 per cent per ingredient) or when ingredients were present in the formulation (0.05 per cent).[45] These benefits were meant to target biodiversity use and conservation projects, as well as cultural projects,[46] with no direct monetary advantages accruing to any member of the association.[47] The settlement also obliged Natura to negotiate a new BS contract with the riverside cooperative COMARU, recognized as an additional case of accessing TK.

I argue that the process that led Natura to accept the mandatory conditions of the settlement was an *agreement without consent*, that is, the company did not yield due to reasonability, but due to a calculation of the risks and benefits involved in the possible outcomes of the dispute.

Which knowledge – legally protected or free disseminated – was accessed?

When the *Protium pallidum* case began in 2001, existing legal provisions left many uncertainties regarding the application of the TK concept. Natura called the accessed knowledge '*diffuse*', seeing it within the public domain in the patent law sense.[48] For the prosecutors taking sides with the *erveiras*, it was equivalent to the definition of TK by the MP 2.186-16/2001. For third interpretations it could be a kind of 'common-pool resource' (Ostrom, 2000) converted to a *de jure* government property or custodianship regime.

What does the Brazilian law say? MP 2.186-16/2001 defines 'access to associated traditional knowledge' as:

> *Acquisition of information on individual or collective knowledge or practice associated to genetic heritage, from an indigenous community or local community, for the purpose of scientific research, technological development or bioprospecting, with a view to its industrial or other application (Article 7, V) [CGEN's official translation].*

The MP's definition of 'associated traditional knowledge' refers additionally to 'information or practice ... with real or potential value' (Article 7, II). Looking at both concepts we find criteria such as the origin of the information, its relevance (value) and its purpose. In the case of breu branco its purpose as an industrial application is obvious. Regarding the origin of the information, the company declared in its application by the CGEN that 'the resins' perfume potential in general has been described in the technical literature',[49] rather than coming from the riverside community of the Iratapuru river. When it stated that the perfume potential was '*disseminated* knowledge *also* identified by the community' [emphasis added],[50] it meant that the associated rights were no longer exclusive. According to Hahn (2004, p350), public domain is the totality of the common knowledge and intangible goods, which cannot be removed from common use by factual or contractual means. For the company the public domain nature of the fragrance was attested not only by literature, but also by the common use of the resin in an extensive region (from Mexico to Brazil).[51] Some of the referred publications predate Brazil's legislation; for example, relating the use of breu branco as an incense for religious purposes (FAO, 1995). The CGEN's Executive Secretariat has considered that the law cannot retroact to cover such cases.[52] This is different, for example, from the Peruvian legislation, which protects TK 20 years retroactively.[53] However, as we will see, for the CGEN the literature records were not able to nullify the legal weight of accessing TK on breu branco.

Concerning the potential value, the central idea at this point is the *shortcut*, which is invoked to justify the claim for benefit sharing with TK providers. The literature shows that a shortcut, as opposed to the random search of 'a needle in a haystack' – the screening of biological material in free nature (WBGU, 2000) – is one decisive means to save time and money in the search for new, commercially valuable chemicals (Kate and Laird, 1999). Simply consulting TK holders can justify ethical and juridical obligations to share benefits as a reward for cooperation, irrespective of whether the expected commercial ends have been achieved. But the scope and amount of the BS, in *fair and equitable* means, should consider the real usefulness of linking the shortcut to market products.

The rights to protect TK are not restricted to economic value, but are rather supposed to protect the culture as a whole in its material, intangible and spiritual facets. Even so, the utility and economic interests related to TK have played a vital role in legitimizing such demands in the political arena. Thus, one voice rose to defend what the herb vendors maintained: 'Without the undue appropriation of the knowledge of these vendors, the Natura Company would not have been able to satisfactorily develop its line of beauty products'.[54] But was the information collected by the *erveiras* actually that important?

Let us assess this reasoning based on the research and development (R&D) of breu branco. Regarding the novelty of the ingredients and the potential function of breu, Natura mentioned known publications about the aromatic qualities of the resin. The processing techniques differ, with the traditional preparation using alcohol and the industrial process based on steam-drag technology.[55] As a result, apparently Natura had accessed widely known TK in free public domain.

However, the decision of the competent regulatory body, the CGEN, was that Natura indeed had accessed TK, by both the herb vendors and the riverside community of Iratapuru. The CGEN changed its initial position, and no longer considered the case as DTK, that is, as TK with special status not included in the MP 2.186-16/2001 and needing a specific legislation. In addition to the pressures of the legal claims, I see a threefold legal justification for this move by the CGEN: there was evidence that Natura had collected information about the uses of breu branco directly from the providers (the shortcut); the traditional uses were considered as having potential commercial value; and the providers were identified as legal holders of TK.

Which side bears the burden of proof? In the current practice, it is borne by the user.[56] Hence, the evidence was not difficult to provide. One of the pieces the CGEN analysed, for example, was Natura's institutional documentary promoting its Ekos line of products. Quotes such as this were used in the documentary: 'To develop this perfume, we recovered an ancient secret for the sensuality of women in native populations, their bath water.'[57] One might question the quality of such evidence, considering the prevalence of aesthetic and symbolic values in marketing, in contrast to precise descriptions. Even rhetorical slip-ups, however, have legal weight a as written record.

Paradoxically, the company recognizes later having accessed TK on breu branco (CBD, 2008, p81).[58] The access of TK was justified as part of the very *marketing conception* of its products:

> *Considering that the said traditional knowledge is the source of inspi-*
> *ration for product development, that is, for creating a line of perfumes*
> *involving the traditions of* banho de cheiro *(aroma bath), ... it is*
> *expressed solely as a source of inspiration of a marketing concept for a*
> *line of perfumes involving the active ingredients breu branco, pripri-*
> *oca and cumaru.*[59]

It is not clear which inspirations Natura refers to here and the confidentiality requirements[60] maintained in the case force us, the observers, to raise plausible explanations. I suggest two possible explanations for the company's paradoxical changing of position: the *hidden shortcut* and the *reframing* thesis.

The first supposition is that the company's interpretation of the story hid a shortcut that in fact took place. The company had seen the information obtained by the interactions with the riversiders and the *erveiras* as freely accessible, and later, through the arguments of the dispute, realized having accessed TK. The hidden shortcut thesis could be described as follows. What the company knew through the literature about the aromatic potential spread by the genus *Protieae* was speculative and abstract, considering that this genus comprises about 135 species[61] 'with high degree of variability of resin composition' (Langenheim, 2003, p358). As Natura had already worked with the Iratapuru riverside community since 1999, it could attest *in loco* to the real commercial potential of the fragrance of one particular species, the *Protium pallidum*, by experiencing empirically its use as incense.

So the company reported in the official anthropological expert opinion about Natura's access to breu in Iratapuru having got a *hint* of the breu's fragrance from the riverside peoples' uses of this resource.[62] The same expert opinion said that the resin is used by the community for many purposes, such as sealing and calking boats, as a mosquito repellent and as incense, but that 'breu had never been used as a fine perfume'.[63] However, it is not exaggerating to consider the use of breu's incense as a shortcut to perfumes, in its socio-legal TK sense. In such a scenario, also information obtained from the *erveiras*, supposed to have already taken place earlier, as by the official interviews in 2003, had 'confirmed the viability of the commercial production'[64] of breu branco, demonstrating it in the form of traditional perfumes.

A different explanation is provided by the *reframing*[65] thesis. In this hypothesis the company's defence was justified, since the idea of the commercial research of breu branco was completely taken from the litera-ture, *before* contacting the providers. However, the company, recognizing that it had lost the case dispute, created a meaningful story explaining the supposed accessed TK, reframing it in favour of its public image. The 'inspiration of a marketing concept' could match the uses of the company's brand names by the Ekos line of traditional designations such as 'aroma bath', launched by the company in its advertising campaigns. Natura thus attaches to the fragrances the double symbolism of a socially and ecologically committed business, along with the exotic attraction of the secrets of the for-est and its native dwellers. But those benefits, although obvious, do not apply to the ABS law, which covers the *information* associated with GRs.

By the hidden shortcut explanation, the CGEN's decision would be legally substantiated. Inversely, by the reframing thesis, it would not be. Between both options, the regulatory body had rightfully decided in favour of the providers, for the reasons explained above, despite the fragilities of the presented evidences.

Independently of this legal and rational dispute, the company's decision to avoid being taken to court can be explained by extra-legal reasons; for example, preventing distorted media coverage of the case, which could dam-age the company's image with sensitive accusations of biopiracy,[66] and using the cooperation with the providers for self-interest reasons. Maybe extra legal motives also played a role in the CGEN's opinion in favour of the riversiders and the herb vendors; for example, the commitment to defend the weak in a strongly asymmetric relationship[67] (see Table 7.1).

Should traders become legitimate TK holders?

This section discusses the concept of legal holders of TK. Natura's lawyers argue that the herb traders at Ver-o-Peso cannot be classified as legal hold-ers because they are not a local community and therefore lack legal standing.[68] The *erveiras* are urban dwellers and their economic activity is trade. Brazilian legislation does not refer to access to TK through com-merce. The MP 2.186-16/2001 defines 'local community' as a:

> *Human group, including descendants of* Quilombo[69] *communities, differentiated by its cultural conditions, which is traditionally organ-ized along successive generations and having its own customs, and preserves its social and economic institutions (Article 7, III).*

A broader version of the concept is provided by Decree 6040/2007, which institutes the National Policy for the Sustainable Development of

Traditional Peoples and Communities, and whose definition is identical to the new draft law on ABS (Article 7, XV):

> *Traditional Peoples and Communities are culturally differentiated groups, who identify themselves as such, possess their own forms of social organization, occupy and use territories and natural resources as a condition for their cultural, social, religious, ancestral and economic reproduction, using knowledge, innovations and practices that are generated and transmitted through tradition (Decree 6040/2007, Article 3, I).*

The question as to whether a particular group is *culturally differentiated*[70] is more difficult to apply the more distant the group is from *ideal types*. Only the latter meet as sharply as possible the attributes of anthropological alterity ('otherness'), contrasting with modern, Western lifestyles. The majority of the Southern hemisphere population could be defined as traditional, as it lives outside the core of modern social systems characterized by the concentration of capital and power and by specialized institutions. At the same time, the classical distinction of modern/traditional is increasingly being

Table 7.1 *The dispute between Natura and herb vendors*

	Natura's vision	*Pro-herb vendors vision*
Source of information	Technical literature	Access to TK
TK accessed	Public domain (free-use sense)	Common rights related to MP 2.186-16/2001
Potential use of the accessed TK	Redundant	Plausible
Is it right to expand rights-holders to include street markets?	No	Yes
Political appeals	Defence of the public domain; business ethos stands in for legal ethos; economic development for common wealth	Strong enforcement of an emerging right; distributive justice; preservation of minority rights

Source: Author's compilation

watered down (Giddens, 1991). The CBD seeks to protect the remaining particular cultural sets against cultural erosion and extinction, especially those fragile in the face of expansive globalization and capitalism. The *erveiras* of Ver-o-Peso live in an urban milieu and work in a popular marketplace in a big city. Narrowly defined, they are not a minority threatened by acculturation. But in a broader sense they can satisfy the legal definition by maintaining forms of *economic reproduction* linked to cultural heritage.

'Self-identification' and 'identification by others' are fundamental aspects of constituting identities in anthropology and in political science. Identification by others is necessary to avoid the undue claims of TK for illicit gain under the law. In this regard, the Ver-o-Peso *erveiras* are icons recognized beyond dispute as part of the Amazon region's traditional culture.[71]

In contrast with the current MP 2.186-16/2001, the new draft law emphasizes the use of territories and natural resources as being essential for traditional peoples' social and cultural reproduction. *Traditional ecological knowledge* is inextricably intertwined with the long-standing link between livelihoods, ancestral territory and smart management of natural resources (Berkes, 1999; Johnson, 1992). The variety of such populations are located in Brazil in terms of professional and ecological categories, including far more groups than the Indians (Diegues and Arruda, 2001; Moura, 2007). This concept does not apply to market traders living in urban settings. On the other hand the *erveiras* are using 'natural resources as a condition for their cultural ... and economic reproduction' (Decree 6040/2007). Consequently, their activities may stimulate the preservation and management of certain species, or may not. In the case of the *erveiras*, the BS contract between Natura and the Ver-as-Ervas Association provides measures to assure the implementation of sustainable practices by suppliers.

Finally, the *erveiras* claim to be promoting the intergenerational preservation of traditions in the use and preparation of the herbs. As was explained by *Beth Cheirosinha*, one of the *erveiras* leaders who was interviewed by Natura: 'Her grandmother, who died at the age of 115, learned everything from the Indians'.[72] That knowledge is maintained and their practices carried out as a livelihood in small commerce.

In short, the strong justification for the inclusion of the herb vendors as legal TK holders is the preservation of recognized intergenerational knowledge of regional biodiversity uses, *as a condition* for their social, cultural and economic reproduction.

The disseminated forms of TK

Nationwide disseminated traditional knowledge

The new Brazilian draft bill defines DTK as 'knowledge disseminated in Brazilian society, freely usable by all, not recognized as being directly associated with the culture of indigenous, Quilombola[73] or traditional communities' (Article 7, XIX). The meaning of 'freely usable by all' here is in a national and not a global sense, since custodianship and BS rights shall be retained by the government, whenever Brazilian DTK is used abroad to develop commercial products.[74] From the definition of the Brazilian draft bill, DTK is better understood as *TK of national custodianship*.

The applicability of the concept reserves some difficulties. Knowledge about something may be *disseminated in society* and at the same time be *directly associated* with traditional communities. How disseminated in society must it be, to no longer be directly associated?

In the case of breu branco, some of its uses in Amazon are broadly shared whilst some are not. Breu as personal perfumery seems not to be of such wide use as other general uses of the plant are. But despite the use of breu branco as incense being a DTK, Natura's track of the riversiders' use of the incense to the company's developed perfume can be justified as accessing TK because this track is a particular source of valuable knowledge. In the case of DTK, a relevant mistake of the Brazilian current law is leaving all knowledge holders, with the exception of the direct providers, outside the sharing of the concerned benefits. Such legal reduction may exclude even the possible inventors of the breu branco traditional uses: for example, considering the traditions of some indigenous peoples:

> *The Tembé and Ka'apor Indians of eastern Amazonia use* breu *as a fire starter, incense, medicine, and for caulking wooden boats. These Indians also rely on the harvest of Breu as one of their main economic products ... For commercial purposes, Tembé Indians recognize two major types:* breu branco, ... *and* breu sarara, ... Protium pallidum *and* P. trifoliatum *are considered to produce breu branco ... (Langenheim, 2003, p358).*

The concept of DTK does not contribute to clear-cut distinctions because TK is usually shared (Dutfield, 2002) and is likely to be disseminated to some degree. For example, the use of the secretion of the *Phyllomedusa bicolor* frog as medicine has been increasingly disseminated by the rubber tappers at least since the 1960s, and today is well known by the local populations of Acre State and the Peru border (Martins, 2006, p60).

But its origin is clearly traceable to a small group of indigenous peoples, who are seen as its legitimate holders (Lima, 2008). It is a crucial fact that dissemination per se does not tell us much about the legal status of specific TK forms.

A new typology for TK types

To better understand the features that DTK associated to GRs can express, I distinguished four types of TK. The typology starts with the knowledge attached to a natural territory and a local culture, and develops to various degrees of detachment. The first two types, the *community-based TK* and the *trade and urban TK*, can or cannot be disseminated to some degree. The last two, *DTK of national custodianship* and *TK in the worldwide public domain*, are always disseminated, but in the former case the related rights can be claimed by a nation state.

Type I

Community-based TK is the guiding model in the legal and political context (WIPO, 2002). Sociologically it is bound to a circumscribed community or a set of communities with particular institutions that are deeply rooted in a territory. They are culturally differentiated from a Western lifestyle in an ethnological sense; for example, the indigenous peoples (Peruvian ABS legislation, Elvin-Lewis, 2006, p85) and the Quilombolas (Brazilian ABS law). In accordance with the concept of traditional ecological knowledge, the ecological criterion links the community reproduction with the management of biodiversity and economic activities of low impact. Additional features are the epistemic differences between TK and the modern scientific system[75] and, concerning right traditions, the correspondence from TK to common property – in contrast to private law (Souza Filho, 2002) as well as to public domain. As Elinor Ostrom (referring to Ciriacy-Wantrup and Bishop) puts it:

> *the difference between property regimes that are* open access, *where no one has the legal right to exclude anyone from using a resource, from* common property, *where the members of a clearly demarked group have a legal right to exclude non-members of that group from using a resource (Ostrom, 2000, pp335–336).*

Type I of TK ranges from locally restricted to transnational disseminated knowledge. The latter is illustrated by the case of hoodia (Wynberg, 2004). Natura's access to breu branco from the riverside dwellers of Iratapuru corresponds to type I of TK.

Type II

Trade and urban TK arises from the empirical recognition that TK can be detached from traditional communities and from territory, as in the urban lifestyle. The economic practices of type II are here usually commerce or services. Examples are popular market vendors, herb medicine traders and urban healers, widespread in Asia, Latin America and Africa (WHO, 2003). Like the *erveiras*, many of them are migrants from traditional communities. They show a weaker connection in managing biodiversity than type I, but have a crucial role in keeping intergenerational practices alive and giving TK value by bringing it to urban consumers. In Africa up to 80 per cent of the population uses traditional medicine for primary health care (WHO, 2003). Immigrants can preserve TK and change practices, adapting to a new social environment (Vandebroek et al, 2007). Type II of holder can concur with type I, as they offer an easier way for users to achieve valuable information on GR, being more accessible and making it possible to arrange legal obligations with individuals in place of communities (Kamau, 2009). Shall the law grant TK protection for these social groups? The case of the *erveiras* (could) answer it positively. It seems that some urban and trade TK holders deserve the rights of legal protection, when practising intergenerational TK associated to biodiversity as a means of their social, cultural and economic reproduction and keeping alive practices that are being extinguished.

Concerning types I and II, the Brazilian draft bill advocates expanding the concept of TK to incorporate modalities of knowledge, innovation and practice expressed outside their original contexts: 'even when made available outside these contexts, such as in data banks, cultural inventories, publications and in commerce' (Article 7, XVIII). Written records are protected since the Brazilian enactment of the ABS legislation. Regarding commerce, this provision includes urban and trade TK holders as legal holders of TK, expanding the original meaning of the concept.

Type III

DTK of national custodianship is knowledge spread in time and space to a degree detaching it from particular groups, but at the same time embedded in national cultural traditions. By the MP 2.186-16/2001 the Federal Government is the contracting party in cases of access to GRs on public land, but not in cases of access to TK.[76] Some states have provided relevant support to the defensive protection of TK; for example, the Indian Traditional Knowledge Digital Library (TKDL)[77] and the Brazilian government in the cupuaçu case (Kleba, 2005). However, the approach is different when the government itself plays the role of TK provider.

As already explained, the new Brazilian draft law is seeking to establish this concept legally. If Natura had been considered to have accessed type III of TK, the company would be free from PIC and BS obligations, while a foreign company would have to negotiate both with the Brazilian regulatory body.

Type IV

Finally there is the *TK in the worldwide public domain*. Here is the particular link to a collective right lost through long-term, cross-cultural and cross-national uses in time and space. It is globally shared common knowledge, like the 'centuries old pharmacopeias of Europe, Greek and Arabic medicines and academic treatises published primarily in the 19th and 20th centuries on African and North American indigenous pharmacopeias' (Elvin-Lewis, 2006, p79).

There are many open questions related to this typology and we shall briefly introduce a few. First of all, the political claims intended to establish type II and III as new legal standards in the national and international arena can be highly controversial. For example, are databases like the TKDL a compendium for type III or rather part of type IV?

Second, the types can overlap. As they may overlap, there are potential conflicts over the legitimacy and scope of true holders. There might be dissent between the interests of different stakeholders. The uses of ayahuasca (*Banisteriopsis caapi*), for instance, rooted in South American indigenous traditions and the object of files against patent applications seen as biopiracy (Hansen and Van Fleet, 2003, p14), became disseminated in Brazilian religious practices and are becoming more and more globally disseminated.[78] Is the legal status of ayahuasca changing from a protected TK into the worldwide public domain?

Third, there is a tension between the options of DTK protected by prior art (public domain) and DTK protected by sui generis law, like the ABS law. The European Patent System includes as prior art 'everything made available to the public by the means of a written or oral description, by use or by any other way, anytime before the date of filing of the patent application' (Hansen and Van Fleet, 2003, p36). Many cases of defensive protection of DTK relies on the identification of DTK with prior art, like the well-known files against patent applications concerning neem (*Azadirachta indica*) and turmeric (*Curcuma longa*) (Hahn, 2004), or the Brazilian cupuaçu (*Theobroma grandiflorum*) chocolate case[79] (Kleba, 2005). By contrast, the positive protection of DTK requires inversely its distinction from prior art. For example, if a 'publication in a highly specialized journal may not constitute evidence that a piece of traditional knowledge has become public domain in a patent law sense' (Seiler et al, 2003, p13), it would not only free concerned scientists from the burden of not publishing sensitive information, but also facilitate access preserving at the same time as PIC and BS obligations.

Finally, DTK presents special challenges regarding PIC and BS. High transaction costs in detecting and consulting a multiplicity of communities can push potential users away. In the case of widely disseminated TK, it makes no sense to quantify and locate all knowledge holders. The establishment of public inventories of TK and the simplification of PIC procedures in cases of DTK shall make access to TK easier; for example, acquiring PIC from only one community can be sufficient for DTK accessed from written sources (Brazilian draft bill, Article 48).

The same concerns apply to BS. In cases of transnational DTK, mutual country agreements and regional common pools (see Winter in this book) should make provisions seeking to avoid a race to the bottom.[80] A special problem concerns legal frameworks such as the current Brazilian law, which does not keep benefits from TK accruing only to direct providers who are favoured by circumstances and fortune. The Brazilian draft bill corrects that imbalance by providing a mode of benefits for both providers and knowledge holders in general (Article 73). While the former are benefited through BS contract with the users, the latter are rewarded through a BS Fund (Article 98 and Article 99 para 2) to be used for environmental and cultural purposes in the region of origin of the TK. Although the benefits may be indirect, they preserve the essential collective quality of TK and bestow greater legitimacy on the process. An additional facilitation to access TK is avoiding the requirement of BS contracts before the access takes place, demanding difficult and often unnecessary negotiations, as the current Brazilian MP 2.186 has done (Article 16 para 4). The draft bill, by contrast, eases access providing that the contracts can be signed just before patenting or commercialization, since PIC allows it (Article 86).

Recommendations

As a result of examining the concept of DTK with the help of empirical cases, the ABS law should be streamlined at the following points:

• The definition of holders and non-holders of TK should be supported by anthropological expert opinions and include the question of the knowledge dissemination. The expert opinions must be made public, in order to be controlled and eventually corrected by the scientific community and the civil society.

- BS funds should be implemented, and the benefits of a particular TK directed to its cultural region of origin. A BS fund guarantees more fairness for the participants in the shared knowledge and decreases the risks of legal claims brought by holders excluded from the rewards.
- The PIC process in cases of DTK should be simplified for users and restricted to the direct providers. Traditional communities should make public which PIC procedure they prefer.
- BS contracts should be made directly before related products are launched or patent applications are applied for, assuming that the providers agree with it and that their rights are assured by the PIC contract. This claim does not absolve the ethical duty of researchers and companies from negotiating more immediate forms of rewards, independently of commercial applications.
- In controversies about accessed TK, both criteria of shortcut (direct consultation of TK holders) and plausibility of commercial uses with potential value should be made evident.
- *Traders and urban people* are potential TK holders and should receive full legal protection, when criteria to be legally established are fulfilled. Those criteria should be clear and narrow potential candidates, avoiding the inflation of unjustified claims and minimizing the transaction costs of users.
- The new legal concept of *DTK of national custodianship* should be, prior to any implementation, carefully evaluated, especially in regard to overlaps with the claims of traditional communities and the claims in favour of open-access, worldwide shared knowledge based on traditional roots.

Notes

1 This chapter is part of a research project funded by the DFG (Deutsche Forschungsgemeinschaft) and coordinated by Prof. Gerd Winter at the Research Centre for European Environmental Law, University of Bremen, Germany. It is also supported by the FAPESP (São Paulo State Research Promotion Foundation). I would like to thank C. M. A. Azevedo, M. Wan, G. Winter, S. A. S. Kishi and E. C. Kamau for their insightful comments on the text.
2 The CBD was signed at the UNCED Conference in Rio de Janeiro in 1992 and ratified by Brazil in 1994.
3 Other sorts of TK are not considered here because they are outside of the CBD legal frame.
4 The first regulation was the MP 2052/2000.
5 The CGEN was created by MP 2186-16 (Article 10) under the Ministry of the Environment, as a decision-making and standard-setting agency, made up of representatives of several federal government agencies.
6 See Kishi in this book.

7 The CGEN itself only began work in April 2002. Its first authorizations for access to genetic patrimony with commercial intent (not to associated TK) were only issued in mid-2004.

8 A CGEN's consultation on PIC and BS took place in 2007. CGEN, consulta pública no 2, Ministério do Meio Ambiente, Brasil.

9 See Kishi (case Oriximiná) in this book.

10 Boletim Interno DPG, CGEN, MMA, March, 2009, http://www.mma.gov.br/estruturas/sbf_dpg/_arquivos/boletim_abril_2009.pdf, accessed 05 May 2009.

11 http://www.planalto.gov.br/ccivil_03/consulta_publica/consulta_biologica.htm, accessed December 2007.

12 See Santilli in this book.

13 Consulted on 15 January 2009, 'Google Academic' had only one entry with a paper referring to 'disseminated traditional knowledge', a paper that does not examine the legal controversies of this concept in relation to ABS, http://icrier.org/pdf/wp141.pdf.

14 The information for this item is mainly based on Kleba, 2008, as well as the CGEN files on the case, hereinafter CGEN file 1.

15 Enciclopédia Povos Indígenas no Brasil, www.socioambiental.org/pib/epienglish/kraho/kraho.shtm, accessed January 2008.

16 CGEN file 1, fl.275ff.

17 25–26 May 2002. A copy of the letter is available in Avila, 2006.

18 Pers. comm. with Cristina Azevedo, CGEN.

19 Interview with the Katukina leader Fernando Katukina, Cruzeiro do Sul, Acre State, 21 July 2008.

20 CGEN file 1, p459ff.

21 CGEN file 1, p459ff.

22 Hultkrantz, Å. (2008) 'Medicine in native North and South America', in Selin, H. (ed) *Encyclopaedia of the History of Science, Technology, and Medicine in Non-Western Cultures*, Berlin, Springer-Verlag, p1567ff.

23 See Kishi in this book.

24 I analysed two files on the *Protium pallidum* case, hereinafter CGEN file 2 (*Processo no 02000.001608/2004-19*) 652pp, and the file of the Federal Public Prosecutor's Office, 330pp, referred to as the MPF file.

25 Natura Annual Report 2007, www.corporateregister.com, p5 (80pp).

26 Natura Annual Report 2006, www.corporateregister.com, p79 (130pp).

27 Law no 0392/1997, Amapá.

28 Created by Law 9985/2000: 'Article 20. A Sustainable Development Reserve is a natural area home to traditional populations whose livelihood is based on sustainable systems of exploitation of natural resources, developed over generations ...'

29 CGEN file 2, pp34–74.

30 MPF file, pp219–225.

31 The community did not have land deeds as required by the law for signing BS contracts for access to GRs, a very common situation in the Amazon region. CGEN file 2, pp249ff.

32 It provided for a lump-sum €2,789 (10,000 Reais) payment, whatever the results of ensuing research. Concerning the net revenues for fiscal year 2003, the corresponding payments amount to €27,727 (R$101,222). CGEN file 2, p28ff.

33 According to CGEN's practice, 'regularization' is the process of bringing already-running projects into conformity with the law.

34 Official notice No 256/2005/CTEC/DPG/SBF/MMA, 19 July 2005. CGEN file 2.

35 MPF file, p160ff.

36 *Folha de São Paulo* (Online) 'Belém renasce com mistura de sabores, cultura popular e história' Busch, A. L. and Vilela, C., 17 June 2005.

37 The private copyright image agreement included payment of €153 (R$500) to each of the participating market vendors. MPF file, pp160ff.

38 www.reporterbrasil.org.br/exibe.php?id=605, accessed June 2008.

39 CGEN file 2, p43.

40 Transcript of testimonies; Minutes of meeting at the Biolaw Commission of the OAB/PA, 18 April 2005. MPF file, pp18–24; p152ff.

41 At 11 November 2005. MPF file, p141.

42 In the settlement, Natura agreed to the demands of the participants to pay legal counsel and provide financial support for the herb traders in the negotiations of a BS contract. MPF file, pp186–189.

43 The community providing priprioca was in Boa Vista do Acará, Pará and that providing cumaru, in Nova Califórnia, Roraima (RECA Project).

44 Today it represents 102 herb traders (90 of them women) working in the Ver-o-Peso market. http://belemhoje.blogspot.com/2008/01/vendedoras.html, accessed June 2008.

45 The percentages are mutually exclusive. MPF file, pp294–310.

46 Today the association provides infrastructure for the manufacture of raw materials, the processing of cosmetic products and the drying of medicinal plants. The association offers courses to the herb vendors for the improvement of the production of phytotherapeutics and cosmetics, on recycling and for suppliers on supporting sustainable practices. www.deputadozegeraldo.com.br/site/news.php?readmore=224, accessed June 2008.

47 MPF file, pp294–310.

48 CGEN file 2, pp387–395.

49 CGEN file 2, p131.

50 CGEN file 2, p129f, pp156–162.

51 Official Letter from Natura, 13 April 2006. CGEN file 2, p392.

52 Cristina Azevedo, CGEN, private communication at the 'IV Seminário de Etnobiologia e Etnoecologia do Sudeste', 8–9 November 2007, UNIFESP, Diadema, São Paulo.

53 Peru Law No 27811 of 2002, Article 13.

54 MPF file, pp79–85.

55 MPF file, p288.

56 Pers. comm. with C. Azevedo, CGEN, Draft bill, article 43 para 2.

57 Author's translation. MPF file, pp230–235. Informational Note 13/2006/CTEC/DPG.

58 In fact, more than once the company changed its statements in recognizing or denying having accessed TK in particular cases, showing unclarity in the correct legal interpretation of the cases; for example, concerning the access to priprioca and also the access to cumaru. CGEN file 2, Official Letter No 415/2006/CTEC/DPG, 23 November 2006, pp528–529; CGEN file 2, Official Letter 257/2006/DPG, p452.

59 Official Letter, Natura, 1 December 2006, MPF file, pp287–289.

60 Documents in the application, such as the expert anthropological opinion on Ver-as-Ervas, are confidential. Personal communication of a technical assistant of the Secretariat of CGEN.

61 Rüdiger, A. L., Siani, A. C. and Veiga Junior, V. F. (2007) 'The chemistry and phar-macology of the South America genus *Protium burm.* f. (Burseraceae)', *Pharmacognosy Reviews*, vol 1, Issue 1, p93.

62 CGEN file II, p72 (expert opinion, p39).

63 Anthropological Expert Opinion by Mary H. Alegretti, 14 July 2004 (p9), CGEN file 2, pp34–74.

64 That is the CGEN's definition of 'potential of commercial use', Orientação Técnica no 6, 28 August 2008.

65 Reframing is here used analogously to its use as a technique in psychotherapy by which, although the concrete facts are unchanged, the perception of those facts is altered in order to make the meaning of a problem more manageable to the patient, Walrond-Skinner, S. (1986) *A Dictionary of Psychotherapy*, London and New York, Routledge & Kegan Paul.

66 MPF file, pp186–189.

67 For example, in the referred settlement the *erveiras* claimed that since Natura began to buy the raw material of breu and other plants, the prices had been raised in an unsustainable way for the traditional perfumary; MPF file, pp186–189.

68 Letter from Natura, 13 April 2006, CGEN file, pp387–395.

69 The hinterland settlements founded by slave fugitives.

70 ILO Convention 169, Article 1, uses a similar criterion.

71 Araújo, A. L. (2006) 'Magia das ervas identifica o Ver-o-Peso', *Jornal O Liberal*, 19 February.

72 Mendes, C. (2006) 'Aroma do Pará gera polêmica', *Jornal O Liberal*, 23 April.

73 Quilombola are the descendants of the Quilombos, the hinterland settlements founded by slave fugitives.

74 Article 48 para 4.

75 TK is usually holistic and spiritual in opposition to analytical and naturalistic science. Johnson, M., 1992.

76 Articles 27, 33. In cases of access to TK by indigenous peoples the FUNAI (official Indian Affairs body) is also party, but in practice not a beneficiary.

77 Presentation on TKDL at the Third Session of Inter-Governmental Committee, *World Intellectual Property Organisation* at Geneva, 17 June 2002, by Gupta, V. K., Director of the National Institute of Science Communication (Council of Scientific & Industrial Research).

78 See Labate, B. (2004) *Ayahuasca Mamancuna merci beaucoup: Diversificação e Internacionalização do Vegetalismo Ayahuasqueiro Peruano.* PhD thesis in Social Sciences, University of Campinas (UNICAMP), Brazil; and *Workshop: The global-ization of Ayahuasca – An Amazonian psychoactive and its users.* University of Heidelberg, Institute of Medical Psychology (Organizers: Jungaberle, H., Weinhold, J., Verres, R., Labate, B.), 16–18 May 2007, www.ritualdynamik.uni-hd.de.

79 It is a special case since it concerns food and agriculture. See Santilli in this book.

80 Like the Auckland Declaration of 2004, Elvin-Lewis, 2006, p78.

References

Avila, T. A. M. (2006) 'Não é do jeito que eles quer, é do jeito que nós quer: Biotecnologia e o acesso aos conhecimentos tradicionais dos Krahô', in Grossi, M. P., Heiborn, M. L. and Machado, L. Z. (eds) *Antropologia e Direitos Humanos 4.* Nova Letra, Blumenau, pp121–183

Berkes, F. (1999) *Sacred Ecology: Traditional Ecological Knowledge and Resource Management*, Philadelphia, Taylor & Francis

CBD (2008) 'Access and benefit-sharing in practice: Trends in partnerships across sectors', *Technical Series No 38,* Montreal, Secretariat of the Convention on Biological Diversity (Laird, S. and Wynberg, R.), 140pp

Diegues, A. C. and Arruda, R. S. V. (2001) 'Saberes tradicionais e biodiversidade no Brasil', *Biodiversidade*, vol 4, São Paulo Ministério do Meio Ambiente, Núcleo de Pesquisas sobre Populações Humanas e Áreas Úmidas do Brasil – NUBAUB, USP

Dutfield, G. (2002) *Protecting Traditional Knowledge and Folklore: A Review of Progress in Diplomacy and Policy Formulation* (Draft), UNCTAD/ICTSD, pp09–19

Elvin-Lewis, M. (2006) 'Evolving concepts related to achieving benefit sharing for custodians of traditional knowledge', *Ethnobotany Research & Applications*, vol 4 (pp75–96), www.ethnobotanyjournal.org/vol4/i1547-3465-04-071.pdf, accessed 30 May 2008

FAO (1995) Memoria – Consulta de expertos sobre productos forestales no madereros para América Latina y el Caribe, *Serie Forestal* no 1, Santiago, Chile, 4–8 July 1994, http://www.fao.org/docrep/t2354s/t2354s08.htm, accessed January 2009

Giddens, A. (1991) *As Conseqüências da Modernidade*, São Paulo, UNESP

Hahn, A. von (2004) 'Traditionelles Wissen indigener und lokaler Gemeinschaften zwischen geistigen Eigentumsrechten und der public domain', *Beitraege zum auslaendischen oeffentlichen Recht und Voelkerrecht* (Max-Planck-Institut), Band 170, Berlin, Springer

Hansen, S. A. and Van Fleet, J. W. (2003) *A Handbook on Issues and Options for Traditional Knowledge Holders in Protecting their Intellectual Property and Maintaining Biological Diversity*, AAAS Science and Human Rights Program, Washington DC, American Association for the Advancement of Science

Johnson, M. (1992) 'Research on traditional environmental knowledge: Its development and its role', in Johnson M., (ed), *Lore: Capturing Traditional Environmental Knowledge*, IDRC, Ottawa

Kamau, E. C. (2009) 'A implementação do Artigo 8j da CDB, o problema do conhecimento tradicional disseminado e a experiência do Quênia', in Kishi, S. A. S. and Kleba, J. B. (eds), *Dilemas do Acesso à Biodiversidade e aos Conhecimentos Tradicionais – Direito, Política e Sociedade*, Belo Horizonte: Ed. Fórum, 2009, forthcoming

Kate, K. T. and Laird, S. A. (1999) *The Commercial Use of Biodiversity*, London, Earthscan

Khor, M. (2002) *Intellectual Property, Biodiversity and Sustainable Development*, Penang, Malaysia, Third World Network/UNEP

Kleba, J. B. (2005) *Propriedade Intelectual e Biopirataria: O caso do Cupuaçu*, XIII Congresso Nacional de Sociólogos, Belém do Pará, 8–11 November

Kleba, J. B. (2008) *Pajés, Etnofarmácia e Direitos Tortuosos – O Caso Krahô/UNIFESP.* VII Jornadas Latinoamericanas de Estudios Sociales de la Ciencia – ESOCITE, Rio de Janeiro, 28–30 May 2008, www.necso.ufrj.br/esocite2008/trabalhos/35972.doc, accessed 10 June 2008

Langenheim, J. H. (2003) *Plant Resins Chemistry, Evolution, Ecology, and Ethnobotany,* Portland; Cambridge, Timber Press

Lima, E. C. (2008) 'As novas formas do kampô: elementos de uma sociologia da disseminação urbana dos saberes nativos', in Lenaerts, M. and Spadafora, A. M. (eds) *Pueblos Indígenas, Plantas y Mercados – Amazonia y Gran Chaco.* V Congreso CEISAL de Latinoamericanistas, Zeta Series in Anthropology & Sociology, vol 3, Argentina, FLACSO, pp169–197

Martins, H. M. (2006) *Os Katukina e o Kampô: Aspectos Etnográficos da Construção de um Projeto de Acesso a Conhecimentos Tradicionais,* MSc thesis, Programa de Pós-Graduação em Antropologia Social da Universidade de Brasília, Brasília, Brazil

Moura, F. B. P. (2007) Conhecimento tradicional e estratégias de sobrevivência de populações brasileiras, Maceió, EDUFAL

Ostrom, E. (2000) 'Private and common property rights', pp332–379, http://encyclo.findlaw.com/2000book.pdf, accessed 10 June 2008

Rodrigues, E. (2001) 'Usos rituais de plantas que indicam acões sobre o Sistema Nervoso Central pelos índios Krahô, com ênfase nas psicoativas', PhD thesis, Universidade Federal de São Paulo, São Paulo, Escola Paulista de Medicina

Rodrigues, E. and Carlini, E. A. (2006) 'Plants with possible psychoactive effects used by the Krahô Indians', *Rev. Bras. Psiquiatria,* vol 28, no 4, pp277–282

Seiler, A., Daele, W. V. and Döbert, R. (2003) *Protection of Traditional Knowledge – Deliberations from a Transnational Stakeholder Dialogue Between Pharmaceutical Companies and Civil Society Organizations,* Discussion Paper No SP IV 2003–102, Berlin, Wissenschaftszentrum Berlin für Sozialforschung

Souza Filho, C. F. M. (2002) Introdução ao Direito Socioambiental, in Lima, A. O. (ed) *Direito para o Brasil Socioambiental* (Instituto Sócio-Ambiental) Porto Alegre, Sergio A. Fabris (ed), pp21–48

Vandebroek, I. et al (2007) 'Use of medicinal plants by Dominican immigrants in New York City for the treatment of common health conditions. A comparative analysis with literature data from the Dominican Republic', in Pieroni, A. and Vandebroek, I. (eds), *Traveling Cultures and Plants: The Ethnobiology and Ethnopharmacy of Human Migrations.* Bergahn Books, New York, pp39–63

WBGU – Wissenschaftlicher Beirat der Bundesregierung Globale Umweltveränderungen (2000) 'Erhaltung und nachhaltige Nutzung der Biosphäre', *Jahresgutachten 1999,* Bremerhaven, Germany, Alfred-Wegener-Institut für Polar- und Meeresforschung

WHO (2003) 'Neurosciences', *WHO fact sheet,* vol 8, no 4, p257

WIPO (2002) 'Traditional knowledge – Operational terms and definitions', Document WIPO/GRTKF/IC/3/9, WIPO, Geneva

Wynberg, R. (2004) 'Rhetoric, realism and benefit-sharing. Use of traditional knowledge of *Hoodia* species in the development of an appetite suppressant', *Journal of World Intellectual Property,* vol 7, no 6, pp851–876

Chapter 8

Protecting TK Amid Disseminated Knowledge – A New Task for ABS Regimes? A Kenyan Legal View[1]

Evanson C. Kamau[2]

Introduction

Different people, groups of people and organizations have perceived TK differently. Previously, it was regarded as barbaric, heathen, devilish and witchcraft (Kamau, 2004, p167; Kihwelo, 2005, p347). However, these perceptions have not been static, but have been transformed with time and by changes in global events. One such event is the emergence of modern biotechnologies, which have enhanced not only the economic, scientific and commercial value of GRs, but also the TK associated with them (the World Intellectual Property Organization, WIPO/TK/CEI/00/ INF.5). Today, TK is known to be a vital lead for inventions in pharmaceutical, cosmetic, agricultural and chemical industries, among others. Hence, many researchers from Western countries travel to developing countries, which are rich in TK, in search of it and the biological resources associated with it. Greater appreciation and respect for TK is continuing to draw international attention to issues related to it (WIPO/TK/CEI/00/INF.5).

Article 8(j) of the CBD urges its contracting parties to:

> [S]ubject to its national legislation, respect, preserve and maintain knowledge, innovations and practices of indigenous and local communities embodying traditional lifestyles relevant for the conservation and sustainable use of biological diversity and promote their wider application with the approval and involvement of the holders of such knowledge, innovations and practices and encourage the equitable sharing of the benefits arising from the utilization of such knowledge, innovations and practices.

Numerous efforts have been undertaken to this effect since the CBD entered into force, but apart from dealing with foreseen challenges, different countries and international fora have constantly encountered new challenges and impediments. Decision IX/12 of the Ninth Convention of the Parties (COP 9) identified a number of components to be further elaborated with the aim of incorporating them in the international regime, as well as components for further consideration. One of the problems identified by the author, which is likely to hinder the effective implementation of any access and benefit-sharing (ABS) regime, is the scattered nature of TK. A lot of TK and innovations based thereon are out of the control of indigenous and local communities. That makes it hard to fortify the prior informed consent system and ensure that indigenous and local communities reap the benefits they deserve from TK utilized by others. The danger created by this scenario is that, after all the legislative work, laws created might at the end just grant them, more or less, artificial rights. The author believes this is imminent as long as all existing loopholes of dissemination and misappropriation of TK exist. This phenomenon needs further consideration.

Focusing on the benefit-sharing (BS) objective of the CBD under Article 8(j), the author advocates for the strengthening of the rights of the indigenous and local communities through abatement and/or regulation of all currently uncontrolled channels of dissemination and misappropriation of TK, as well as the redirection of a share of benefits reaped by privately utilizing such knowledge back to such communities. First, he asserts the right of the communities for benefits from TK based on the value and ownership and the obligation of CBD contracting parties to equitably share benefits. He then classifies 'disseminated traditional knowledge' (DTK) by conceptualizing the term and developing criteria for differentiating DTK from core TK. Further, he questions whether communities have a right to also share benefits from DTK and looks at the problems DTK presents to effective protection of TK. Finally, he proposes how benefits from DTK should be shared and makes recommendations on how ABS regimes could eliminate or minimize the effects caused by DTK.

Asserting the right of benefits for TK

The arguments made below shall be based on the manner of acquisition, transmission, accumulation, storage and dissemination of TK and the obligation of Article 8(j) of the CBD.

Acquisition, transmission, accumulation, storage and dissemination

TK is usually acquired intuitively in a traditional setting through observations and experiments. Existing studies by most ethnographers, for example, tend to depict the transmission of skills through the social and cultural forms of learning as a disorganized, unstructured and highly individualistic process (Ruddle, 1993, p17f). However, there are studies that show that learning in those general forms is structured and culturally specific (Mead, 1930; Read, 1960; Ruddle, 1993).[3]

Observations by researchers from diverse disciplines produce remarkably consistent generalizations about certain structural and processual characteristics of transmission of TK. These characteristics were exemplified for instance in a case study of a mixed peasant economy in the Orinoco Delta, Venezuela, featured in Ruddle and Chesterfield (1977). They are summarized by Ruddle (1993, p18) as follows:

1 There exist specific age divisions for task training in *most* activities [emphasis added].
2 Different tasks are taught by adults in a similar and systematic manner.
3 Within a particular task complex ... individual tasks are taught in a sequence ranging from simple to complex.
4 Tasks are gender and age specific and are taught by members of the appropriate sex.
5 Tasks are site specific and are taught in the types of locations where they are to be performed.
6 Fixed periods are specifically set aside for teaching.
7 Tasks are taught by particular kinsfolk, usually one of the learners' parents.
8 A form of reward or punishment is associated with certain tasks or task complexes.

The term 'traditional' is often misleading. It is frequently interpreted to mean primitive, untechnical, outdated or archaic, inferior and so on. That gives the impression that the knowledge acquisition process in a traditional setting is also poor and lax. In contrast (Njoroge and Bussmann, 2006, p332),[4] it is organized (highly structured), systematic, disciplined, demanding and living or evolutionary. 'Traditional' only indicates that it is tradition-based (Hansen and VanFleet, 2003, p3). It is created, preserved and disseminated in a way that reflects the traditions of the communities who created it (Hansen and VanFleet, 2003).

Training in herbal medicine is rigorous, demanding keen plant identification, collection and (composition) preparation ability, proper acquisition of diagnostic skills and medicine administration know-how (Kamau, 2004, p166). This demands substantial time and sacrifice, which makes a family the most convenient training institution (Kamau, 2004). Most herbalists, for example, receive their first lessons from very close and often older relatives, mostly parents or grandparents (Kamau, 2004; Ohmagari and Berkes, 1977, p209ff; Ruddle, 1993, p22; Tabuti et al, 2003a, p122f; Takako, 2003, p114). In exceptional cases, folk healers receive training not only from their kinsfolk, but also from traditional expert healers outside their families (Takako, 2003).[5]

Training in herbal medicine often begins at a very early age. Some herbalists claim to have been introduced to the vocation at an age of only 12 (Kamau, 2004, p166, note 737). According to Valentine Nde Fru[6] from Cameroon, his 63-year-old grandmother, Mama Lum Sonia Neh, started practising at the age of 13.[7] That indicates herbal medicine training can commence even much earlier than the age of 12.

Learning in herbal medicine takes the form of on-the-job training, that is, learning by doing (or apprenticeship) (Ohmagari and Berkes, 1977, p122ff; Ruddle, 1993, p20; Tabuti et al, 2003a, p122ff; Tabuti et al, 2003b, p20). Generally, transmission of TK starts by familiarizing the learner verbally and visually with the physical elements of the appropriate location (Ruddle, 1993). Trainees in herbal medicine accompany their parents or grandparents in collection of herbs (Kamau, 2004, p166).[8] In the field they learn how to identify the plants, their names and the best methods of harvesting relevant parts (Njoroge and Bussmann, 2006, p334; Tabuti et al, 2003b, p21; Tabuti et al, 2003c, pp279, 284).[9] For small herbaceous plants, whole plants are usually collected (Njoroge and Bussmann, 2006; cf. Tabuti et al, 2003b, p21). Use of some herbs is determined by the sheer nature of the texture of the bark, shape of the leaves or fruit (Kamau, 2004, p167). Herbalists have the capability of grouping herbs according to specific body-part ailments (Kamau, 2004). At home trainees learn how to prepare herbs and store formulations (concoctions) (Tabuti et al, 2003b, p40f).[10] Most of the concoctions are a combination of extracts from a number of different herbs (Kamau, 2004, p167; Tabuti et al, 2003b, p40).

Patients are usually treated at the herbalists' private homes (Kamau, 2004). In the process, the trainees learn how to diagnose sicknesses and administer medicine (Kamau, 2004).

At every level of learning the old lessons are revised (repeated) before a new lesson is introduced (Ruddle, 1993). Learning thus proceeds additively and sequentially (Figure 8.1) from simple to complicated tasks until an entire task is mastered (Ruddle, 1993). When trainees are competent enough, they are allowed to assist the instructor in performing the task, as well as independently experiment and use personal initiative (Ruddle, 1993). Later they are entrusted with the treatment of simple illnesses even in the absence of the main practitioners (Ruddle, 1993). Serious illnesses require a high proficiency level and hence more thorough and longer training.[11]

Generally, herbalists accumulate their knowledge by constantly and continually gathering information from the general community (O'Connor, 2003, p678) around their vicinity (Otieno-Odek, 1994, pp79–103). According to a study conducted by Chege (see Kamau, 2004, note 735),

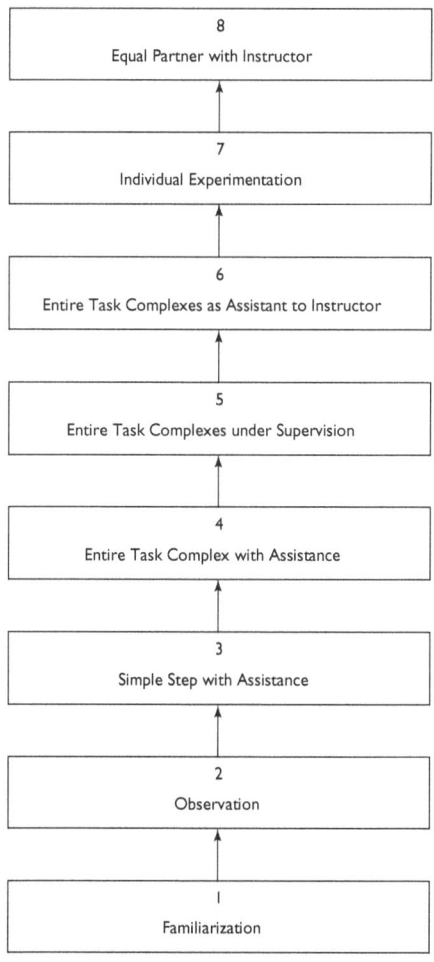

Source: Ohmagari and Berkes (1997, p205), after Ruddle and Chesterfield (1977)

Figure 8.1 *Identified learning sequence for traditional skills and knowledge*

up to 95 per cent of herbalists appropriate knowledge that is gathered and generated by the general community. Nonetheless, some build on the community knowledge by continuous innovation (Kamau, 2004, p175), as well as acquiring information outside the local environs (Kamau, 2004, p166). For that purpose they travel to other parts of the country and at times beyond country borders (Kamau, 2004). Some herbalists are not dogmatic in applying existing knowledge. They subject it to extensive experimentation in their practice, initially by trial and error and experiences gathered over time, until they discover new uses and formulas not contained in the general-community knowledge (Kamau, 2004, p167). Therefore, although TK is often referred to as 'common' or 'collective' knowledge, it is not always common or commonly distributed in a community. Among Kadazan/Dusun people of Sabah in east Malaysia, the efficacy and usage of particular medicinal plants are known to everyone (Takako, 2003, p110). Specific plants together with the specialized skills and knowledge required in treatment of patients, however, are occasionally known exceptionally to the *Bobohizan*, high priestesses.[12] At times they are even confidential and their practice restricted to individual experts (Tabuti et al, 2003a, p123; Takako, 2003).

Dissemination of traditional medicine knowledge (TMK) is determined by the level at which it is held. Knowledge held at general community level is disseminated freely through exchange between individuals within a community, or neighbouring communities (Kamau, 2004, p167f). Knowledge held at an individual level is considered secret and hence exchanged between close confidants (Kamau, 2004, p167; Tabuti et al, 2003a, p123). This pattern of diffusion of TMK is maintained down the generations as the older members of the community informally (mostly orally) pass it over to the younger ones (Evans, 2002; Kamau, 2004, p166f; Njoroge and Bussmann, 2006, p333; O'Connor, 2003, p678; Tabuti et al, 2003c, p279).

From the information above, I make the following conclusions:

- The greatest part of TK on medicinal properties of biological resources is a property of the general communities, whether tribal or regional.
- A part of herbal medicine knowledge, no matter how marginal, is an innovation of individual herbalists.
- General community medicinal knowledge acts as a trailblazer for formation and administration of new knowledge held by individuals.
- Dissemination of TK by any member of a specific community for non-traditional usage without the prior informed consent (PIC) of the community is a violation against community rights.

These facts point to the obvious truth based on morals and justice that the general community should share in benefits derived directly or indirectly from common (traditional) knowledge. All users of TK for commercial purposes should respect this credo irrespective of whether they are foreigners or locals. It is interesting to note that many TK holders believe that, whether commonly or individually accumulated, TK cannot be owned by anyone as all knowledge and resources come from God (Swiderska, 2006, p12). Thus, they never consider themselves as exclusive owners of TK. Tabuti et al see traditional medicine practitioners (TMPs) as (vital) depositories of TK of healing (Tabuti et al, 2003a, p119). The communities benefit from the individual knowledge of local experts or healers and they reciprocate by giving them special reverence (Vivekanandan et al, date unknown). Nonetheless, experts may also extract a fee for the services they offer, which they retain as their private property, not necessarily because they themselves developed the knowledge or held it secret from other community members, but rather because they, and not the whole community, fulfil certain spiritual and cultural requirements (Koopman, 2003), as well as deliver certain services. In addition, as already mentioned, they innovate on collective knowledge, giving it new and unique qualities, and therefore deserve some reward. Therefore, it is immoral and unjust for anyone to monopolize benefits from TK in total exclusion of the communities to whom the knowledge belongs.

Be it in the area of medicine, cosmetics or food, TK is known as traditional mainly due to the characteristics examined above. They prove that: (1) indigenous and local communities are the genuine source of TK; (2) TK cannot be separated from its legitimate owners as it is tightly intertwined with their culture; (3) these communities have invested much labour in creation of TK; (4) creation of TK involves a genuine and vigorous inventive activity; (5) TK is dynamic; and (6) as TK is based on the tradition of sharing knowledge, no single person is entitled to exclusive monopolization of TK and benefits thereof. (Points three, four and five also advocate for ample rewarding of the indigenous and local communities.)

BS obligation under Article 8(j)

Under Article 8(j), the CBD requires that '*Each Contracting Party shall ... encourage the equitable sharing of the benefits arising from the utilization*' of TK with its holders. Indigenous and local communities are the legitimate holders of TK (see below) and therefore, whereas it is their right to receive benefits from the utilization of their knowledge, it is the obligation of users of such knowledge to share the benefits therefrom with them.

Classifying DTK

Conceptualizing DTK

The concept of DTK and the ideas used to formulate it are mostly inspired by the current state of distribution of TK in Kenya. However, the situation described below might also be familiar to other countries, as Kleba in this book shows.

As a result of free and unrestricted dissemination of TK over a long period of time, a tremendously large portion of TK is no longer under the control of indigenous and local communities. There are three identifiable levels at which TK is currently held. First, intact TK still held by indigenous and local communities. Second, TK disseminated nationally. This is held by traditional herbal medicinal practitioners and their assistants, freelance traditional herbalists, researchers and research institutions, academia and parataxonomists, for example. In addition, there are plenty of freely obtainable publications, some of which not only list medicinal plants and the diseases they are used to treat (Kagombe et al, 2006; Mukonyi and Gachathi, 2004; Tabuti et al, 2003c), but also show their chemical structures (Mukonyi and Gachathi, 2004), as well as describe how traditional medicine is prepared from the plants and administered (Tabuti et al, 2003c, pp281–283). Third, worldwide-disseminated TK. TK under second and third levels is held privately by numerous entities that are estranged from indigenous and local communities and hence is referred to as DTK. In this study, focus is limited only to the second level and its impact on effective protection of the rights of indigenous and local communities, as well as BS.

The CBD and the concept of DTK

The CBD does not address the issue of DTK. However, the use of the term 'holders' in Article 8(j) may help in construing a likely view of the CBD in this issue.

At first sight, the term seems to cause some sort of confusion as to whom it refers, depending on how one interprets it. If it is interpreted broadly, it would include all persons who have some TK in their possession and by doing so acknowledge them as legitimate owners of the TK they hold with the right to dispose it at their will, as well as monopolize benefits from its commercialization. Such an interpretation would consequently acknowledge DTK as a separate and independent body of knowledge from the core TK. A narrower interpretation, on the other hand, is, in my opinion, more consistent with what the CBD had in mind.

The term 'holder' cannot be plugged out of the provision (Article 8(j)) and interpreted in isolation. It should also be noted that the CBD refers to TK as the 'knowledge *of* indigenous and local communities'. 'Of' undoubtedly indicates ownership, which implies that the term 'holders' as used in the CBD is a synonym of 'owners'. There is no disagreement that the indigenous and local communities are the legitimate owners of TK. So by asking the contracting parties to promote the application of TK with the approval and involvement of the holders of TK, the CBD is obviously referring to the indigenous and local communities.

WIPO is clearer on this issue. It uses the term *'traditional knowledge holder'* and defines it as all persons who create, originate, develop and practise TK in a traditional setting and context (WIPO/GRTKF/IC/3/9, §37(ii)(b)). This affirms the view of the narrower interpretation. Therefore, any TK that is not created, originated, developed and practised in a traditional setting and context is DTK.

From the concept of DTK and the analysis of the term 'holders', an attempted definition of DTK reads: Traditional knowledge held privately by entities isolated or estranged from the indigenous and local communities who use TK to earn gains privately without recognition of indigenous and local communities and in their total exclusion in sharing benefits.

Criteria demarcating DTK from non-DTK

From the analysis above, any TK falling under the following criteria is classified as DTK:

- TK held by entities that are not connected to any indigenous/local community
- TK not owned collectively
- TK practised and developed in a non-traditional setting and context
- TK held by entities that are not occupants/indigenous of a specific territory, which is related to the usage of the TK
- TK held by entities that do not necessarily live close to nature
- TK held by entities that do not see it as their assignment to pass it on to the next generation
- TK practise which has no direct relationship to conservation and sustainable use activities of environments from where the raw materials are fetched
- TK practise which has no proof of proficiency of learning/training in a traditional setting.

Should indigenous communities share benefits from DTK?

In exclusion or with inclusion? The puzzle

A strong argument exists for protection and rewarding of TK. It is not clear though whether this should proceed separately for the two *quasi-national* blocks of TK.[13] If yes, the following question arises: Is it just to allow holders of disseminated knowledge to derive benefits from TK in exclusion of indigenous communities? Existing definitions and descriptions, the process of creation of TK, Article 8(j), among others, agree on vital issues such as the value, ownership and dynamism of TK. As already argued, any TK whether intact or in public domain originates from the indigenous/local communities. In any case, customary law does not demand non-disclosure (novelty) as a conditio sine quo non for claim of rights or ownership over TK. Excluding the custodians of TK fully from the benefits earned by disseminated (traditional) knowledge holders is hence unfair and unjust: it is equivalent to freeriding. It encourages and allows biopiracy at the expense of indigenous/local communities. Likewise, it contradicts and weakens ongoing efforts to regulate ABS. On this basis, the author concludes that indigenous and local communities have a right to share in benefits derived from the use of TK and GR associated with it in any of the three blocks of TK. It is also vital to abate the current trend of extreme dissemination of TK without which any endeavour to reward the indigenous/local communities for their contribution (to science/innovation) will be gravely undercut.

Hurdles created by exclusion of DTK from regulation

Disseminated knowledge has and is likely to continue disadvantaging the true custodians of TK, as well as the general ABS process due to the following reasons:

- Disseminated knowledge holders are ready to earn quick and easy money. Many of them make and sell products based on TK, as well as land-based biological resources as a way of earning a living.[14] They neither have a vision for a long-term enterprise nor the interests of the rightful custodians and the environment at heart. For the user, it is likewise easier to imagine a reward for a single person or a few persons. There is also a general tendency of a community (representative) demanding benefits based on the size or needs of the community that scares off interested or potential users (parties), giving advantage to disseminated knowledge holders.[15]

- Most likely, users might prefer approaching disseminated knowledge holders so as to avoid cumbersome access procedures.[16] It is easier to deal with the provider one-to-one and thus escape complicated PIC, mutually agreed terms (MAT) and MTA (material transfer agreement) procedures and/or complex contracts (Kamau, 2004, p169f, note 746). This makes DTK a preferred source of knowledge.
- A user who has already accessed DTK would most likely also get associated (biological) resources from the same source or some of the many local parataxonomists/'suppliers' (biopirates). This gravely weakens the PIC system, as it undermines it and denies it a chance for learning and transformation.
- DTK holders are able to locate the market where demand is high for TK with more ease. They are not restricted to remote areas, but are in a better position to expose and advertise their products. Most genuine owners of TK are located in remote areas and only depend either on the same patients or (rarely) on new patients recommended by old and constant patients.
- The more the true custodians of TK are overshadowed and cut off from potential markets for their products and knowledge, the less lucrative the practice of TK becomes. This discourages younger generations from taking up and continuing TK and practice. This, of course, causes further erosion and hence gradual disappearance of TK. The death of an elder is at times compared to the burning down of a library, yet the former is even more consequential as a permanent and total loss is suffered: it might still be possible to trace copies of burnt materials in other libraries, but a dead-elder's knowledge can never be recovered.
- The more the misappropriation of TK continues, the less the indigenous/local communities would have interest to further evolutionalize and revolutionalize TK. If TK becomes dormant in growth, then there's a danger that it might eventually vanish.
- TK practitioners who develop new methods, drugs and/or products might resolve to keep their knowledge secret, thus hindering dissemination and growth of knowledge. Likewise, their knowledge vanishes with their death, and thus problems that were once solved are left in favour of seeking fresh solutions.
- Many DTK holders are not properly trained and lack the proficiency necessary for good and safe practice.[17] A lot of new scepticism has been expressed concerning the efficacy and health-safety of TK based on experiences made, for example, with 'mobile-jerrican clinics (therapists)'.[18] They are deemed to mar the reputation of TK and initiate a retreat therefrom, except for desperate patients who lack other alternatives. There is a real danger that with numerous publications of plants

and their usage, more self-professed TMPs will mushroom to the detriment of both the patients and the legitimate custodians of TK.[19]

In brief, the injustice, unfair competition, ineffective protection as a result of a weakened ABS procedure, deteriorating reputation of TK caused by doubtable efficacy and safety, and eroding and gradually disappearing knowledge caused by the existence of disseminated knowledge, are new challenges that ABS regimes ought to address in order to save the situation.

New tasks for **ABS** regimes and possible antidotes

In the midst of all the threats that hinder effective protection, the following questions arise: Is there any intellectual property protection (IPP) that can be granted to TK? If not, what regime would be befitting for TK? Are there other possible ways of ameliorating as well as abating the negative effects caused by DTK?

Trade secret protection?

Among the forms of intellectual property (IP) that exist, trade secrets seem to be closer to TK. What are trade secrets? Trade secrets derive their definition from three main characteristics. According to Cornish and Llewelyn (2007) these are: (1) valuable information; (2) not generally known to the relevant portion of the public; and (3) being subject of reasonable efforts to maintain their secrecy (Cornish and Llewelyn, 2007, p308; Francis et al, 2007, p30f). Thus certain factors would determine what a 'trade secret is and what is not'. These include: the extent to which the information is known outside the business; the extent to which it is known to those inside the business, that is, by the employees; the precautions taken by the holder of the trade secret to guard the secrecy of the information; the savings effected and the value to the holder in having the information as against competitors; the amount of effort or money expended in obtaining and developing the information; and the amount of time and expense it would take for others to acquire and duplicate the information.[20] Therefore, a trade secret is any valuable business information that is not generally known and is subject to reasonable efforts to preserve confidentiality.[21]

Patent law does not offer any protection for trade secrets because they do not fulfil one of the conditions of patent protection, that is, disclosure. In addition, some of the trade secrets do not meet the patentability criterion of novelty, as they require only a minimal level of inventiveness (Francis et al,

2007, pp10, 22).[22] They are hence maintained through secrecy and transferred under an oath of no breach of confidence.

For a breach of confidence (in respect of technical, commercial, personal and other information) to be actionable in the American and English courts, the following conditions are necessary:

1 The information itself must have the necessary quality of confidence about it (Cornish and Llewelyn, 2007, §8-09). No right of action would exist if the object is available in the open market and an obtainer is able to analyse it so as to find out its secret content (see Francis et al, 2007, p9). This applies also if it contains partly public and partly private information, or if the information has been made freely and entirely public either before it was imparted to the defendant in confidence, or in the interval between that time and the trial of the action if the defendant's breach of confidence is not the cause (Cornish and Llewelyn, 2007, §§8-10–8-12).

2 That information must have been given in circumstances indicating an obligation of confidence (Cornish and Llewelyn, 2007; Francis et al, 2007, p10).[23] The supplier gives it to the acquirer on condition that the latter will keep it secret. The circumstances under which the information is supplied and acquired may also give rise to an obligation to keep it secret;[24] for example, in employment[25] or commissioning.

3 There must be an unauthorized use of the information to the detriment of the party communicating it (Cornish and Llewelyn, 2007, §§8-10–8-12).

Restraint of breach of confidence spawns from equity, which most often grants injunctions as a remedy for breach.[26] Enjoined may be both actual and threatened misuse of trade secrets, but that does not preclude development of technology by fair and honest means (Francis et al, 2007, p8), as noted below. Thus a court may issue an injunctive order either restraining a person who obtained a trade secret by improper means from disclosing it (Francis et al, 2007, p10), or instructing a person who has been using or disseminating the trade secret information to desist from doing so. Damages and other remedies may be granted in certain circumstances.

Generally, the type of remedy will depend on whether the obligation of secrecy is imposed on another party involuntarily by operation of law (tortuous, which will result in injunction) or voluntarily assumed (contractual, which will result in damages) (Francis et al, 2007, p19). But even criminal liability may be imposed in some countries; for example, in the USA (Francis et al, 2007, p8). This is the case under the statutes of many US

States, as well as the Federal Economic Espionage Act (FEEA) of 1996 (18 U.S.C., §1831ff). It has never been clear under which circumstances direct and indirect recipients of information would have liability imposed upon them (Cornish and Llewelyn, 2007, §8-06).

Trade secret law does not offer protection against discovery by fair and honest means; for example, by independent invention, accidental disclosure or reverse engineering,[27] that is, working the process used to develop a product backwards by using that product.

In order to find out whether trade secret protection can be given to TK, the question we need to ask first is: Does TK qualify for it? To answer that question, it is imminent to find out whether TK possesses characteristics that qualify information as a trade secret. As mentioned previously, TK is held at different levels. Hence, a further question would be: If it does qualify, at what levels?

To briefly reflect on the criteria qualifying information as trade secrets, first, the information must be valuable. This is true of TK at all levels.[28] Second, the information must not be generally known to the relevant part of the public. Here we start having doubts concerning almost all levels of TK due to its widely disseminated nature. However, as stated before, some of the traditional skills and knowledge are still intact, being restricted to individual experts. Such knowledge would qualify, but all the DTK is excluded. Nonetheless, how far and where it should be disseminated for it to be considered as generally known to the relevant part of the public is debatable. Apart from specific skills and knowledge that are restricted to experts, TK is generally community knowledge. Therefore, in my opinion, any TK that is still restricted to a particular community cannot be termed as DTK and should hence qualify. Third, the information must have been subject of reasonable efforts to maintain its secrecy. Again, this will depend on the level the knowledge is held, and presumably only that which is held by individuals (experts) and exchanged only between close confidants would qualify (Kamau, 2004, p167; Tabuti et al, 2003a, p123). Community knowledge is exchanged freely within the community, as well as between communities. It is often not shielded against the risk of disseminating beyond those communities and therefore does not pass the third criterion in many cases.

Some literature seems to suggest that it is possible to give TK trade secret protection (see Axt et al (eds), 1993, p62f; Dutfield, 2000, p88ff; Gollin, 1993, pp159–197). Some TMPs are even pushing for it.[29] It is true that trade secrets are favourable for TK as the information does not have to be novel. They also offer perpetual protection as long as secrecy is maintained. In addition, they afford the TK holder an opportunity to impart information that has not attained patentability level to an entrepreneur with the hope that the latter will help to develop it without fear of undesired disclosure. But that alone does not guarantee effective protection. So the deciding question is: Does trade secret law offer effective protection for TK?

I conclude tentatively that intact TK qualifies for trade secret protection, but express doubt whether trade secret law can effectively protect it. One, the rights of indigenous/local communities over TK are still considered perpetual and, although trade secret law allows perpetual ownership of rights, it guarantees that only as long as secrecy is maintained. Two, trade secret protection does not prevent reverse engineering and offers no remedy against accidental disclosure. These first two arguments show that it is barely possible to guarantee perpetual rights over TK under trade secret law. Third, it is extremely hard for indigenous communities to establish breach of confidence, especially after the knowledge has left their territory. That is exacerbated by the possible existence of similar knowledge. Fourth, the remedies granted by trade secret law are not permanent: in practice, injunctions have a limited duration; damages cannot replace the value of rights of indigenous/local communities over TK. Also, litigation in cases involving breach of confidential information is complicated and would definitely be too expensive for such communities.

My final conclusion hence is that, although a specific category of TK qualifies for trade secret protection, trade secret law is not appropriate for protection of TK as a whole. If used to protect intact TK, it has to be applied very cautiously. It would, however, imply that its conditions are acceptable by the indigenous/local communities, thus demanding a re-construing of (their) rights over TK. Unless that question is resolved, a tailor-cut regime (sui generis), taking into consideration the specificities of TK, would still be needed to ensure effective protection. Nonetheless, the concept of trade secret protection is useful in advising national legislation on development of a similar approach, as well as introduction of claims of injunction and damages. It can also be used to alert the ABS regime (PIC/MAT/MTA) to the fact that DTK has its origin in localizable knowledge, and therefore DTK holders have an obligation to share benefits from transfer and use of such knowledge with the indigenous and local communities. Finally, it can be used to inform customary law on ways of communicating TK.

Ideas for a sui generis intellectual property rights regime

Criteria of TK to be protected

In order to be protected, TK must possess a certain level of novelty (having real or potential value) with the ability to examine according to the principle of complete identity of technical solution. It should also involve an inventive step with prominent substantive features and notable progress as compared to the existing technology. In addition, it should be practically applicable: its product(s) must have a medical effect. Also, it should be possible to carry out/exploit its methods, as well as realize its use industrially.

TK to be protected should belong to an identifiable indigenous or local community, or a number of identifiable indigenous or local communities, or individual expert(s) living within the said community(ies). There has to be proof that the TK was created, developed, held or preserved by such communities or individuals if claim of ownership is disputed.

The TK should not be generally known to the relevant part of the public. It must have been the subject of reasonable efforts to maintain its secrecy either through written or unwritten norms of customary law or codes of conduct. Finally, it must be associated with GRs.

There are currently no requirements given for protecting TK with patents, but a national sui generis form of IP could be created.

Access procedure, BS and litigation

In contrast to concepts establishing TK as property rights, regulatory concepts requiring consent and contracting are more in use. They could be designed as follows:

- Step 1: The potential user must obtain the PIC of indigenous/local community(ies) or individual(s) duly identified as owners of the TK to be accessed. The PIC must be in the form of a written triune document with names and addresses/locations of parties on both sides and affiliation(s) of applicant, where this applies. Where a community possesses administrative structures or where the knowledge concerned belongs to more than one community, a council of elders representing such a community or communities, respectively, shall give the PIC. It shall contain the names of all the council elders and the community(ies) they belong to. One representative from each side shall then sign it.
- Step 2: The parties or representatives of the parties (provider and user) take the PIC to the access authorizing office at the nearest environmental department, which plays a dual role of examining the PIC and processing the official access permit on the one hand, and witnessing between both parties by placing its seal on the PIC on the other. The office collects the administrative fee, faxes a copy of the permit and PIC to the central office and files one copy of the originals. The parties keep a copy each of the originals.
- Step 3: Finally, an agreement/contract based on mutually agreed terms is made. The contents of the contract may include:
  - provider's obligation to disclose the TK to the user on exclusive or

non-exclusive terms – a short description of the TK to be accessed based on present uses; user's obligation to pay an agreed, up-front payment (75 per cent of a normal patent licence remuneration for exclusive rights and 50 per cent for non-exclusive)
- user's obligation to disclose source during patent application
- provider's right to co-own intellectual property rights (IPRs)
- user's obligation not to licence third parties without the consent of the provider
- provider's right to share commercial and non-commercial benefits from utilization, including from third party licences
- provider's right to share benefits from all proceeds of utilization of TK even after the patent has expired
- provider's right to continue using the TK in a traditional context and obligation to maintain secrecy
- penalties for violation (such as damages and criminal sanctions) and choice of law (preferably the law of the provider country).

Ways of sharing benefits from utilized TK

The user shall surrender 25 per cent of the monetary benefits; 75 per cent (of the 25 per cent) shall go to the providing community and 25 per cent to a national trust fund for environmental conservation and restoration. If the provider is an individual living in the community, 25 per cent (of 25 per cent) of the benefits shall go to the individual, 50 per cent to the community fund and 25 per cent to the national trust fund. The user shall share results of further research with the provider so as to assist in improving traditional usage. The user shall transfer technology utilizing the TK to the provider so as to assist in acquisition of modern ways of manufacture. Lastly, the user shall assist in capacity building.

Tapping benefits from DTK: Types of funds

DTK holders have an obligation to share benefits. Whereas they may give PIC for usage of such knowledge prior to an access authorization by the ABS regulating body, the BS contract should acknowledge the communities from whom the TK stems (if known) as co-parties and the obligation of the providing party to share benefits with them.

Funds are proposed as an appropriate way of sharing benefits from utilization of DTK with indigenous and local communities. Is there a sensible way of doing so? In my opinion, there is. But the first thing would be to categorize TK depending on whether it is possible to establish its source or not.

The following types of funds and how they should function are suggested:

- Fund 1 ('X Community Fund') This fund should be used for benefits from utilization of TK originating from a specific indigenous/local community. Practitioners who provide information concerning the concrete source of the TK – based on the orthodox manner of acquisition – should pay 75 per cent of the benefits for the community into this fund and 25 per cent into the national trust fund.
- Fund 2 ('Y and Z Communities Fund') Into this fund, users should pay benefits from TK originating from a number of indigenous/local communities. Practitioners who use TK ownership which is claimed by more than one community should pay 75 per cent of the benefits for the communities into this fund and 25 per cent into the national trust fund.
- Fund 3 ('National Trust Fund') Into the national trust fund should come all benefits from utilization of TK, the ownership or source of which is impossible to establish. The fund should be administered by the government for environmental conservation and restoration, as well as community projects such as clinics, clean water, education and infrastructure.

Benefits paid into the funds from utilization of DTK should include a percentage of the holders' monthly earnings from treatment of patients and any other benefits,[30] including benefits from contracts of usage by third parties. Although the benefit-sharing formula proposed above should generally apply, for contracts with foreign users this may be adjusted to match the size of benefits, as well as the scope and terms of the contract. Nonetheless, benefits should reflect the input of individual innovation on collective TK and value addition: the more individual innovation and value addition, the less the benefits into the fund.[31] These conditions should likewise apply to local institutions and/or industries that manufacture products based on TK.

Revision of the definition of TK

Present definitions of TK leave out knowledge held by either individuals or entities that have no relationship with the indigenous/local communities. That seems to remove such knowledge from their custody and suggests they have no claim of any rights over it. It is proposed that these concerns should be integrated in a new definition, which could read as follows:

> *Traditional knowledge is a collective intellectual property of a society based on a systematic and coherent body or stock of culture-specific knowledge of the indigenous and local communities occupying a specific geographical territory about the relationship of living beings with*

*one another and with their environment. It also includes their inno-
vations and creations resulting from intellectual activity in the
industrial, scientific, literary or artistic fields and also their practices
and beliefs based on orally preserved past experiences and observations of older generations. This knowledge is held, constantly evolved
and enlarged over time through use and adaptation to new demands,
and culturally transmitted down through generations by the
indigenous and local communities. It includes such tradition-based
knowledge held by entities that are alienated from the indigenous/
local communities and that use TK to earn gains privately without
recognition of indigenous and local communities and in their total
exclusion in sharing benefits.*

Auxiliary measures

TK-friendlier intellectual property law

The current Kenya Industrial Property Act (IPA)[32] is very insensitive to
TK. Even where the invention is based on GRs and TK, it does not require
disclosure of source and proof of PIC during patent application. I suggest
its amendment to include such a requirement by inserting a subsection to
this effect under section 34, which defines what a patent application shall
contain. The subsection could read: 'Where the invention is based on
genetic resources and/or traditional knowledge, the applicant shall disclose
the source of such genetic resources and/or traditional knowledge in application for a patent and produce a valid proof of prior informed consent
authorizing access'.

The IPA should also have a section or subsection on withdrawal and/or
transfer of rights for post-grant challenges to rights in case a violation of disclosure requirement was committed.

Likewise, the IPA should construe the doctrine of novelty and define
categories for immunity[33] under section 23 (novelty) such that the public
domain for TK is shrunk or narrowed. Why should this be important if TK
cannot be patented? First, there are numerous cases when TK is accessed
from what would be called 'public domain', which at times is a culturally
normal area for indigenous and local communities to share knowledge.
Such traditions cannot be altered. Second, indigenous peoples' knowledge
accessed freely and used in inventions is valuable knowledge without which
the invention could not have attained its stature. Once a patent has been
granted for such an invention, the TK is also indirectly patented and knowledge that was and still is in the 'cultural knowledge exchange area'

monopolized. Third, there is still a lot of secret, as well as new, constantly evolving TK, which can be upgraded to the level of conventional inventions with the help of modern technology and patented. Would the issue of perpetual rights over TK then not be an impediment?

I think the debate on sharing in benefits from utilized TK has strayed from the right path if perpetual rights are to be claimed over TK on the basis of which a patented invention has been made. According to patent law, the invention is free for utilization by the public after a fixed duration of time has expired which, in other words, implies that the invention is henceforth public domain. In these circumstances, what would have been left of indigenous peoples' rights over TK? In my opinion, once the indigenous peoples have accepted patenting of inventions based on their TK, the claim of perpetual rights over it loses strength. It is more reasonable to broaden their claim of ownership of TK, including that which is said to be in public domain as a result of the usual traditional way of exchanging knowledge.

It is imaginable that the law could be amended to include a subsection under section 23, which gives a longer 'period of grace' exceptionally to TK. According to the IPA, immunity for all categories of prior disclosure lasts 12 months. My proposal is that TK is granted a 20-year immunity, within which any successful claims made by the rightful indigenous/local communities will see the full rights and damages being transferred to them. Any successful claims made thereafter should regulate how future rights would be shared between the person who applied for protection and the legitimate communities, but without demanding damages from the former. This will enable the indigenous peoples to continue using TK freely within the traditional context and also give them the prospect of reclaiming their rights where their IP has been abusively disclosed, or used in an invention without their PIC or without them sharing in benefits. It will also in turn encourage sharing of knowledge (disclosure) and innovation.

Measures of a more voluntary character
In order to strengthen the rights of the indigenous and local communities further and minimize the effects of DTK, other measures of a more voluntary character may in addition be undertaken. Local institutions and industries should be obliged to produce PIC for products manufactured based on TK. This will help to hinder cheap commercialization of TK and re-divert any benefits to the rightful beneficiaries.

The state should establish standards of efficacy and safety. This will yield validation, confidence, value addition and increased potential for BS. It will also encourage plant conservation.

Traditional medicine practitioners practising in a more or less secular business setting must register their businesses as required by law, get licences and pay taxes. This will help the government to improve their ser vices, which include acquisition of more laboratory equipments where herbal practitioners can test their drugs and doses, offer corresponding capacity building, etc,. Traditional medicinal healers practising under traditional conditions, and still attached and recognized by corresponding indigenous/local communities, should be exempted from registration. The registration process should also encompass proof of proficiency. This could be a means of sieving out dangerous and unhealthy practices, as well as drawing a line between legal and illegal health care services. It might also assist in abating local biopiracy[34] and overharvesting of biological resources/environmental degradation.

A Code of Conduct (CoC) should be established for basic researchers, including those working for or affiliated to local public institutions, restraining them from bringing into public domain any information that has not yet been captured in databases. Indigenous/local communities should be sensitized on the need to demand an official permit signed and duly sealed with the stamp of the pertinent licensing authority, in this case the National Council of Science and Technology (NCST), before they provide any information.

Conclusion

TK is valuable knowledge. Most of it involves a rigorous inventive activity. Like conventional inventions, there are ample justifications why it should be protected and rewarded.

It is widely agreed now that indigenous and local communities deserve a reward from users of TK for their effort in creating it. Available definitions and descriptions of TK, the CBD and many (working) documents of international and national organizations, institutions and non-governmental organizations identify the indigenous and local communities as the true custodians of TK. They also acknowledge the value of TK. Notwithstanding, the issue as to who should have custodianship over DTK and whether the indigenous and local communities should share in benefits from its utilization has not yet been broached.

The status of DTK and how it relates to the core TK needs to be determined. There are two likely approaches. The first would be to place all knowledge based on TK under the common umbrella of TK and hence create either a direct or indirect link with indigenous and local communities. The second would be to de-link it from the latter, thus acknowledging the existence of a separate body of knowledge belonging to the nation and ignoring any effects that it might have on the rights of the indigenous and local peoples, including the right to share in benefits generated by DTK. The argument that it is difficult to trace the true origin of DTK and thus to classify it under any indigenous or local community speaks more for de-linking it. However, as seen above, there are stronger arguments, which speak for linking it. My opinion is that DTK should be linked to TK as far as it is traceable to particular indigenous/local communities. DTK, which cannot be attributed to any of these communities – either because it cannot be clearly traced to any such communities or based on its wide national application having become known to the relevant part of the public – should be considered as a national heritage (public domain) freely usable by all citizens.

Substantive criteria and a legal basis upon which these rights may be claimed are lacking. The WIPO working definition of the term 'traditional knowledge holder', for example, implicitly ignores or leaves out issues concerning all TK held and practised out of a traditional setting and context. This loophole seems to exclude DTK from the scope of TK requiring regulation, and by doing so risks the possibility of denying indigenous and local communities custodianship over it and also the right to share in benefits arising from its utilization. The definitions and pertinent terms such as 'holder' must be augmented and cautiously interpreted, respectively, to address this issue.

Efforts to effectively protect and reward TK are likely to be seriously hampered by the existence of DTK. There are many factors that place DTK holders at an advantage over the legitimate custodians of TK. The former are hence likely to undercut the latter as far as rewarding of TK is concerned. DTK holders are also likely to water down protection measures under ABS regimes because they act as loose dissemination channels of TK. Under these and other circumstances, it is important for ABS regimes to address problems created by DTK. In my opinion, the indigenous and local communities should not be totally excluded from sharing in benefits derived from the full or partial use of the knowledge of indigenous and local communities by DTK holders.

In summary, ABS regimes need to carry out the necessary alterations and inclusions so as to aid the ABS process against a backdrop of challenges caused by DTK. Possible approaches of regulating this conflicting area might vary and stretch from revising existing definitions to include: intellectual property maxims, doctrines and laws; creating a sui generis IPR for TK; categorizing TK and TK holders; creating benefit funds according to categories of TK; finding a concrete method of calculating the benefits to be paid into community funds; introducing CoC for access and usage of

TK by local public institutions, which include recognition of indigenous and local communities in publication of materials containing information on TK; and revision of the current definition of TK.

Notes

1 This chapter has been written parallel to an ongoing bioprospecting research project at the Forschungsstelle für Europäisches Umweltrecht (FEU), University of Bremen, Germany, with the title 'Law and practice in access to genetic resources with the example of Brazil, Kenya and Germany'. I would therefore like to thank the sponsor of the project, the German Research Foundation (DFG), from whom this study has tremendously benefited. Many thanks to Professor Gerd Winter (project supervisor) for his invaluable comments, critique and suggestions. I also owe thanks to my Brazilian colleagues, Professor John Kleba (Instituto Technológico de Aeronáutica, São José dos Campos-SP) and Ms Sandra Kishi (Prosecutor, São Paulo), for their comments. Of course I cannot forget to thank a dear person, Brendan Tobin (Asociacion para la Defensa de los Derechos Naturales (ADN), Peru), who encouraged and motivated me to research further on the issues discussed here after my first attempt to raise them at a Bremen ABS workshop in the winter of 2008.
2 The views expressed herein are solely those of the author.
3 Ruddle (1993) quotes among others Raum, O. F. (1940) *Chaga Childhood, A Description of Indigenous Education in an East African Tribe*, London, OUP.
4 Note that the use of traditional plants for management of diseases both in animals and humans is not haphazard.
5 According to Takako (2003), such a practice is upheld in Tebing Tinnggi in North Samatra, Indonesia.
6 Interview with Nde Fru at the Forschungsstelle für Europaesches Umweltrecht, University of Bremen, 23 November 2007. Nde Fru was a Masters degree candidate in law at the time.
7 Nde Fru, ibid.
8 Also Nde Fru, ibid.
9 According to Tabuti et al (2003b, p21), a significant proportion of the concoctions are made using leaves (37.3 per cent) and roots (34.3 per cent). Use of other parts is generally below 5 per cent. In some cases the whole plant is used (8.2 per cent). Overall, use of perennial parts and reproductive parts (flowers, fruit, seeds) is substantial at 42.4 and 6.8 per cent, respectively.
10 Nde Fru, note 6.
11 Nde Fru, note 6: Herbal compositions against such sicknesses as malaria (e.g. from eucalyptus and pawpaw leaves) and typhoid (from eucalyptus) are very concentrated (strong) and require careful administration. They likewise often have side effects. Anti-malaria herbal medicine from pawpaw leaves, for example, causes extreme itching, especially in the early stages of the sickness. Some mixtures cause high fever. These side effects are counteracted using either non-conventional methods or, at times, conventional medicine. High fever is suppressed, for example, by covering the patient with a thick blanket to induce sweating so as to lower the body temperature. Alternatively, conventional medicine such as piriton or chloroquine is administered.

12 Takako (2003) describes the *Bobohizan* as ritual specialists and practitioners of traditional medicine.

13 For the three blocks of TK identified in this chapter, see above under 'Conceptualizing DTK'.

14 Due to economic frustration, even the true custodians of TK are often forced to succumb to the pressures of basic (life's) needs and hence surrender their knowledge to users for a small price. Nde Fru (note 6) says, 'as a traditional healer, my grandmother possesses certain capabilities that are individually developed and not widely known in the community. However, if Bayer or a researcher wishing to develop a product based on her (traditional) knowledge offered her a goat, pig or cow, she would give out the secret'.

15 That does not imply that DTK holders should be abated to allow indigenous/local communities to extract exorbitant gains for the knowledge. They, or their representatives, need to be schooled on how to carry out constructive bargain, as well as demand rewards that have long-lasting benefits.

16 Possible access restraints created in Kenya after the new ABS regulations (2006) and possible ways of abating them are discussed by Kamau (2009).

17 Interviewee, anonymous (April 2007, Kiambu-Kenya): According to the interviewee, many who claim to be traditional medicinal healers access scanty and shallow knowledge concerning plants and their usage either from the general community or genuine practitioners, and then use humans through trial and error as laboratories in a quest to create themselves a profession. The interviewee gave an example of the usage of aloes for treatment of intestinal worms and cited examples of numerous cases of overdose, resulting in various side effects, viz. stomach cancer. The interviewee admitted that, although he too treated patients, he neither originates from a family with a pertinent background, nor went through any traditional training. He graduated from the University as a chemist and was employed in a long-existing herbal clinic, where he helped establish the right doses for drugs. Although the owner of the clinic is very secretive as far as methods of mixing concoctions is concerned, the interviewee managed to learn the plants used in the clinic and their usage. He also claimed that the owner of the clinic does not collect the herbs himself, but employs locals to harvest them. It is also very likely that the collectors also do some small-scale practising to substitute their living. The interviewee justified himself by claiming that, being a University graduate and also conversant with the right methods of establishing doses, it is unlikely for his patients to suffer any side effects. Nonetheless, whether the efficacy and safety of the drugs is established, this scenario clearly depicts how TK leaves the domain of its genuine custodians and disseminates from one person to the other, being used for personal gains without any benefits for the former. Even efficacy and safety cannot justify such an injustice seeing that even a long-existing clinic still has problems fixing the right dosage, hence little or no added value.

18 Mobile-jerrican clinics: This term makes reference to vendors of traditional medicine who transport it mostly in transparent plastic containers (jerricans).

19 See note 17.

20 See http://law.freeadvice.com/intellectual_property/trade_secrets/, accessed 3 June 2008.

21 See US Trade Secrets Act (UTSA), Section 1(4), which defines a trade secret broadly as information, including a formula, pattern, compilation, program, device, method, technique, or process, that: (1) derives independent economic value, actual

or potential, from not being generally known to and not being readily ascertainable by proper means by other persons who can obtain economic value from its disclosure or use; and (2) is the subject of efforts that are reasonable under the circumstances to maintain its secrecy.

22 According to Cornish and Llewelyn (§8-10), a budwood of a new plant variety, which is a physical object, for example, may embody secrets.

23 *Coco v Clark* [1969] R.P.C. 41 at 47 approved and was relied upon in subsequent cases.

24 In *Att-Gen v Guardian Newspapers* (No 2) [1990] 1 A.C. 109, 281, it was stated that ' duty of confidence arises when confidential information comes to the knowledge of a person ... in circumstances where he has notice, or is held to have agreed, that the information is confidential, with the effect that it would be just in all the circumstances that he should be precluded from disclosing the information'. It might hence be enough to show that the person ought to have known that the information was confident. In other words, the restriction of non-disclosure or non-use might be express or implied.

25 Although the practice between countries varies, the overwhelmingly dominating rule holds that an employee's duty of fidelity, whether expressed or implied in a contract of employment, requires that the employee acts in the employer's best interests at all times in the course of employment. This includes the protection of trade and commercial secrets, as well as any information given to him by the employer and that which he generates in the course of employment. In certain circumstances, the employee would be restrained from deliberately or secretly entering into competitive work with his employer or another employer. An employer can stop the employee from extracting information, for example, by copying it out or deliberately memorizing it with a view to taking it away on departure, but information that would naturally be remembered may be taken away. If no valid express covenant between the employer and the employee exists, restraining the latter from usage of such information within a legally permissible scope (types of business, duration and area of operation) after employment, the employer cannot receive any legal redress for its usage by the ex-employee in a manner that is detrimental to his interests. For a comparative analysis of practices in 13 countries, see Lagesse, P. and Norrbom, M. (eds) (2006) *Restrictive Covenants in Employment Contracts and Other Mechanisms for Protection of Corporate Confidential Information*, Kluwer Law International and International Bar Association. In English law, infractions of duty by fiduciaries might invoke liability upon third parties if the latter knowingly assisted in the breach of trust or equitable obligation (Cornish and Llewelyn, §8-33). But a third party who overhears when information is being imparted upon another is under no legal obligation to preserve the confidence (Cornish and Llewelyn, §8-34).

26 According to Cornish and Llewelyn (§8-06), the scope of the modern law started forming in the 1850s with two cases, one of which involved a recipe for a medicine (*Morison v Moat* (1851) 9 Hare 241)), where injunctions were granted against indirect recipients of the confidential information and jurisdiction said to arise by virtue of property, agreement, confidence, trust and bailment.

27 That was stated, for example, in a ruling at a US court in *National Tube Co. v Eastern Tube Co.*, 3 Ohio Cir.Ct.R., N.S., 459, 462 (1902), aff'd, 69 Ohio St. 560, 70 N.E. 1127 (1903), as quoted in Francis et al (2007, p10).

28 See 'Asserting the right for benefits'.
29 Interviews carried with TMPs in Kenya, July/August 2008. Confirmed in interviews at the Kenya Industrial Property Institute (August 2008).
30 For that, TMP need to learn how to manage records of their patients, as well as book keeping.
31 Practitioners who have developed special abilities by constantly innovating TK will thus have the opportunity to gain for their intellectual contribution that is above and beyond collective knowledge. Practitioners with genuine TK proficiency, but void of individual abilities, will hence act more or less as marketing managers: they earn for their services, but re-direct a big chunk of the gains to community funds. This might help to fight against 'everyone-is-a-traditional-healer' attitude and make it profitable for clinics based on proficiency and legally established, as well as encourage innovation. It will also assist in 'netting' benefits for communities.
32 E-copy available online at http://www.kipi.go.ke/patents/ipa/ipact2001.pdf, accessed 1 October 2008.
33 Patent law knows generally three cases under which disclosure made prior to (application for) protection does not damage novelty. First, disclosure made by the inventor or his predecessor in title within a limited 'period of grace' foreseen by law. Second, disclosure made by the inventor or his predecessor in title at exhibitions. Third, disclosure made through an evident abuse in relation to the applicant or his predecessor in title. (For details and how some legal systems deal with this issue, see Kamau, 2004, p40ff. Note that the new section is No 23 according to the amended version of the IPA.)
34 According to the interviewee quoted in note 17, some local biopirates harvest plants secretly even at night, including from private lands. Another interviewee, Mr Bernard Kamondo of the KEFRI, Muguga Regional Research Centre, in a post-interview tour of the forest area under the institution, showed some freshly rehabilitated species of trees, including *Prunus africana*, which had been degraded as a result of theft and overharvesting by locals.

References

Axt, J. R., Corn, M. L., Lee, M. and Ackerman, D. M. (eds) (1993) 'Biotechnology, indigenous peoples and intellectual property rights', *Congressional Research Service*, Washington DC, 16 April, http://www.ipmall.fplc.edu/hosted_resources/crs/93-478.pdf, p62f, accessed 3 November 2008
Cornish, W. and Llewelyn, D. (2007) *Intellectual Property: Patents, Copyright, Trade Marks and Allied Rights*, London, Sweet & Maxwell
Dutfield, G. (2000) *Intellectual Property Rights, Trade and Biodiversity*, London, IUCN, Earthscan
Evans, S. (2002) 'Chilean stories: Exploring herbal medicine in South America', *Australian Journal of Medical Herbalism*, 14(1), Commentary
Francis, W. H., Collins, R. C., Stevens, J. D., Grove, A. M. and Schmidt, M. J. (2007) *Cases and Materials on Patent Law Including Trade Secrets-Copyrights-Trademarks*, sixth edition, American Casebook Series, West Publishing Co
Gollin, M. A. (1993) 'An intellectual property rights framework for biodiversity prospecting', in Reid, W. V. et al (eds), *Biodiversity Prospecting: Using Genetic*

Resources for Sustainable Development, Washington DC, WRI, INBio, Rainforest Alliance, ACTS, pp159–197

Hansen, S. A. and VanFleet, J. W. (2003) *Traditional Knowledge and Intellectual Property. A Handbook on Issues and Options for Traditional Knowledge Holders in Protecting their Intellectual Property and Maintaining Biological Diversity*, Washington DC, AAAS

Holmberg, A. (1950) 'Nomads of the long bow', *Smithsonian Social Anthropology Publication* 10, Washington DC, Smithsonian Institution

Kagombe, J. K., Kariuki, J. G. and Luvanda, A. M. (June 2006) 'Socio-economic and natural resources baseline survey in Mukogondo landscape, Laikipia District', *FOR-REMS-KEFRI Project*, Report No. 6

Kamau, E. C. (2004) *A Hard Patent System: An Impediment to Technological (Economic) Development in Less Developed Countries*, Baden-Baden, Nomos Verlag

Kihwelo, P. F. (2005) 'Indigenous knowledge: What is it? How and why do we protect it?', *JWIP*, vol 8, no 3, pp345–359

Koopman, J. (December 2003) 'Biotechnology, patent law and piracy: Mirroring the interests in resources of life and culture', *Electronic Journal of Comparative Law*, vol 7.5, http://www.ejcl.org/ejcl/75/art75-7.html, accessed 21 January 2008

Mead, M. (1930) *Growing Up in New Guinea: A Comparative Study of Primitive Education*, New York, William Morrow and Co.

Mukonyi, K. W. and Gachathi, N. F. (October 2004) 'Survey on utilization of non-wood forest products in Mukogondo district', *FORREMS-KEFRI Project*, Report No 4

Njoroge, G. N. and Bussmann, R. W. (2006) 'Herbal usage and informant consensus in ethnoveterinary management of cattle diseases among the Kikuyus (Central Kenya)', *Journal of Ethnopharmacology*, vol 108, pp332–339

O'Connor, B. (2003) 'Protecting traditional knowledge. An overview of a developing area of intellectual property law', *JWIP*, vol 6, no 5, pp677–698

Ohmagari, K. and Berkes, F. (1977) 'Transmission of indigenous knowledge and bush skills among the western James Bay Cree women of subarctic Canada', *Human Ecology*, vol 25, no 2, pp197–222

Otieno-Odek, J. (1994) 'The Kenya patent law: Promoting local inventiveness or promoting foreign patentees?', *Journal of African Law*, vol 38, pp79–103

Read, M. (1960) *Children of Their Fathers, Growing Up Among the Ngoni of Nyasaland*, New Haven, Yale University Press

Ruddle, K. (1993) 'The transmission of traditional ecological knowledge', in Inglis, J. T., (ed) *Traditional Ecological Knowledge. Concepts and Cases*, International Program on Traditional Ecological Knowledge and International Development Research Centre, pp17–31. Also available online at http://www.idrc.ca/openebooks/683-6/

Ruddle, K. and Chesterfield, R. (1977) *Education for Traditional Food Procurement in the Orinoco Delta*, Ibero-Americana 53, Berkeley, University of California Press

Swiderska, K. (2006) *Banishing the Biopirates: A New Approach to Protecting Traditional Knowledge*, Gatekeeper Series 129, London, IIED

Tabuti, J. R. S., Dhillion, S. S. and Lye, K. A. (2003a) 'Traditional medicine in Bulamogi county, Uganda: Its practitioners, users and viability', *Journal of Ethnopharmacology*, vol 85, pp119–129

Tabuti, J. R. S., Lye, K. A. and Dhillion, S. S. (2003b) 'Traditional herbal drugs of Bulamogi, Uganda: Plants, use and administration', *Journal of Ethnopharmacology*, vol 88, pp19–44

Tabuti, J. R. S., Dhillion, S. S. and Lye, K. A. (2003c) 'Ethnoveterinary medicines for cattle (*Bos indicus*) in Bulamogi county, Uganda: Plants species and mode of use', *Journal of Ethnopharmacology*, vol 88, pp279–286

Takako, Haruyama (2003) 'Transmission mechanism of traditional ecological knowledge', *Policy Science*, vol 11, no 2, pp109–118

Vivekanandan, P. et al (date unknown) 'Protecting traditional knowledge of small scattered and disadvantaged grassroots innovators and traditional knowledge holders: Honey bee perspective. Agenda for policy and institutional change', available online at http://r0.unctad.org/trade_env/test1/meetings/tk2/honeybee.pdf, accessed 21 January 2008

WIPO/TK/CEI/00/INF.5, Intellectual Property and Genetic Resources, Traditional Knowledge and Folklore – Background Document, www.wipo.int/edocs/mdocs/tk/en/wipo_tk_cei_00/wipo_tk_cei_00_inf_5.doc, accessed 22 May 2009

PART THREE

RECENT DEVELOPMENTS IN EXEMPLARY COUNTRIES

Chapter 9

The Law-Making Process of Access and Benefit-Sharing Regulations – The Case of Kenya

Anne N. Angwenyi[1]

Introduction

The issue of access to GRs and benefit sharing (ABS) was one of the central themes in the negotiation of the CBD. For decades, GRs had been generally regarded as the common heritage of humankind – openly and freely accessible – without the authorization of the country in which they were found and without any obligation to share benefits from their exploitation. An important goal in the negotiations of the Convention was to redefine the conditions under which the benefits arising from the use of GRs would be shared with the countries of origin of such resources. The third objective of the Convention therefore focuses on the 'fair and equitable sharing of the benefits arising out of the utilization of genetic resources'.

The Convention establishes an elaborate framework of objectives, principles and obligations relating to ABS. These provisions need to be translated into national legal requirements for effective implementation. The Bonn Guidelines on Access to Genetic Resources and Benefit Sharing provide some guidelines to Parties on how to implement their ABS obligations. Since the entry into force of the Convention, Kenya has developed both framework legislation that domesticates the provisions of the Convention and specific regulations to implement the relevant provisions of the framework legislation. This chapter examines the key legislation governing ABS in Kenya and makes recommendations for the enhancement of enforcement of the law on ABS.

General legislation and policies with relevance to access to GRs in Kenya

The constitution of Kenya

Kenya, like many other states, has yet to address the legal status of GRs. However, it approaches the concept of property based on the English common law system, except for the instances where this is superseded by the Constitution or statute. The understanding of real property includes land and whatever is erected or growing upon or affixed to land, to the extent that it is considered immovable by law (Lettington, 2001, p151).

The provisions of the 1992 Constitution of Kenya addressing property are introduced by the umbrella provisions of section 70, which provides every citizen with 'protection for the privacy of his home and other property and from deprivation of property without compensation'. The question as to whether private landowners also have ownership and control of the GRs found on their property is not expressly provided for in the laws. However, an interpretation of the Registered Lands Act (Chapter 300 of the Laws of Kenya) and the Transfer of Property Act of India 1882, which govern individual ownership of land and the nature of interests one gets under these laws, indicates absolute ownership of land, together with all rights and privileges belonging or appurtenance thereto subject to enumerated overriding interests. Further, Kenya inherited the Anglo-Saxon common law tradition pursuant to which ownership is said to extend to everything that is found beneath and above someone's private property (Lettington, 2001, p151). The exceptions to these broad categories of property generally involve specifically identified assets, such as minerals or mineral oils, and rights of way, such as through airspace, over which the government retains control for the benefit of the public.

The 1992 Constitution of Kenya refers to the environment only in the context of governmental powers for the purposes of conservation. Consequently, it does not directly refer to the ownership of, access to or benefit sharing of GRs. However, certain provisions can have direct impacts on these questions. In particular, the Constitution's provisions regarding personal property and trust land may be relevant. Some of the provisions of the Constitution of Kenya most relevant to GRs are contained in chapter IX, dealing with the status of trust land in the country. Section 115 places the principal responsibility for trust land in county councils. Subsection (2) of section 115 obliges county councils to hold trust land for the benefit of the ordinary residents of the land, and to 'give effect to such rights, interests or other benefits in respect of the land as may, under the African customary law for the time being in force and applicable thereto, be vested in any tribe, group, family or individual'.

The constitutional issue which arises thus is whether GRs are an asset controlled by the state or whether they are the subject of private ownership and control. There is little case law to provide guidance on this issue. However, given Kenya's common law tradition, it is arguable that GRs constitute part of the rights that make up real property, in that they invariably are growing upon or are affixed to land in some manner. This approach is supported by a decision of the High Court, which held 'that according to common law and/or customary law of the inhabitants of this country, those entitled to the use of land are also entitled to the fruits thereof which include the fauna and flora unless this has been negated by law'.[2] However, practice on the ground with various on-going projects show no consistency on the subject of control and/or ownership of GRs. For instance, the Royal Botanical Gardens, Kew, project on collection of plants and seeds provides for benefits to go to national agencies rather than to the traditional occupants of the trust land where the activities are carried out, thus implying governmental control of the GRs involved (Lettington, 2001, p151). The activities of the Kenya Wildlife Service (KWS) and International Centre of Insect Physiology and Ecology (ICIPE) collaboration sheds no light as the collection of micro-organisms are from soils collected in national parks which are government land. The question thus remains as to whether the government owns nationally applicable rights to GRs, has limited powers to regulate the use of these rights, or only has powers over GRs on government-controlled land. It is suggested that only a clear statutory/constitutional or regulatory regime may clear up this issue, since litigation thus far has not been on a constitutional basis.

One of the practical challenges that may arise if GRs are defined as state owned is that it precludes, unless otherwise provided for in law, private contracts or transactions where a landowner can enter into private transactions with a bioprospector. While individuals can deal with their property as they see fit, if the use of biological material falls into the category of GRs, however defined, then state-stipulated procedures will kick in and have to be observed. The clarification of the legal status of GRs in national legislation is therefore crucial to the implementation of Article 15 of the Convention, it being essential in defining access requirements, procedures, rules and rights over these resources.

The Wildlife (Conservation and Management) Act (amended in 1989)

There are several operative elements of the Wildlife (Conservation and Management) Act that either directly or in a manner of interpretation govern the management of GRs in protected areas. Section 3A provides the

KWS with the mandate of formulating policies regarding the conservation, management and utilization of all types of fauna (not being domestic) and flora. Further, Section 3A(g) allows KWS to 'conduct and co-ordinate research activities in the field of wildlife conservation and management', whose clause has been used to provide for the Service's GRs activities. Bioprospecting activities fall within the ambit of these sections.

The Crop Production and Livestock Act (Revised Edition 1977)

The Crop Production and Livestock Act provides for the control and improvement of crop production and livestock. Of specific relevance to access to GRs, Section 4(1)(b) of the Act gives broad powers to the Minister for Agriculture to regulate the methods for the production of any crop, while Section 4(1)(c) provides powers regarding the improvement of the quality of any agricultural produce. The rules promulgated under the Act on African Produce (1964) regulate the improvement and inspection of legumes, sorghum, potatoes, rice, bulrush, millet, finger millet, wheat, fresh fruit and onions grown by Africa. However, these rules have not been used to govern access to GRs for food and agriculture or research based on these resources (Lettington, 2001, p151).

Policy and legislation specific to access to GRs

In any discussion on ABS in Kenya, it is important to make the 1992 UNCED the reference point. This is because, other than the fact that Kenya signed and ratified the CBD, it was from this time forward that national awareness on the potential opportunities for commercial exploitation and the need to regulate access started growing.

Therefore, consequently, national laws relating to GRs fall into two broad categories: the pre-UNCED laws and the post-UNCED laws. Prior to UNCED, ABS issues did not attract national attention. For this reason, laws enacted in this period did not include any specific provisions on access to GRs or sharing of benefits arising from the exploitation or utilization of such resources. Most of the laws therefore tend to go back to the pre-colonial times.

The CBD adopted at UNCED in 1992 increased the political and economic significance of these GRs. Consequently, laws in Kenya enacted post-UNCED in the area of natural resources contain provisions on access to GRs. Thus the rationale behind the current regulatory regime on natural resources in Kenya is multi-faceted. The laws therefore encompass conservation, as well as enhancement of research, development capacities and revenue among others. It is increasingly becoming clear to the Kenya Government that biological resources are an important and integral part of Kenya's natural resources. Thus the Government is making serious attempts to consolidate these resources with the understanding that they will contribute to national economic development, as is reflected in the Access to Genetic Resources Regulations that have recently been gazetted.

To date, however, there has been minimal policy debate on the issue of access to genetic and biological resources. Prior to the publication and entry into force of the Environmental Management and Coordination Act, several initiatives were undertaken to develop regulations governing GRs in Kenya. This issue first appeared in discussion during the development of the First National Biodiversity Strategy and Action Plan of 1999, which made mention of the need to regulate Kenya's biological resources through policy and legislative measures. In 2000, under the Chairmanship of the National Council on Science and Technology (which is mandated to deal with research in Kenya), an expert group made up of experts from the NGO sector, relevant government ministries and departments worked on developing a draft regulatory system with an accompanying committee to oversee the implementation of the future regulatory system. At the same time, although with no statutory mandate, a second initiative was undertaken by the National Museums of Kenya, and pursuant to its activities of collecting plant and seed varieties with the Royal Botanic Gardens at Kew, a drafting sub-committee including government ministries and the Attorney General developed a preliminary draft of possible regulations. Unfortunately, both initiatives stalled and have yet to be reinstated. Notwithstanding the failure of these initiatives, the entry into force of the Environmental Management and Co-ordination Act (1999; EMCA)[3] had a significant impact on the initiatives as the Act provided a statutory mandate to the National Environment Management Authority to develop legislative measures governing access to GRs in Kenya.

The EMCA (1999)

The EMCA is Kenya's framework legislation coordinating all environmental management activities in the country. As such, it constitutes the primary implementing legislation for the CBD. A number of the provisions of the Act have either direct or indirect potential impacts on the issue of access to GRs. Section 42(3) provides the Minister for Environment with broad powers to issue orders, regulations or standards for the management of riverbanks, lakeshores, wetlands and coastal zones. In particular sub-section (g), (h) and (j) are relevant to GR management. Section 42(3)(g) further provides that the Minister may regulate the 'harvesting of aquatic living and non-living resources to ensure optimum sustainable yield'. Section 42(3)(h) provides for 'special guidelines for access to and exploitation of living and non-living resources in the continental shelf, territorial sea and the Exclusive Economic Zone'. Section 42(3)(j) provides for the management of biological resources. This application supersedes previous legislation on fisheries GRs as this legislation does not make specific mention of these resources, while EMCA makes a particular case for GRs.

EMCA further elaborates the issue on GRs more explicitly through Section 53, which mandates the National Environment Management Authority (NEMA) to 'issue guidelines and prescribe measures for the sustainable management and utilisation of GRs of Kenya for the benefit of the people of Kenya'. Accordingly, the provisions of any guidelines issued, or measures prescribed, shall include:

- appropriate arrangements for access to GRs of Kenya, including the issue of licenses and fees to be paid for that access
- measures for regulating the import or export of germplasm
- the sharing of benefits derived from GRs of Kenya
- any other matter that the Authority considers necessary for the better management of the GRs of Kenya.

Pursuant to these provisions, NEMA has issued the relevant regulations, namely the Environmental Management and Co-ordination (Conservation of Biological Diversity and Resources, Access to Genetic Resources and Benefit Sharing) Regulations 2006. The scope of the regulations is fairly broad. Except for a list of things that they do not apply to, all access to and use of GRs is covered by the regulations. All bioprospectors are required to obtain a research clearance certificate, prior informed consent (PIC) from the community and/or property owners, and enter into a MTA that includes the sharing of monetary and non-monetary benefits. What this implies in effect is that, although GRs may be privately owned by virtue of common law principles or constitutional rights, all access to GRs as defined must be granted only with the permission of the relevant state authority. In other words, even though a land owner may allow a bioprospector to obtain the resources from his land and enters into an agreement based on mutually agreed terms, only NEMA can issue an access permit which would allow the bioprospector to utilize the resources for the purpose indicated in the application for such permit. Thereafter an MTA is entered into between the relevant lead agency or community and those seeking access in order to transfer the resources out of the country.

Sub-section 50(f) of the EMCA provides that any measure for the conservation of biological diversity 'shall protect indigenous property rights of local communities in respect of biological diversity'. The term 'indigenous property rights of local communities' is not defined by the Act, but given Kenya's historical recognition of customary law in various fields, it would seem, at a minimum, to indicate an intention to recognize customary rights over natural resources. Such an interpretation would seem to be consistent with other references to community rights in the Act.

Section 43 provides that 'the Minister may, by notice in the Gazette, declare the traditional interests of local communities customarily resident within or around a lake shore, wetland, coastal zone or river bank or forest to be protected interests'. Sub-section 48.2 provides that the Director-General of NEMA 'shall not take any action, in respect of any forest or mountain area, which is prejudicial to the traditional interests of the local communities customarily resident within or around such forest or mountain area'. There is, however, no definition of the term 'traditional interests' in the Act and it may or may not include interests in GRs.

The making of Kenya's ABS regulations

The EMCA requires that, in developing regulations, NEMA must do so in consultation with relevant government ministries and departments represented in the Standards and Enforcement Review Committee. The Regulations were drafted over a period stretching from 2004 to 2006 and were developed through a consultative process involving key government agencies and civil organizations.

The Environmental Management and Co-ordination (Conservation of Biological Diversity and Resources, Access to Genetic Resources and Benefit Sharing) Regulations 2006 Legal Notice 160 of 2006 presents the most comprehensive attempt by the government to date to put in place a regulatory framework for ABS. The Regulations do not specifically define GRs, which are defined in the parent Act, the EMCA, as any genetic material of actual or potential value. The Regulations are set out in five parts. Part I addresses preliminary issues, defines the key terms, sets out the objectives of the regulations and sets the scope of application. Access is defined as 'obtaining, possessing and using GRs conserved, whether derived products, and, where applicable, intangible components, for the purposes of research, bio-prospecting, conservation, industrial application or commercial use'. Although not specifically stipulated, protection of traditional and community knowledge is provided for in the mention of 'intangible components' in the definition element of the Regulations. One of the reasons for non-specificity on protection of TK is that at the time of gazetting these Regulations the Attorney General had constituted a taskforce[4] to look into matters of inter alia TK and folklore, and make recommendations as to their protection either through the enhancement of intellectual property laws or the development of a sui generis system. Further, Regulation 3 specifically precludes the exchange of GRs, their derivative products or intangible components associated with genetic resources carried out by members of local Kenyan communities amongst themselves and for their own consumption from the requirements of the Regulations. Further, the Regulations exempt access to GRs derived from plant breeders in accordance with the Seeds and Plant Varieties Act Cap 326, human GRs and approved research activities intended for educational purposes within recognized Kenyan academic and research institutions from the requirements therein.

Part II of the Regulations provides for the conservation of biological diversity through the requirement of an environmental impact assessment for persons who engage in activities that may potentially have an adverse impact on the environment, propose to introduce exotic species in Kenya or unsustainable use of natural resources. Further, this part provides for the conservation of threatened species, the inventorying of biological diversity by NEMA and relevant lead agencies, monitoring the status of biological diversity in Kenya and protection of environmentally significant areas.

Part III of the Regulations lays out the institutional framework for the management of GRs. It designates NEMA as the competent authority for all matters relating to access to GRs. This part also attempts to demarcate the responsibilities between NEMA and other lead agencies such as the National Council for Science and Technology, which is mandated through the Science and Technology Act to issue clearance for research undertaken in Kenya through a research permit. Regulations 9–17 provide for the application of an access permit, the requirement for PIC of the government or local community, the determination of such a permit, validity, terms and conditions of the permit, and the requirement for a register of all access permits granted by NEMA. Further, Regulation 18 specifically provides that there shall be no transfer of GRs outside Kenya without an access permit and MTA.

Part IV of the Regulations in particular provides for the principles that apply to benefit sharing and sets out the generic benefits to be shared (monetary and non-monetary), including the participation of Kenyan citizens and institutions in any activities being conducted with the GRs, joint ownership of patents, payment of access fees and royalties. Kenya being a party to the CBD, the principles encapsulated in this Convention, as well as those in the Regulations on benefit sharing, constitutes the terms and condition for access. It is noteworthy that, even with an access permit, without an MTA one will not be able to export any genetic materials collected.

It is important to note that the benefit mentioned in the Regulations are not exhaustive, but are rather an indication of what any MTA may contain. At the time of drafting the Regulations, the drafters were of the opinion that benefit-sharing mechanisms should not be curtailed, but instead the provisions on benefit sharing should remain open-ended to allow for innovative benefit-sharing regimes to be incorporated into the MTA. It is hoped that the legal experts drafting such MTA would refer to the various global guidelines such as the Bonn Guidelines, the Africa Model Law on the Protection of the Rights of Local Communities, Farmers and Breeders, and for the Regulations of Access to Biological Resources.[5]

Part IV of the Regulations finally provides for the holding of information as confidential on application by the applicant for a permit as may be determined by NEMA, in addition to penalties for contravention with the Regulations.

Recommendations and conclusion

In as much as the Regulations came into force six months after their gazettement, the actual operationalization of the same only started recently, in January 2008, with NEMA calling for potential and actual bioprospectors to bring their activities in line with the requirements of the Regulations. It is unclear how much biodiversity may have left the country and the loss of benefits along with these losses. For GRs that have already left the country through agreements entered into prior to the coming into force of the Regulations, Kenya will rely on the reporting requirements on users to bring their activities in line with the requirements of the Regulations, in order to track compliance with the terms and conditions of access. It is therefore in this context that the country has pushed vigorously at international negotiations for an international regime on ABS that caters for an international certificate of origin/source/legal provenance to be considered. Such a certificate accompanying GRs would ensure transparency and traceability and provide a guarantee that the legal requirements in the country have been fulfilled.

As Kenya is a country rich in biodiversity and recognized as such by belonging to mega-diverse groups of countries, there is a need for NEMA to tighten its coordinative role between the various actors involved in the access regime, such as the National Council for Science and Technology, the KWS, Kenya Forest Service and local government authorities to avoid future loss in terms of benefit sharing to the country. This is both through the advantages and experiences gained in negotiating benefits, and in terms of the national advantages that coordination in benefit-sharing policy can bring in areas such as capacity building and infrastructure development.

Further, in order to implement the requirements of the Regulations, intensive capacity building and awareness raising is a priority requirement to develop the necessary skills required of the policy makers and administrators, legal experts, scientists and researchers, as well as local communities and local administrative authorities. In addition, it is necessary that the national focal point on ABS, NEMA, is clear on *its* mandate, scope, roles and responsibilities. Expertise in the scientific, commercial and legal areas that make up ABS should be found within NEMA. The process for granting access should be transparent, minimally bureaucratic and should promote communication and collaboration rather than suspicion and frustration. In addition, capacity building should include the ability to:

- assist communities, private landowners, other organs of the state in the negotiation of ABS agreements
- analyse benefit-sharing agreements and understand the provisions they contain
- develop an understanding of the opportunities and risk associated with bioprospecting
- gain knowledge of the spectrum of benefits that can be included in benefit-sharing agreements
- acquire expertise in international and regional law and policy and national law and policy. It is important that there is consistency and negotiating expertise in the membership of the teams working on the complex issues of the international regime on ABS, and the World Trade Organization Agreement on Trade-Related Aspects of Intellectual Property Rights (WTO-TRIPS) team on patents and intellectual property rights, as the global strategy now is to build synergies between these two international regimes.

A concerted effort must be made by the government of Kenya to build ABS capacity and raise awareness about ABS issues at a variety of levels: from assisting government with analysing agreements, developing negotiating and legal drafting skills and permit database management, monitoring skills to track the collections and use of resources to determine whether the collected resources are being used in compliance with the requirements of the benefit-sharing agreements, through to improving awareness amongst the research community about the importance of PIC.

It must further be remembered that Article 8(j) of the CBD recognizes the important role that local communities play in maintaining the ecosystems and individual species. In addition, communities that are aware of the question of access to GRs are ones that can assist the government in enforcement. It is extremely difficult for an outsider to conduct activities in a rural African community without that community knowing something about it. Lastly, benefits negotiated out of an ABS MTA based on mutually agreed terms may make an immediate and direct difference to local communities in the name of, for example, dispensaries, refurbishments of schools, cattle dips and boreholes, which when accumulated can make a major difference in poverty alleviation and environmental sustainability in the area.

In undertaking research on this subject, the author noted a number of issues that may potentially arise in Kenya's operationalization of the Regulations and wishes to point these out for further consideration by the enforcers of the Regulations and policy makers.

Continued and effective participation of Kenya in the development of an internationally binding legal regime on ABS is vital. Despite the fact that Kenya has gazetted Regulations, this is a rapidly evolving and complex field that requires due diligence and follow-up on the issue as an international legally binding regime on ABS is being worked on by state parties to the Convention. Inevitably, once the international regime has been finalized, Kenya will be required to harmonize/domesticate its legislations in line with the international regime. This will therefore entail that participation in the meeting on the international regime on ABS be consistent and effective, with an appropriate level of involvement of legal and scientific/technical experts.

Different sectors (state corporations, private commercial users, research and academia) use genetic and biological resources in vastly different ways, and adopt a diversity of approaches and tools for ABS associated with these resources. It is important that the dramatic differences in the ways genetic and biological resources are used by the various sectors are incorporated into policy deliberations by Kenya's policy makers. The Regulations that have been gazetted broadly provide for uniformity of principles and consistency in approach. This generic framework should further be elaborated in different and flexible ways for different sectors, types of research (e.g. academic versus commercial, discovery versus development and commercialization) and scales of such type of activity.

There is a need to build capacity in many provider countries and amongst intermediary institutions to ensure that *potential negotiating and other inequalities between parties are reduced*; knowledge of business, law and advances in science and technology is significant; and opportunities for long-term, mutually beneficial relationships need to be enhanced.

Studies on experience from other countries (e.g. South and Central American countries) indicate that there is widespread frustration in seeking

PIC, negotiating mutually agreed terms and sharing of benefits associated with the use of TK as the same remain unclear. Because of these difficulties, many companies have adopted a 'hands off' approach to the use of TK, whilst others have little awareness of the need to enter into ABS arrangements when using TK. In cases where TK is used, there is typically strong reliance by companies on the use of intermediary institutions, such as research institutions, non-governmental organizations (NGOs) or governments, to resolve difficult issues. As such, Kenya must ensure that communities and national authorities are sufficiently aware that their role in ABS pursuant to the regulatory regime is not to stifle access, but ensure that access is regulated pursuant to the laws of the land if such access is indeed provided. Further, in sharing of benefits, and spreading the benefits to other conservation priorities in Kenya, Kenya may consider establishing a Biodiversity Trust Fund into which a pre-determined percentage of monetary benefits accrued from the agreement may be deposited. These monies can thereafter be distributed in other parts of the country requiring resources to promote biodiversity conservation, and not particularly the area in which the resource was derived. This Trust Fund may have similar objectives as that of the National Environment Trust Fund as established by Section 26 of the EMCA.[6] Further, in designing benefit-sharing schemes, it is imperative that the knowledge related to the resources is recognized and protected. Kameri-Mbote (2008, p412) suggests that the guiding principle should be the recognition that all forms of knowledge are equally important to Kenya's development and there is need for the co-existence of TK and other forms of intellectual protection. More specifically, NEMA needs to put in place a framework for protection of TK related to GRs as a sui generis system rather than bringing it under an existing intellectual property rights regime. Such a regime ought to be based on facilitation of exchange rather than on monopoly of rights.

Legal certainty and clarity of rights to material is vital to promote and protect industry investment in research and development and commercialization. As such, it is necessary that Kenya's laws on intellectual property laws need to be reviewed and brought in line with, subject to national sovereignty, international regimes on ABS under the CBD, and intellectual property rights (IPR) under the World Trade Organization rules.

Problems of genetic identification, combined with capacity constraints and the sheer complexity of designing a *monitoring and tracking system* that suits different types of genetic material and sectors, pose significant challenges for the development of a compliance system that is both cost effective and effectual. These difficulties point to the need for provider country institutions and companies to enter into ABS arrangements and partnerships, and to build trust and collaboration over time. Increasingly, it appears unlikely that countries can effectively and comprehensively regulate, or groups can adequately track and monitor, the use of resources they provide to users. This points to the importance of building monitoring capacity in Kenya, to ensure her commitment to agreements and to promote transparent and fair transactions, and establishing on-going and long-term partnerships. Such approaches are vital to ensure that the use of material can be monitored and consequential benefits assured.

The Environmental Management and Co-ordination (Conservation of Biological Diversity and Resources, Access to Genetic Resources and Benefit Sharing) Regulations 2006 Legal Notice 160 represent a first attempt by Kenya to establish a comprehensive and well-coordinated regulatory and administrative system on ABS, which intends to address the letter and spirit of the CBD. However, it must be noted that this system will only work in an environment of well-sensitized communities and policy makers. As NEMA in Kenya has only just started operationalizing these Regulations, one must wait and see the impact NEMA will have in regulating ABS activities in the country.

Notes

1 At the time of writing the chapter the author was the Head of Legal Services, NEMA, Kenya. She is currently a Programme Officer at the Royal Danish Embassy, Nairobi. The views expressed in this chapter are those of the author and do not necessarily represent those of NEMA or the Danish Embassy.
2 *Abdikadir Sheikh Hassan and 4 Others v. Kenya Wildlife Service*, Civil Case No 2959 (High Court of Kenya, 1996), cited in Lettington, R.
3 Environment Management and Coordination Act (1999), No 8 of 1999, entered into force January 2000.
4 Taskforce on the Development of Laws for the Protection of Traditional Knowledge, Genetic Resources and Folklore – Gazette Notice No 1415, 22 February 2006.
5 African Model Law on the Protection of the Rights of Local Communities, Farmers and Breeders, and for the Regulation of Access to Biological Resources endorsed by the Organization of African Unity, www.grain.org/docs/oua-modellaw-2000-en-pdf, accessed 24 December 2007.
6 The aim of the National Environment Trust Fund is to facilitate research intended to further the requirements of environmental management, capacity building, environmental awards, environmental publications, scholarships and grants.

References

Kameri-Mbote, P. (2008) 'Access to genetic resources and benefit sharing in Kenya', in Okidi, C. O., Kameri-Mbote, P. and Akech, M. (eds) *Environmental Governance in Kenya: Implementing the Framework Law*, Nairobi, East Africa Educational Publishers, Chapter 17

Lettington, R. (2001) 'Access to genetic resources in Kenya', in Nnadozie, K. et al (eds) *African Perspectives on Genetic Resources. A Handbook on Laws, Policies and Institutions*, Washington, Environmental Law Institute, Chapter 11

Chapter 10

Brazil's Experience in Implementing its ABS Regime – Suggestions for Reform and the Relationship with the International Treaty on Plant Genetic Resources for Food and Agriculture

Juliana Santilli

Introduction

Brazil was one of the first megadiverse countries to enact national legislation[1] on ABS, aimed at implementing the CBD at the national level. Provisional Measure no 2186-16/2001[2] regulates access to GRs, to associated TK, benefit sharing derived from their use, and the transfer of technology for the conservation and use of biological diversity. More recently, Brazil has ratified and published the International Treaty on Plant Genetic Resources for Food and Agriculture (ITPGRFA),[3] which provides a differentiated legal regime for the plant GRs that come under its Annex I, kept in ex situ collections and in the public domain, as long as their use is intended for food and agriculture. The ABS regime set up under MP 2186-16/2001 was conceived above all for wild GRs, particularly for their chemical, pharmaceutical or industrial use, with no consideration for specificities of plant GR's for food and agricultural uses. MP 2186-16/2001 does apply, however, to both wild and domesticated GRs, and makes no distinction between the two in terms of ABS. We shall thus look first at the general provisions of MP 2186-16/2001, and then analyse its application to plant GRs for food and agriculture.

The Genetic Patrimony Management Council (CGEN) is responsible for GR's management policies. The Council was established in 2002 and is made up of representatives of several government agencies, under the coordination of the Ministry of the Environment (specifically, the Biodiversity and Forests Secretariat).[4] Decree 3945/2001 established the Council's membership based only on representatives of federal agencies, leaving out representatives of any other sectors, such as biotechnology companies, researchers in scientific institutions or indigenous and traditional communities. Since 2003, representatives of these sectors have been attending CGEN meetings, with the right to speak but not to vote. In 2007, Decree 6159 stipulated: 'In order to contribute to decision-making, the Management Council may decide to invite experts or representatives of various sectors of society involved with the subject matter'. The participation of all stakeholders in the Council, able not only to speak but to vote as well, is fundamental for it to be able to act as a body that can mediate potentially conflicting interests, as well as to achieve effective social control over its work. Other councils have long had this kind of participation; for example, the National Environment Council (Conama) and the National Water Resource Council, with representatives from several sectors of society, as well as official agencies. Such participation respects the constitutional precept of social and democratic participation in the management of environmental resources.

The CGEN publishes standards for enforcement of Provisional Measure 2186-16/2001 and rules on the granting of authorizations for access to GRs and to associated TK, following the acquiescence of concerned indigenous and traditional communities. Scientific research is defined as research that a priori has not identified any potential for the economic use of its results. Bioprospecting is defined as an activity aimed at identifying components of the country's GRs and information on associated TK, when there is a potential for their commercial use. When there is a prospect for commercial use, benefit-sharing contracts must be signed between the providers and users of the GRs, and these contracts must be approved by the Council to assure they are in line with the law.

ABS legislation today: The main instruments

Provisional Measure 2186-16/2001 created a legal regime based on two main instruments: authorization of access to GRs and associated TK, and the benefit-sharing contract. IBAMA, the Brazilian Institute of Environment and Natural Resources, which is in charge of the administrative implementation of federal environmental laws, is the federal agency responsible for authorizing access to GRs for the purposes of scientific research with no potential for economic use and which do not involve access to associated TK. When access to GRs is aimed at research with the potential for economic use (bioprospecting or technological development), or if it involves access to associated TK, CGEN is responsible for issuing the legal authorization for access.

Even so, access to GRs can only be granted following the previous acquiescence of the indigenous peoples (when access occurs in indigenous territories), of an environmental agency (when access occurs in a protected area) or of the owner of private land. When access takes place in waters under Brazilian jurisdiction, on the continental platform or in the exclusive economic zone, the previous acquiescence of the maritime authority must be obtained, or even of the National Defence Council, if it involves an 'area that is indispensable to national security' (e.g. military or national border areas).

If access to TK held by indigenous and traditional communities is involved, the Authorization of Access[5] depends on their previous acquiescence, without which the CGEN cannot grant authorization. When there is a prospect of commercial use, a benefit-sharing contract must be signed with the indigenous and traditional communities providing for benefits such as profit sharing, payment of royalties, access to and transfer of technology, no-cost licensing of products and processes, and training of human resources. The Provisional Measure also stipulates the need for a material transfer agreement to cover the sending of any sample of GRs, indicating whether there was access to associated TK.

Access to and collecting of biological material

A number of questions have been raised over the past eight years as the legislation has been applied to specific cases. In response to the need for clarity on which activities are covered by the law, CGEN published an official 'technical orientation' to make it clear that access is different from the collecting of biological material. According to CGEN, access is 'the activity carried out with GRs with the objective of isolating, identifying or using information of genetic origin or molecules and substances arising from the metabolism of living beings and of extracts obtained from such organisms'. The activity only requires authorization from CGEN when it fits this definition. The collecting of biological material in protected areas requires another type of official permit, but not an authorization from CGEN.

Genetic resources and TK shared by multiple indigenous or traditional communities

Another shortcoming in the access legislation is its focus on bilateral contracts between the providers and users of GRs and TK, thus ignoring situations in which resources and knowledge are shared by diverse traditional communities. This is probably one of the law's most serious loopholes, due to countless situations in which knowledge on the characteristics, properties and uses of biological resources are held and/or produced by various traditional peoples. When TK is shared by more than one traditional people, the exercise of rights by one or more knowledge holders must not restrict the rights of other peoples and communities who also hold that knowledge. Otherwise, access legislation may give rise to disputes among the communities regarding rights over resources and knowledge, and may prejudice the free circulation of biological objects and exchanges among communities' traditional practices that are fundamental for the maintenance of biological diversity. This situation was clearly exemplified by the Krahô case, where bioprospecting was frustrated due to lack of representation of some villages in the agreement for access.[6] So the question is: who has legitimacy to represent indigenous and traditional peoples when authorizing access to GRs and in benefit-sharing contracts?

Brazil's enormous socio-diversity holds back adoption of homogeneous standards or a single criterion for representation. After all, there are many indigenous peoples, *Quilombolas* and traditional communities, with tremendous ethnic and cultural differences from each other and living in different ecosystems. Some indigenous peoples, for example, have their chiefs represent them, based on variable attributes that qualify them for the exercise of power, such as age, experience, being good warriors, good shamans, skilled hunters, fishermen or farmers. Other indigenous peoples, meanwhile, grant political decision-making power to their Councils of Elders. Official Brazilian law must restrict itself to recognizing and conferring legal validity to these forms of representation.

Indigenous peoples, *Quilombolas* and traditional communities in Brazil

According to the Instituto Socioambiental,[7] today in Brazil there are 227 indigenous peoples, speaking 180 different languages, with a population of 600,000, or 0.2 per cent of the Brazilian population. The Brazilian Constitution (article 231)[8] recognizes their social organization, customs, languages, beliefs and traditions, and indigenous peoples have the right to the exclusive use of the natural resources located in their traditional lands. The *Quilombolas* are communities of Afro-descendents originally founded as *Quilombos*, or villages of runaway slaves, and also have territorial rights assured by the Constitution. Both indigenous peoples and *Quilombolas* need to authorize access to GRs located in their territories, as well as to their TK, and to participate in benefit sharing. According to the Palmares Cultural Foundation,[9] 1,000 *Quilombola* communities have been officially identified, with a population estimated at approximately 2 million people, most of them located in the States of Maranhão and Bahia.

Other traditional communities in Brazil include artisan fisherfolk, nut gatherers (*castanheiros*), rubber tappers (*seringueiros*) and other extractive communities, adding up to approximately 4.5 million people, according to the Ministry of the Environment. Decree 6040/2007 officially defines traditional communities as 'culturally differentiated groups that recognize themselves as such, having their own forms of social organization, occupying and using territories and natural resources as a condition for the cultural, social, religious, ancestral and economic reproduction, using knowledge, innovations and practices generated and transmitted through tradition'. Access to the resources and TK held by these communities also depends on their previous acquiescence. Some environmental conservation units created by law,[10] such as extractive reserves and sustainable-development reserves, allow for the presence of traditional communities and seek to reconcile the conservation of both biological and cultural diversity. Management plans on these reserves must ally scientific research with the use of TK about the management of natural resources.

Collective benefits

Since many resources and much TK are shared by several communities, indigenous organizations have proposed that, in such cases, the forms of benefit sharing should be collective, through the creation of Benefit-Sharing Funds. In this manner, all communities sharing a given resource or TK would have access to the money deposited in the Funds, which would be divided into ecological and ethnographic regions and managed by the communities themselves.

Benefit-sharing mechanisms

Under the current law, benefit-sharing contracts are only required when authorization is requested for access to GRs and TK for purposes of bioprospecting (with prospects for commercial use). In the event of access to a GR located in a federal-domain protected area (parks, ecological stations, etc.), the Union must be a party to the benefit-sharing contract. In other situations where the Union is not a party, it is entitled only to a share of the benefits.[11]

If, for example, the benefit-sharing contract is signed with a private landowner (on whose land the resources are to be accessed), the benefits will not necessarily go to conserving biodiversity. There is a proposal, presented by the Ministry of Environment, to invert that logic, so that benefits are distributed to public funds and used in biodiversity conservation projects, while the private landowners would only receive a share. The proposal also provides for the creation of a new fee to be used as a benefit-sharing mechanism, which would be levied on the sale of products derived from the access to GRs and to TK (1 per cent) and on royalties earned through the exercise of patents and breeders' rights (2 per cent).

Provisional Measure 2186-16/2001 provides that authorization of access to GRs for bioprospecting purposes (with prospects for commercial use) requires the prior signing of a benefit-sharing contract. Therefore, even before research activities begin to inform anyone about their outcome (whether they would actually lead to potentially valuable results with prospects for commercial use), the Provisional Measure requires signing of a benefit-sharing contract. For both the providers and the users of resources and knowledge, it is hard to stipulate the value of benefits as long as the results of the work are still unforeseeable. A Presidential decree (Decree 6159/2007) actually stipulated that, if the provider agrees, the benefit-sharing contract could be drawn up and signed at a later date, as long as it is prior to the development of any new commercial product and to any filing for patent protection. This could make the results of research clearer to the party when they finally negotiate the terms of their benefit-sharing contract.

So far, the only contract for the use of GRs and benefit sharing for the purpose of bioprospecting, to which traditional communities are parties, was the contract signed between the Federal University of Rio de Janeiro and the *Quilombola Association* in the municipality of Oriximiná, in Pará State, in Brazil's Amazon region.[12] That bioprospecting was aimed at searching for bioactive substances in medicinal plants based on the TK of *Quilombola* communities in Oriximiná, and the contract is valid for 18 months, starting on the day its authorization was issued (2 January 2008) by the CGEN.

The contract between the University and the *Quilombola Association* provides that the benefits will be negotiated once the results of the research allow for the development of a commercially viable product, and will define a share of the profits for the *Quilombola* communities (a percentage will be set, as well as a formula for calculations and transfers to the communities). It also provides that possible intellectual property rights (IPRs) over the products or processes developed during this work will belong to the university and to the communities.

The CGEN has granted other authorizations of access to traditional associated knowledge, for the purposes of scientific research (with no commercial interest), with the prior acquiescence of the stakeholder communities. Benefit sharing is only required, however, once a possibility of economic exploitation has been identified. Researchers are obliged to report the origin of the associated TK every time the results of their research are published. They must also include a warning, in all media used to disseminate the research, that the uses of the results of the research for commercial use depends on the prior acquiescence and on a benefit-sharing contract signed with the participating communities.

Establishment of different legal regimes for scientific research and for bioprospecting

Another controversial issue is the differentiation of rules on access for scientific research and for bioprospecting, one of the researchers' primary demands. In August 2006, CGEN approved Resolution 21, which exempts four kinds of research and scientific activities from the need for authorization for access: (1) research aimed at assessing or revealing the history of evolution of a species or taxonomic group, the relations of living beings with each other or with the environment, or the genetic diversity of populations; (2) tests to identify parents or sex and karyotype, or deoxyribonucleic acid (DNA) analysis, aimed at identifying a species or specimen; (3) epidemiological research or research aimed at identifying etiological agents for diseases, as well as measurement of the concentration of known substances whose volumes, inside the organism, are indicative of disease or physiological state; and (4) research aimed at constituting collections of DNA, tissues, germ plasm, blood or serum. These lines of scientific research were exempted from the need for authorization of access because the isolation, identification or use of genes, biomolecules or extracts (all acts of access to GRs) in such activities is done circumstantially, as a methodological tool, in contrast to projects whose objectives are directly related to access to GRs.

In one of the draft bills presented by the government, with the support of researchers, access to GRs for the purpose of scientific research (with no prospects for commercial use) would not require authorization, except when foreign institutions or profit-making Brazilian institutions are involved in the research activities. If a scientific research project acquires commercial objectives, it would have to follow the rules for this kind of research. Commercially oriented research projects would depend on authorization of access to GRs and to TK. Most indigenous organizations do not accept this distinction, but hold that any access to genetic resources and TK must be subject to authorization by the agency responsible for GRs and by the communities holding such resources or knowledge.

Proof of origin of resources at the patent office

Provisional Measure 2186-16/2001 (Article 31) stipulates that applicants for patents or other forms of intellectual property must inform the origin of GRs and TK used in the course of the processes or in the products they wish to protect. Applicants must also sign a declaration that they have complied with all the requirements of Provisional Measure 2186-16/2001 and provide the patent office with the number and date of the Authorization of Access issued by the National Genetic Resources Council, following acquiescence by indigenous and traditional communities.

Although the proof-of-origin requirement has been law since 2001, it only began to be enforced after the publication of CGEN Resolution 23, in November 2006. A study published by the Instituto Socioambiental (Novion and Baptista, 2006)[13] in March 2006 showed that until that date fewer than 10 per cent of the patent applications filed at the Brazilian patent office, the INPI, identified the origin of the genetic material or of the associated TK, and that no patent application filed at the INPI had attached an Authorization of Access issued by the CGEN. Since Provisional Measure 2186-16/2001 can only be enforced inside Brazil, it is important that other countries using GRs also adopt similar laws, obliging applicants for patents or other IPRs to declare the origin of GRs and of the TK used in developing processes and products they intend to patent or otherwise protect.

The establishment of a binding international benefit-sharing regime is another important step in this direction, and it must attend both to the rules of the CBD and to the multilateral system created by the ITPGRFA.

Plant GRs for food and agriculture and the new International Treaty: options for implementation at the national level

It is very difficult to apply bilateral regimes governing TK of wild species that is shared by several local communities. The difficulties created by bilateral regimes on plant GRs for food and agriculture are even more serious and insoluble.

All varieties of cultivated plants are the result of selection and breeding work done over generations by farmers, and agrobiodiversity is the fruit of the complex and dynamic management of crops carried out by farmers. As people accustomed to sharing and promoting exchanges of their genetic material, knowledge and farming experience through social networks that are regulated by their own local rules, how do local farmers decide who will authorize access to these plant GRs and be eligible for benefits derived from their use? Countless exchanges that take place among different countries and farmers have allowed for the creation of new varieties, based on the combination of genetic materials from such diverse origins, that it is often difficult to attribute a single origin to a newly obtained variety or even to identify the various regions of origin of the materials used to develop and/or breed a given variety. Generally, many varieties are used in selection and breeding processes to obtain new varieties, both by farmers in the field and by institutional breeders and researchers. Local farming systems are neither closed nor static, and farmers constantly try out new varieties – whether brought in by other farmers or by agricultural research centres – and incorporate new material into their own stock.

Brazil's MP 2186-16/2001 and most other access and benefit-sharing laws require contract-based relations between 'providers' and 'users', while creating 'direct' benefit-sharing mechanisms through which farmers may be compensated for genetic material accessed 'on-farm' and used to obtain new plant varieties. This system is not appropriate for regulating ABS among local communities that jointly hold resources and knowledge associated with their agrobiodiversity. In Brazil, no contract for the use of GRs and economic benefit sharing has ever been signed between bioprospectors and local farmers, under the provisions of MP 2186-16/2001, to provide concrete benefits for farmers or for agrobiodiversity.

Benefits arising from the use of plant GRs (for food and agriculture) must be shared in a collective manner, directly linked to recognition of farmers' rights, which are essentially collective. Farmers must enjoy rights such as to save, use, exchange, produce and sell their seeds, free from any legal obstacles and restrictions that are inappropriate for local productive systems, as well as to share in benefits arising from the use of agrobiodiversity through collective mechanisms and public policies that enhance and strengthen local and traditional farming systems, and finally to participate in national, regional and local decision-making on agriculture, land tenure, environmental and other policies that affect the conservation and sustainable use of agrobiodiversity. Instead of singling out holders of plant GRs with whom to share benefits, the law must create legal grounds that allow farmers to continue to dynamically conserve and manage agrobiodiversity resources. Otherwise, greater restrictions will be imposed on access and on the free circulation of plant GRs.

These issues must be faced when a country draws up legislation on ABS for plant GRs for food and agriculture and tries to implement the ITP-GRFA. The treaty has an entire chapter on farmers' rights, which recognizes their contribution to the conservation of agrobiodiversity and to food and agricultural production. The responsibility for implementing farmers' rights lies with countries, which must draw up national laws that recognize and make them effective. The recognition of these rights must cover the whole gamut of local agriculture, including not just indigenous and traditional farming, but all forms of family, agroecological and peasant farming, all of which play important roles in the conservation of agrobiodiversity. Farmers' rights, as key and fundamental components in any legislation regarding the management, conservation and sustainable use of agrobiodiversity, must be recognized and covered by legislation on access to plant GRs.

There are other aspects to be considered as well, to implement the Treaty in Brazil. The multilateral ABS mechanism created by the Treaty applies only to plant GRs that come under its Annex I[14] and are in the public domain, as long as their use is intended for research, breeding and training in the field of food and agriculture. The multilateral system does not apply to access to plant GRs found in situ[15] (i.e. in their natural environments), and its rules do not cover the collection of or access to plant GRs carried out *domestically*, inside their own countries of origin. When research institutions or private companies intend to access in situ plant GRs, they must therefore obey national laws, since the Treaty does not refer to such cases.

Article 19 para 2 of MP 2186-16/2001 provides that 'shipments of samples of a genetic heritage components *of species that have facilitated exchange in international agreements*, including on food safety, of which Brazil is signatory, shall be carried out according to the conditions defined therein'. Since Brazil has already joined the ITPGRFA, crops covered by its Annex I must be shipped based on the rules of the multilateral system created by that international agreement.

While the Treaty is intended to regulate shipments and exchanges of genetic material between different countries, shipments and exchanges between domestic institutions and researchers must also be regulated, once the Treaty enters into force in each country, following the rules of the multilateral system (exclusively for Annex I crops, kept in ex situ collections, in the public domain, when access is for use in research, breeding and training in the fields of food and agriculture). After all, it makes no sense to facilitate access to ex situ collections for overseas institutions and researchers, through the multilateral system, while domestic institutions and researchers must submit to MP 2186-16/2001. In addition, it is important for not only federal but also State institutions to make their collections of plant GRs available through the multilateral ABS system. Another matter Brazil will have to deal with in its new legislation is whether or not to include in the multilateral ABS system its in situ plant genetic resource collections located on public domain lands. This option has already been considered by some of the Treaty's member countries.

As mentioned above, access to in situ plant GRs depends on domestic law and is not ruled by the Treaty. In Brazil, access to in situ GRs is regulated by MP 2186-16/2001 and is therefore subject to the bilateral ABS regime, which is why it is not legally possible today to include in situ plant GRs from public lands in the multilateral ABS regime. As new legislation is being drafted, however, legislators should bear in mind the precedents of special legal regimes for indigenous lands, *Quilombolas* and sustainable-use conservation units, as well as for extractive and sustainable-development reserves, all of which allow for the presence of traditional populations. Any entry or collection of biological material on land occupied by indigenous peoples, *Quilombolas* or traditional populations depends on their consent. At the same time, the targeting of resources into benefit-sharing funds must consider the objectives of the National Policy for the Development of Traditional Peoples and Communities, as established by Decree 6040/2007, according to which such funds must be administered with the participation of the National Commission on the Sustainable Development of Traditional Peoples and Communities, which includes representatives of traditional communities.

It is up to each country to decide whether or not it will place its ex situ, public-domain, non-Annex I agricultural crops under the multilateral system. When such (non-Annex I) crops are included in the multilateral system, domestic institutions and researchers should also enjoy facilitated access. To include them, however, Brazil would have to change its MP 2186-16/2001, since para 2 of Article 19 only allows the shipment of GRs from '*species that have facilitated exchange in international agreement*', but the multilateral system created by the Treaty only covers the species listed in Annex I. We believe it is too early for Brazil to include other crops in the multilateral system beyond those listed in Annex I, since it must monitor the effectiveness of implementation of the Treaty's benefit-sharing mechanisms, such as exchange of information, access to and transfer of technology, capacity building and the sharing of economic benefits arising from the sale of products. A new (national) legal regime should, however, set rules for ABS for all plant GRs in the fields of food and agriculture, whether they are found in situ or ex situ.

Brazil should also defend the position, internationally, that payments into the Treaty's benefit-sharing fund should be a fixed percentage of all sales of products derived from genetic material accessed through the multilateral system, *whether or not it is available with no restrictions to third parties for research and breeding*, since this possibility is expressly provided for in the Treaty.[16] In other words, benefit sharing should be de-linked from the existence or absence of protection through intellectual property rights over products derived from genetic material accessed under the multilateral system. This will be the only way to produce enough funds to implement plans and programmes for the conservation and sustainable use of agrobiodiversity in developing countries. Since benefit sharing is not compulsory when products are protected by plant breeders' rights (because these rights do not limit access for research and development), but only when patents are granted, it is easy to conclude that there will be little money coming out of the multilateral system's compulsory benefit sharing. Today, the only countries that allow patents on plant varieties are the USA, Japan, Australia and New Zealand. Of the four countries, only Australia has ratified the Treaty. In addition, a new plant variety averages ten years to be developed, meaning that it will still take a long time before any percentage of sales find their way into the Treaty's benefit-sharing fund.

The Treaty, however, does not regulate access to privately-held ex situ collections or to plant GRs found in situ. Both must be regulated by domestic law. It would be important for a new law on access to plant GRs to include a provision that genetic material collected (in situ) on land in the public domain, even when kept in privately held collections, must necessarily be accessible to public institutions and to farmers. Access by private institutions to public collections must also be conditioned to reciprocity vis-à-vis those institutions' collections. In other words, to gain access to public collections, private institutions would have to allow public institutions and farmers to access their own collections. Even though this condition cannot be imposed on collections for crops under the multilateral system because of Brazil's obligations under the Treaty, which do not allow any unilateral changes in the rules of the multilateral system, Brazil's other crops and in situ resources can still be covered by the country's own domestic rules.

After all, GRs are public-interest goods, whether they are found in the public or private domain, and their use and access should be determined by the public interest. When genetic material is collected on land in the public domain, even if it is kept in private ex situ collections, it is even clearer that it must be accessible for public institutions and interested farmers. The Treaty (Article 11.4) provides that within two years of the entry into force of the Treaty (it came into force internationally on 29 June 2004), the Governing Body shall assess whether natural and legal persons (holders of ex situ collections) who have not included their plant GRs in the multi-lateral system will continue to merit facilitated access, or whether other 'measures as it deems appropriate' are to be taken. The Treaty thus allows for the possibility of blocking access by institutions who do not make their own collections available to third parties.

The discussion is still underway on a proposal by which Brazil's access law would levy a fixed percentage fee on the sale of all products developed from plant genetic material accessed in public ex situ collections or collected in situ, whether or not those products are available with no restrictions to third parties for purposes of research and breeding. A simpler option, however, would be to charge the fee on all seed sales in the country, thus removing the need to determine the origin and genetic composition of new products. This solution was adopted in Norway, which earmarks 0.1 per cent of the value of all seed sales in the country to the Treaty's benefit-sharing fund, in order to support agrobiodiversity conservation and management initiatives.[17] Brazil could set up a national benefit-sharing fund with the same purpose. This would be one way to materialize the 'user pays' principle set forth in the National Environment Policy,[18] which obliges users of environmental resources to pay for their economic use. This principle is also present in other Brazilian laws, such as Law 9433/97, on the National Water Resource Policy, which creates a fee for the use of water resources. The 'user pays' principle is aimed at internalizing the environmental costs of economic activities, and the users of plant GRs should therefore contribute to activities intended to promote their conservation. A percentage of the sales of seeds in the country should therefore go into a national benefit-sharing fund administered with the participation of representatives of local, family and traditional farmers, to support plans and programmes focusing on in situ and on-farm conservation of agrobiodiversity, and to implement farmers' rights. This form of benefit sharing is more coherent with the nature of plant GRs than, for example, trying to identify the 'providers' of those resources.

Conclusion

The revision of Brazil's ABS legislation (MP 2186-16/2001) is still underway, and several draft bills have been discussed by stakeholders in the process. MP 2186-16/2001 was conceived mainly to deal with wild GRs, for chemical, pharmaceutical or industrial uses. It is very difficult, therefore, to apply it to plant GRs. With the entry into force in Brazil of the ITP-GRFA, a new multilateral ABS system must be implemented for certain farm crops. Farmers' rights are a fundamental component in any law regarding the management, conservation and sustainable use of agrobiodiversity, and must be covered in the new national legislation.

Notes

1   The Brazilian States of Acre and Amapá, located in the Amazon region, had in 1997 approved state laws to regulate access to GRs and TK.
2   Provisional Measures are a form of legislation issued by the President of Brazil. They enter into effect as laws upon publication and may later be approved, amended or rejected by the national Congress. Provisional Measure 2186-16/2001 was first published in June 2000 and has never been voted for by Congress. However, it became a legally binding instrument due to a constitutional amendment.
3   Presidential Decree 6.476/2008 approved the International Treaty in Brazil, which came into force internationally on 29 June 2004.
4   Decree 3945/2001 established the composition of CGEN and its operating rules. It has since been amended by Decree 5439/2005, by Decree 4946/2003 and by Decree 6159/2007.
5   Authorizations for Access may only be requested by Brazilian legal persons, public or private institutions, active in biological or related research. When a foreign legal person is involved in research that includes access to or the remittance of GRs or access to associated TK, the research must be carried out under the coordination of a publicly owned Brazilian institution.
6   See Kleba and Kishi in this book for a detailed discussion.
7   http://www.socioambiental.org.
8   The International Labor Organization's (ILO) Convention 169, on Indigenous and Tribal Peoples, is now in effect in Brazil. It provides that the fundamental criterion to determine which (indigenous or tribal) groups are subject to its provisions is the awareness of the indigenous or tribal identity (in other words, their own self-identification). The self-identification criterion is also adopted to identify the *Quilombolas*.
9   http://www.palmares.gov.br. The Palmares Cultural Foundation is the federal agency responsible for policy making for *Quilombola* communities.
10 Law 9985/2000 created the National System of Nature Conservation Units.
11 Under the present Law, benefits received by the Union (essentially, the federal government) are deposited in the National Environmental Fund, the Naval Fund and the National Scientific and Technological Development Fund.
12 See Kishi in this book for a detailed discussion.
13 That study came out of the Andean–Amazon Initiative to Prevent Biopiracy, a South American network of non-governmental organizations and research institutions.
14 Of the crops listed in Annex I and included in the Treaty's multilateral system, the only crop species of Brazilian origin is manioc (cassava, or *Manihot esculent*), exclusive of its wild relatives. During the negotiations, there was strong pressure to include peanuts (groundnuts) as well in the multilateral system, but this Brazilian species was left out.
15 Article 12.3(h) in the Treaty provides that access to plant GRs for food and agriculture found in situ will be provided according to national legislation.

16 Article 13.2(d)(ii) of the Treaty provides that the Governing Body may, from time to time, review the levels of payment with a view to achieving fair and equitable sharing of benefits, and it *may also assess, within a period of five years from the entry into force of this Treaty* [it came into force internationally on 29 June 2004], *whether the mandatory payment requirement in the MTA shall apply also in cases where such commercialized products are available without restriction to others for further research and breeding.*

17 According to Norway's Minister of Agriculture, Terje Riis-Johansen, if all of the Treaty's member countries paid in the same percentage of domestic seed sales, the Treaty's benefit-sharing fund would raise approximately US$20 million per year, to support farmers who conserve diversity. Source: 'Norway announces annual contribution to the benefit-sharing fund of the International Treaty', available at: http://www.planttreaty.org, accessed 17 October 2008.

18 Law 6938/81 creates the National Environment Policy and establishes the 'user pays' principle in its Article 4 para VII.

Reference

Novion, H. P. D. and Baptista, F. M. (2006) 'O certificado de procedência legal no Brasil: estado da arte da implementação da legislação', available at: http://www.socioambiental.org/inst/docs/download/estudocertificadopatentes.pdf

Finding a Path Through the ABS Maze – Challenges of Regulating Access and Benefit Sharing in South Africa[1]

Rachel Wynberg and Mandy Taylor

Introduction

Like many other countries, the regulation of ABS is relatively new in South Africa, despite the fact that the country has been a party to the Convention on Biological Diversity (CBD) since 1995 and actively engaged in bioprospecting for decades (see, for example, Laird and Wynberg, 1996). A central reason for the high levels of interest in bioprospecting in South Africa is the country's extraordinarily rich and unique biodiversity, placing it as the third most biologically diverse country on earth (World Conservation Monitoring Centre, 2002). This is largely due to extremely high levels of endemism in the plant kingdom, with the Cape Floristic Region alone holding one of the six most significant concentrations of plant diversity in the world (Cowling and Richardson, 1995). Intraspecific genetic diversity is also unusually high, adding to the potential for developing new medicines, crops, cosmetics, ornamental plants and other useful products. Moreover, South Africa's technological and scientific research capacity, combined with its well-developed infrastructure and institutional capacity, place it as a leader in the African region for bioprospecting and biotechnology development, and also provide an extremely attractive environment and springboard for foreign companies wishing to bioprospect in southern Africa.

Until recently, however, the commercial development of South Africa's biological resources took place in a legislative vacuum. Now that vacuum is being filled by a specific regulatory ABS framework, articulated through a chapter of the National Environmental Management Biodiversity Act (10 of 2004) (the Biodiversity Act) and the regulations passed under that Act, which came into effect in April 2008.[2] The purpose of this chapter is to examine the new legislative framework regulating bioprospecting and ABS in South Africa and to explore some of the challenges posed in implementing this legislation.

From ad hoc-ism to regulation

Before the Biodiversity Act came into effect, an approach to bioprospecting emerged that was characterized initially by a relatively ad hoc and minimalistic response. This subsequently evolved into a more creative focus on developing national and regional collaborative platforms for bioprospecting, an example being a national consortium comprising major South African scientific research institutions and universities, focused on the discovery of drugs for malaria and tuberculosis (TB) from indigenous plants (Crouch et al, 2008). Most commonly, however, smaller bioprospecting initiatives slipped 'under the radar', whereas larger initiatives were characterized by bilateral contracts between those desiring access to GRs (typically a foreign company or foreign research institute) and those providing that access (typically represented by a local research institute). These contracts filled a necessary gap but were developed outside of any legal framework, the absence of which is widely considered to have been a major reason for the failure of bioprospecting to deliver optimal benefits for South Africa over the past 15 years (Kidd and Mayet, 2003; Wynberg, 2004a).

Examples of this failure include:

- a 1999 research and licensing agreement to develop South African plant species as ornamental and horticultural crops, between the then National Botanical Institute (now the South African National Biodiversity Institute) and the US-based multinational, Ball Horticulture. This agreement was roundly criticized for not enabling effective technology transfer to South Africa and for inadequate consideration of intellectual property issues (Glazewski et al, 2001; Henne and Fakir, 1999; Wynberg, 2004a);
- use of the indigenous San people's TK of the succulent plant *Hoodia* in the development of anti-obesity products, and the patenting by the South African-based Council for Scientific and Industrial Research (CSIR) of a process to extract active constituents of the plant. The CSIR captured international attention as it failed to obtain the prior informed consent (PIC) of the San before lodging a patent based on the San's knowledge, thereby limiting the benefits the San could obtain from the patent (Chennells, 2003; Wynberg, 2004b; Wynberg and Chennells, 2009);

- an agreement between the University of the Free State (UFS) and the New York Botanical Garden (NYBG), which provided the pharmaceutical company Merck with biological extracts from South Africa. This agreement was criticized for having extremely weak benefit-sharing requirements and a lack of clarity as to scope and definitions (Wynberg, 2003). This was highlighted by the discovery of platensimycin, a previously unknown class of antibiotics produced by a strain of the bacteria *Streptomyces platensis*, isolated from soil samples from plants collected as part of the NYBG-UFS programme (Wang et al, 2006). Although plants were covered by the bioprospecting agreement, it was not clear if terms of the agreement also extended to associated microbial species (Wynberg et al, 2007).

Many of the cases described were accompanied by controversy and public concern that the natural and cultural heritage of South Africa was being 'sold for a song', without proper controls and oversight (see Glazewski et al, 2001; Gosling, 2001). This, combined with ongoing bioprospecting activities and South Africa's ratification of the CBD, led to the initiation in 1995 of a two-year period of public consultation, linked in part to a broader post-apartheid law reform initiative to develop a biodiversity policy that represented the interests of all South African citizens (Wynberg and Swiderska, 2001). In 1997, this culminated in the publication of a Biodiversity White Paper, the first national policy to incorporate ABS and explicitly prioritize the need for legislative and administrative mechanisms to control access to South Africa's GRs and ensure fair benefit sharing (Department of Environmental Affairs and Tourism, 1997). Following on from the Biodiversity White Paper, and seven years in the making, South Africa's Biodiversity Act was finally promulgated in 2004, which, together with the regulations passed under that Act in 2008, signalled the development of a new era for bioprospecting in South Africa.

The Biodiversity Act

The Biodiversity Act and ABS regulations comprise the primary legislative means for regulating ABS in South Africa, with the Biodiversity Act providing a broad framework, regulating all aspects of biodiversity conservation and use. An important rationale for the Act was to resolve the fragmented nature of biodiversity-related legislation; to facilitate cooperation between the different levels of government (national, provincial and local); and to give effect to constitutionally protected environmental rights.

One of the objectives of the Act is to provide for 'the fair and equitable sharing among stakeholders of benefits arising from bioprospecting involving indigenous biological resources' (section 2 of the Act), and one of the ten chapters of the Act, entitled 'Bioprospecting, access and benefit-sharing', seeks to give effect to this objective. This chapter sets a fairly sparse legislative framework, leaving the detail to be dealt with in subordinate national legislation – the ABS regulations.

The purpose of the ABS chapter of the Biodiversity Act is:

- to regulate bioprospecting involving indigenous biological resources
- to regulate the export from South Africa of indigenous biological resources for the purposes of bioprospecting or any other kind of research
- to provide for a fair and equitable sharing by stakeholders in benefits arising from bioprospecting involving indigenous biological resources (section 80 of the Act).

In contrast to the narrow definition of GRs embraced by the CBD, the Biodiversity Act defines 'indigenous biological resource' broadly in relation to bioprospecting to include any living or dead organism of an indigenous species, any genetic material or derivatives of such organisms, or any chemical compounds and products obtained through use of biotechnology. As discussed below, the breadth of this definition has significant implications as to the nature of activities regulated. Material of human origin is excluded from the ambit of the law, as are exotic organisms and indigenous biological resources listed in terms of the International Treaty on Plant Genetic Resources for Food Agriculture (ITPGRFA) (section 1 read with section 80(2) of the Act). The term 'bioprospecting' is defined broadly to include 'any research on, or development or application of, indigenous biological resources for commercial or industrial exploitation' (section 1 of the Act).

The framework established by the chapter, and illustrated in Figure 11.1, is as follows:

- A permit is required before anyone can carry out bioprospecting involving indigenous biological resources or before anyone can export indigenous biological resources for the purposes of bioprospecting or other research.
- A permit will only be issued if there has been material disclosure to stakeholders and if their consent to the bioprospecting has been obtained.

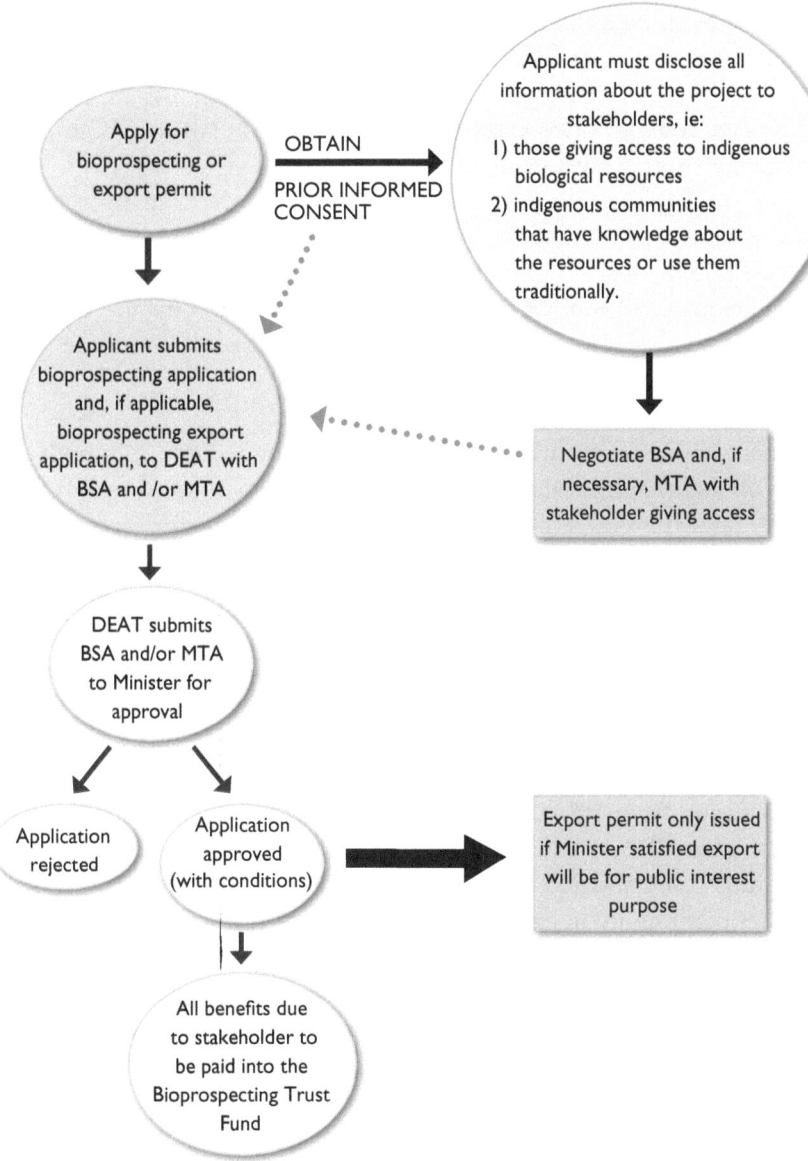

Source: Wynberg, R. and Taylor, M.

Note that the figure excludes the procedures to be followed to obtain a research export permit. DEAT Department of Environmental Affairs and Tourism

Figure 11.1 *Process prescribed by the ABS regulations to obtain a bioprospecting permit or bioprospecting export permit*

- Consent must be reflected in a benefit-sharing agreement that allows for sharing by the stakeholder in any future benefits that may result from the bioprospecting or research.

The Act envisages two categories of stakeholders whose PIC to a bio-prospecting project must be obtained. They are:

- those who give access to the indigenous biological resources – for example, a land owner
- indigenous communities whose knowledge or traditional use of indigenous biological resources has contributed to, or may contribute to, the bioprospecting.

Benefit-sharing agreements (BSAs) must be entered into with both categories of stakeholders and, in addition, a material transfer agreement (MTA) must be entered into with stakeholders who give access to the indigenous biological resources. BSAs and MTAs must be approved by the national minister responsible for the environment and the minister may require the authority responsible for issuing permits to take steps to ensure that the negotiations around the agreement take place on an equal footing, and that the resultant agreement is fair and equitable (section 82(4)(b) and (c) of the Act). The Act sets out what must be included in BSAs and MTAs.

The Act also establishes a Bioprospecting Trust Fund, into which all money arising from BSAs must be paid, and from which all payments to stakeholders will be made (section 85 of the Act). The initial thinking behind this Fund was to establish a mechanism to enable benefit flows to the wider community in cases where it was not possible to identify specific knowledge or resource holders. However, this innovative approach has not been adopted in the Act, which instead sets up the Fund to simply channel money due to identified stakeholders in accordance with the provisions of the relevant BSAs, with the accounting officer for the Fund having no discretion as to the allocation of funds.

ABS regulations

ABS regulations to give effect to the Act were gazetted in March 2007 for public comment, following a lengthy consultative process, with a revised and final version promulgated in February 2008. These regulations came into effect on 1 April 2008. No fewer than 14 drafts of the regulations were produced prior to their promulgation, indicating the complexity of the issues being dealt with in the regulations.

The most important aspects of the ABS regulations are:

- They make a distinction between the 'discovery phase' of a bioprospecting project and the 'commercialization phase', which is reinforced in recent amendments to the Act.
- They provide that the national minister responsible for the environment will be the issuing authority for bioprospecting permits and export permits if the export is for bioprospecting purposes.
- They make provincial ministers the issuing authorities for export permits if the export does not involve bioprospecting.
- They provide that foreign people or companies may only apply for bioprospecting or export permits if they apply jointly with South African people or companies.
- They require applicant holders to report annually to the national minister.
- They make provision for integrated export and bioprospecting permits.
- They provide that an export permit will only be issued if the issuing authority is satisfied that the export will be for a purpose that is in the public interest.
- They provide that the national minister may refuse to approve a BSA if no provision is made in the agreement for enhancing scientific knowledge and technical capacity to conserve, use and develop indigenous biological resources, or for any other activity that promotes conservation of the resources.

The regulations also provide a pro forma BSA, which lists possible monetary and non-monetary benefits that may be appropriate.

Issued with the regulations is a notice exempting the following activities from the application of Chapter 6 of the Act:

- research other than bioprospecting if the research is conducted in South Africa and is not for commercial purposes
- the export of ex situ indigenous biological resources if the export is for research other than bioprospecting
- the trade of commercial products purchased from a bioprospector, provided the bioprospector has complied with the regulations
- keeping, breeding and trading in wildlife not directed at bioprospecting
- the collection, use and trade of indigenous biological resources for domestic use or subsistence purposes
- the artificial propagation and cultivation of flora species for the cut flower and ornamental plant markets
- aquaculture and mariculture activities for consumption purposes.[3]

Challenges raised by the Biodiversity Act and the regulations

While the new Act is a tremendous step forward in terms of ABS regulation in South Africa, there are some critical problems with the approach adopted. The regulations sought to address some of these problems, but as subordinate legislation there was only so much they could do. This means that some of the problems can only be resolved by way of legislative amendments to the Act. This has been acknowledged by government and in May 2009 amendments to the Act (the National Environmental Laws Amendment Bill) were gazetted which will go a long way to resolving some of the problems identified.

Problems identified with the approach adopted in the Act are set out below, together with an analysis of the extent to which these problems have been resolved by the regulations or may be resolved by the recent legislative amendments.

The property clause in the constitution

All South African legislation, including the Biodiversity Act, must be read in accordance with the country's Constitution. Amongst other things, the Constitution gives concurrent executive and legislative responsibility for environmental matters to national and provincial government, and it protects various environmental rights. It also includes a property clause which protects people from being arbitrarily deprived of their property.

A central issue is that the Biodiversity Act does not vest ownership of GRs in the state. This has the effect that the state has no right to benefit in bioprospecting unless the collection of the resources occurs on state land, in which case the state will be a stakeholder as defined in the Act. Legislators elected not to vest ownership of GRs in the state as there was a concern that to do so may infringe constitutionally protected property rights. In terms of the common law, a landowner owns everything beneath and above the land, and the argument was that to vest ownership of GRs in the state may amount to a deprivation of an owner's right to use and dispose of the resources. The Constitution provides that no one may be deprived of property except in terms of a law of general application and no law may permit arbitrary deprivation. It also provides that property may only be expropriated for a public purpose or in the public interest, and on payment of just and equitable compensation.

A strong case, however, can be made against this argument, which presents an outdated and erroneous perspective about the nature of biological and GRs. The common law concept of ownership was developed prior to any understanding as to the genetic make-up of plants and other organisms and it is therefore silent on the issue of ownership over GRs. In any event, even if the common law concept of ownership over biological resources on one's land extends to ownership over the genetic components of those resources, the common law can be changed by legislation as long as the legislation is not contrary to the Constitution. It is our view that legislation that vests owner-ship of GRs in the state will not be contrary to the Constitution because:

- It is reasonable to distinguish between physical organisms and the genetic material in those organisms, given that GRs are intangible and information-based resources and are not unique to the resources on a particular landowner's property, but are common to those resources wherever they are found.
- This distinction provides a sufficient basis for the law to provide that, while landowners enjoy property rights over the physical organisms found on their land, they do not enjoy property rights to the genetic material in those organisms.
- To the extent that this distinction may deprive landowners of some of the normal incidents of ownership such as the right to use the resources, dispose of the resources and refuse access to the resources, this depriva-tion will be justifiable if it is contained in legislation that applies to everyone.

This was not something that the regulations could address and it is not something that is dealt with in the proposed amendments to the Act.

Cooperative governance

As national and provincial government have concurrent responsibility for the environment in South Africa, the possibility that one of the nation's nine provincial governments will pass legislation that is in conflict with the ABS chapter of the Biodiversity Act cannot be ruled out. If this happens, the national Act will prevail over the provincial Act if it can be shown that:

- the national legislation is necessary to protect the environment; or
- the regulation of bioprospecting requires uniformity across provinces and the national legislation provides that uniformity through norms and standards, frameworks or national policies.

If a dispute of this nature ever arises, national government will have to argue that the Biodiversity Act should prevail over conflicting provincial legislation because of:

- its international law obligations with regard to the management and con-servation of biodiversity (the Bonn Guidelines, for example, stress the need for a central focal point)
- the need for uniformity between provinces in the granting of bio-prospecting permits
- the need to protect the country's genetic heritage.

It would have helped in a dispute of this nature if the legislation had vested ownership over GRs in the state, which it fails to do.

Will the Act hinder academic research?

A concern repeatedly expressed by scientists has been that over-regulation of the pre-commercial or 'discovery' phase of a bioprospecting project may hamper academic research (Crouch et al, 2008; Wynberg, 2005). Prior to the amendments, the problems posed by the Act in this regard were as follows:

- The Act throws the net wide in firstly requiring that a permit be issued before any bioprospecting of indigenous biological resources can take place, and secondly by defining bioprospecting and indigenous biologi-cal resources very broadly.
- The Act did not distinguish between the different phases of a bio-prospecting project and provided that a permit would not be issued even for the early exploratory stages of a project unless the necessary MTAs and BSAs were in place.

The difficulty with this approach is that it failed to recognize that bene-fit-sharing agreements are typically only developed once research and development is further advanced (Ten Kate and Laird, 1999). At the early stage of a bioprospecting project the commercial outputs are unknown or unclear, and it is therefore difficult for parties to enter into a benefit-sharing agreement as envisaged by the Act. It also meant that the permitting requirements of the Act were unduly onerous and complex for applicants engaged in the exploratory or research phases of a project.

The regulations seek to lessen the bureaucratic burden involved in applying for a permit by making provision for an integrated export and bio-prospecting permit. An earlier version of the regulations distinguished between the commercialization and discovery phases of a bioprospecting project and required a simple notification procedure for the discovery phase. The notification provision were left out the of the regulations but have now been gazetted as amendments to the Act. This resolves an anomaly in the regulations which bizarrely distinguish between the discovery and commercialization phases, but then regulate the two phases identically. There clearly is no point in differentiating between the two phases if one is going to regulate them identically.

A further problem with the regulations is the fact that they exempt 'research other than bioprospecting, provided that the research is conducted within the borders of South Africa and the research is conducted for the purposes of commercial or industrial exploitation'. This may lead to huge problems of definition as to what falls within the regulatory net and what falls outside the net – that is: When is research part of the discovery phase of a bioprospecting project (which is regulated) and when is research 'research other than bioprospecting', which is exempt?

Recognizing these problems, the amendments to the Act seek to resolve them by:

- inserting definitions of the commercial and discovery phases into the Act
- providing for a simple notification procedure for the discovery phase
- allowing the relevant national minister to exempt categories of research from the application of the relevant provisions of the Act and regulations.

Bioprospecting, biotrade and the complexities of definitions

A further challenge with respect to the scope of the Act lies in its broad definition of 'bioprospecting' and 'indigenous biological resources', both of which go beyond research involving genetic material or biochemical material and the development of that material for commercial purposes. The CBD by contrast focused narrowly on GRs – defined as 'genetic material of actual or potential value' – although interpretation of this definition has been a matter of some dispute (see International Federation of Pharmaceutical Manufacturers and Associations, 2006; Rosenberg, 2006) as bioprospecting entails the commercial use not only of genetic material, but also of chemical compounds found within the organism as well as derivatives and products from the genetic material. Excluding derivatives, biochemicals or metabolic extracts from international and national laws thus significantly curtails benefit-sharing opportunities, and as a result biodiversity-rich countries such as South Africa are increasingly drafting ABS laws to go beyond the CBD. However, poorly defining what constitutes derivatives, biochemicals or metabolic extracts can lead to legal confusion, and has created concern on the part of industries that use these resources in research and development (Crouch et al, 2008; Laird and Wynberg, 2008).

For example, by including the word 'application' in the definition of bio-prospecting and by extending the scope of the Act to include all indigenous biological resources, there is an argument that in some circumstances even trade in biological resources, commonly referred to as 'biotrade', is covered by the definition. This broader category includes GRs, but also organisms or parts thereof, populations, or any other biotic component of ecosystems with actual or potential use or value for humanity (see Union for Ethical Biotrade, 2007). These might include, for example, non-timber products harvested from the wild; medicinal, food, cosmetic and other plants grown on farms and sold as commodities in international trade; and even raw materials grown in bulk to supply the manufacture of pharmaceuticals. Historically, many of these natural, biodiversity-based products have entered commodity markets similar to those for agricultural products, but distinctions between the categories of genetic and biological resources are becoming more difficult to determine, more especially when TK holders are involved.

For example, the succulent plant *Hoodia* spp. is being developed both as a GR, to be included in patented extracts, and as a herbal medicine, where the raw material is simply dried, cut and incorporated into products (Wynberg and Chennells, 2009). Devil's claw (*Harpagophytum* spp.), a plant traded for more than 50 years, is exported as a commodity to herbal markets throughout the world and nationally regulated as such, yet its bio-chemical constituents are extracted through a patented process and sold as derivatives in a variety of products (Stewart and Cole, 2005). *Pelargonium sidoides* is another example of a plant widely traded as a commodity, with its extracts incorporated into patented formulations throughout the world (Brendler and van Wyk, 2008). A recent BSA (Institute of Biodiversity Conservation, 2004) to develop the cereal crop *Eragrostis tef* or Teff, the staple diet of Ethiopia, as a gluten-free food reveals a similar 'grey area' between what constitutes a GR and a food product (Laird and Wynberg, 2008). These and many other cases illustrate well the overlapping and sometimes artificial boundaries that exist between trade in GRs and biolog-ical organisms, and the difficulties of prescribing legislation under such circumstances. This is made all the more complex by TK claims for many of the species traded. Indeed, TK is far more commonly associated with resources involved in biotrade than those involved in bioprospecting initia-tives (Laird and Wynberg, 2008; PhytoTrade Africa, 2007). Moreover, companies are increasingly developing BSAs associated with biotrade, see-ing this as an extension of ethical trading, good business practice and social responsibility, rather than for purposes of compliance with the CBD (Laird, 2008; Laird and Wynberg, 2008; Union for Ethical Biotrade, 2007).[4]

Broadening the CBD concepts of ABS to these categories of products may well be where the real economic benefits of biodiversity lie, although the regulation of such varied activities and products presents major challenges. This is well illustrated by the broad approach adopted in the Biodiversity Act, which presents specific legal problems in that even trade in rooibos tea, a crop indigenous to South Africa, and other major agricultural commodities could be interpreted to fall within its ambit. This problem can be dealt with in one of two ways. The first is to rely on a purposive interpretation of the definition, reading the word 'application' as being limited by the preceding words – 'research on or development'. In other words, insofar as 'application' must be given a specific meaning, that meaning must be understood in the context of the research and development of genetic or biochemical material. It cannot stand alone and be understood as 'the application of indigenous biological resources for commercial or industrial exploitation'. The alternative, and more cautious, approach is for the national minister for the environment to exempt certain activities, such as trade, from the application of the ABS chapter of the Act (section 86 of the Act). This, however, would have to be done in such a way so as not to blur the distinction between biotrade and bioprospecting, and to provide a loophole for benefit sharing to be avoided in situations where derivatives or biochemicals are traded. Moreover, the inclusion of certain activities in the exemptions raises the question as to whether a purposive interpretation of the general definition will now be harder to apply.

Ensuring that benefits from bioprospecting serve national interests

The goal of ensuring that national interests are served by the use and development of South Africa's GRs, articulated in the Biodiversity White Paper (DEAT, 1997), is largely unrealized in the Biodiversity Act in that the Act fails to deal expressly with the issue of ownership over GRs. As indicated above, the Act's approach is to ensure that benefits flow to stakeholders, who are defined as people providing access to the particular resources, and indigenous communities with knowledge of, or who have traditionally used, the resources.

If the state owns the land, the state is a stakeholder and must be a party to the BSA that must be entered into with all stakeholders. In this capacity, it can play a significant role in ensuring an appropriate sharing of benefits in the public interest. However, if the resources are collected on private land, the state will have no involvement as a stakeholder, other than in the requirement that all MTAs and BSAs must be approved by the minister. The only beneficiary of a BSA in these circumstances will be the private landowner.

The requirement that BSAs must be entered into with private landowners if the resources are being collected on their land elevates the rights of private landowners in a way that does not reflect current practice in South Africa, or current expectations of landowners, and will probably be difficult to implement. It also entrenches existing inequalities in land ownership in South Africa, distorted through decades of colonialism and apartheid, and thus goes against the principle of equitable benefit sharing. In the past, landowners have generally been happy to sign a simple MTA that allows researchers to collect samples from their land and that accords the landowner no right to future benefits.

In summary, the problem is this: as the Act does not vest ownership in GRs in the state, the state is not a stakeholder unless also a landowner giving access to the resource. This means that bioprospecting on private land will benefit private landowners rather than the state or the wider community.

The regulations sought to address this problem by:

• providing that a permit may only be issued to a non-South African person or company if that person or company works in collaboration with a South African person or institution
• providing that a BSA must make some provision for:
  – enhancing the scientific knowledge and technical capacity of persons, organs of state or indigenous communities to conserve, use and develop indigenous biological resources
  – any other activity that promotes the conservation, sustainable use and development of the relevant indigenous biological resources.

The prescribed format for a BSA provides a list of possible non-monetary, monetary and in-kind benefits that could be included in the agreement. The list includes: support for conservation; lodging voucher specimens with national institutions; participation of South Africans in research; access to international collections by South Africans; scientific capacity development; technology transfers; fees, upfront payments, milestone payments or royalties; and the provision of equipment and infrastructure.

The Bioprospecting Trust Fund

As indicated above, even without vesting ownership of genetic and biochemical material in the state, the Act could have ensured that some benefit flows from bioprospecting to the wider community. There are various ways this could have been done, including requiring bioprospecting agreements to make a financial or in-kind contribution to promote conservation in the area of collection; requiring that there be some transfer of skills to South African research institutions; or requiring that research results be shared with South African institutions. An alternative way would have been to require that all BSAs incorporate a benefit that promotes the national interest or to provide that a portion of any financial benefit gets paid into the Bioprospecting Trust Fund, with the Fund being given responsibility for distributing surplus money in specified public interest ways.

This is not, however, how the Fund has been set up. Only one section of the Act deals with the Fund and it does little more than establish the Fund; give the director general of the relevant department responsibility for its management; and provide that all money due to stakeholders in terms of BSAs be paid into the Fund (section 85 of the Act). The Act also makes provision for regulations dealing with the administration of the Fund. What this means is that in its current form the Fund cannot be more than a conduit for money due to stakeholders.

The one way the regulations sought to address this problem was to include a provision that if there is for any reason surplus money in the Fund the money must be distributed to:

* conserve indigenous biological resources
* build capacity amongst indigenous communities
* enhance the scientific knowledge or technical capacity of South African people and institutions.

However, as currently structured, it is unlikely that there will ever be significant surplus funds in the Trust Fund. To date, no money has been deposited into the Fund.

Identification of indigenous communities

Indigenous communities whose knowledge or traditional use of indigenous biological resources has contributed to, or is likely to contribute to, bioprospecting are stakeholders in terms of the Act. The term 'indigenous communities' is not defined, and it was clear from the public consultation process held prior to the drafting of the regulations that in many cases applicants will have great difficulty in identifying relevant indigenous communities (Wynberg, 2005).

The following scenarios were raised during the consultation process to develop the ABS regulations and indicate some of the problems that may be encountered in identifying the correct communities (Wynberg, 2005):

- If the knowledge resides in a small sector of a community (e.g. amongst traditional healers, individuals or particular families) rather than in the community as a whole, should the community as a whole benefit?[5] It was suggested that stakeholders be defined in a limited sense as the holder of the knowledge rather than the community at large.
- If the knowledge or use is held widely across a number of communities, must an applicant for a permit get the agreement of all those communities? What protocols should apply where knowledge or resources straddle political boundaries?[6] It was suggested that if a plant is endemic in an area, then it will be relatively easy to identify stakeholders, but if the plant occurs widely through the region – as is the case in *Hoodia* spp., for example – it will be much more difficult. Prior research is needed within the community to find out who are actually the bearers of TK, and importantly, of accurate and relevant knowledge.
- If there is already published research on the knowledge or traditional use held by an indigenous community, must a subsequent researcher still enter into a BSA with that community? Here it was recommended that the existence of published research should not preclude the need for a BSA, especially since a great deal of knowledge of South African plants has already been recorded.
- Is it possible to enter into an agreement with a community that is not formally organized (e.g. as a cooperative or a communal property association)?

An earlier draft of the regulations inserted principles to assist in the identification of communities, but these were excised from the final regulations which provide no assistance as to how one must identify the correct indigenous community. The amendments to the Act also provide no clarity except to expand the provision to provide that an indigenous community or an indigenous individual may be the holder of TK and therefore a stakeholder in terms of the Act.

Prior informed consent

The notion of PIC is incorporated in the Act by:

- requiring full disclosure to stakeholders of all material information
- requiring that a BSA be entered into with stakeholders prior to the granting of a bioprospecting permit
- providing that an issuing authority may facilitate negotiations between an applicant for a permit and a stakeholder to ensure that those negotiations are conducted on an equal footing and that the resultant agreement is fair and equitable

- providing that a BSA must make provision for a regular review of the agreement.

Key issues raised during the consultation process (Wynberg, 2005) focused on:

- whether PIC should be required for knowledge already in the public domain, given the difficulty of identifying the holders of such knowledge
- the need to ensure that holders of knowledge are properly informed in local languages, that meetings are widely advertised, that the value of their knowledge is described, and that all risks, rights and responsibilities are clearly spelled out. Researchers also need to be mindful of the correct procedures to be followed when working in communities. There is also a need to publish the intent to negotiate with certain parties to ensure that negotiations take place with the right party
- the fact that the type of consent may vary according to the scale and nature of the project.

Obtaining PIC is clearly a complex, challenging and community-specific process and is unlikely to be a one-off event, but rather one that progresses over time and is focused on relationship building. These are challenging questions to consider in the context of a relatively rigid legal framework.

Institutional landscape and permitting

Finally, a number of national departments administer legislation that has an impact on bioprospecting, but the Act does not provide a central focal point as proposed by the Bonn Guidelines. Instead the Act provides for the designation of issuing authorities via regulations. An issuing authority may be the national minister responsible for the environment or an organ of state in the national, provincial or local sphere of government.

This is partly resolved in the regulations, which designate the national minister responsible for the environment as the issuing authority for all bioprospecting, and relevant provincial ministers as issuing authorities for export of indigenous resources for a research purpose that is not bioprospecting.

Towards implementation

In many ways, South Africa is an international leader in ABS. A number of pioneering bioprospecting case studies have emerged from this country, and its legal framework has developed over many years of policy development and public consultation. While it is still too early to judge effectiveness, it is fair to say that the legislative gaps and inconsistencies described above present significant constraints towards achievement of a coherent legal framework that facilitates access to genetic and biological resources, and ensures the fair and equitable sharing of benefits derived from their use. The Act's failure to deliver wider community benefits from bioprospecting is a particular weakness requiring redress through an amendment to vest ownership of GRs in the state. The confusion created by the Act in regulating research is another area requiring urgent clarity if natural product development is to be stimulated and not impeded in South Africa.

There are also major implementation challenges, including insufficient capacity, political will and a lack of awareness as to the rights, roles and responsibilities of different interest groups and constituencies. Capacity development is especially crucial, within both national and provincial departments. Within the national department of environmental affairs, capacity needs to be developed to ensure greater leadership and strategic direction; to provide relevant information (e.g. corporate policies, standardized contracts, information on ABS measures); and to give technical assistance, where necessary, to provincial government departments, researchers, communities and companies. Because of the complexity of the issues, such expertise is neither easy nor quick to develop, and it will be necessary for the national department either to ensure the inclusion of individuals with relevant scientific, commercial and other expertise as part of its staff, or to seek technical advice to ensure that decisions made are informed and evidence-based. An expert ABS task team has recently been established to advise the national department with respect to implementation of the ABS regulations, but its confinement to government representatives is too narrow to reflect the diversity of stakeholders involved in bioprospecting. The development of South African positions for the negotiation of an international regime on ABS similarly requires wider consultation and transparency than is currently practised. Capacity development is also needed within provincial departments to alert them to the implications of the ABS regulations and their roles and responsibilities, as well as within communities and interest groups (e.g. researchers, traditional healers, NGOs) affected by the ABS regulations. This could include the development of a package of materials geared specifically towards the interests and rights of different user groups, translated and simplified where appropriate.

Information management is a crucial aspect of implementation because of the range of laws that impact on or are impacted by the Biodiversity Act, and the variety of permits required by these laws. The development of a single electronic database for permit applications for biodiversity use is thus recommended, which could cover ABS, the Convention on International Trade in Endangered Species of Wild Fauna and Flora (CITES), legislation relating to threatened and protected species, and provincial research permits. This could include information about the application, its status and existing permits granted.

Finally, major gaps remain with respect to the interface between the ABS legal framework in South Africa and intellectual property rights. Although the Biodiversity Act covers TK held by farmers for indigenous agricultural GRs, and the requirement for benefit sharing with holders of knowledge, this does not include non-indigenous GRs that farmers may have improved and developed. The protection of farmers' rights thus remains a key legislative gap in South Africa. This issue will become more critical with South Africa's signature and ratification of the International Treaty on Plant Genetic Resources for Food and Agriculture (IRPGRFA). Similarly, although legislative steps have been taken in South Africa requiring applicants to furnish information relating to the use of indigenous biological resources or TK in an invention, a broader strategy is required to ensure that the intractable issues associated with TK use and protection are adequately incorporated into a workable ABS framework. Indeed, by its nature, ABS regulations exist at the juncture of many interlacing bodies of law that 'criss-cross' the same biological material, bringing together a complex mix of scientific, conservation, trade and legal elements that fit uneasily into a regulatory whole. While no single law is ever likely to address these collectively, bringing TK, innovation, science, biodiversity conservation, economic development, technology and equity into an overall coherent strategy remains the greatest challenge of all.

Notes

1   This chapter draws in part from Taylor and Wynberg (2008), published in the *South African Journal of Environmental Law and Policy*, vol 15(2). Permission to reproduce parts of this article is gratefully acknowledged.
2   The regulations were published for comment in the Republic of South Africa, Government Gazette, 16 March 2007, vol 501, no 29711, Notice 329 of 2007. The final regulations were published in the Government Gazette, 8 February 2008, no 20739, No.R. 149.

3   Government Gazette, 8 February 2008, no 20739, No.R. 149.
4   See Kleba, this book.
5   See Kleba and Kamau, this book, for a discussion on the problem of disseminated traditional knowledge and benefit sharing.
6   See Winter ('common pools') in this book for a discussion on possible solutions.

References

Brendler, T. and van Wyk, B. E. (2008) 'A historical, scientific and commercial perspective on the medicinal use of *Pelargonium sidoides* (Geraniaceae)', *Journal of Ethnopharmacology* 119, pp420–433

Chennells, R. (2003) 'Ethics and practice in ethnobiology, and prior informed consent with indigenous peoples regarding genetic resources', Paper presented at a conference on biodiversity, biotechnology and the protection of TK, St Louis, 4–6 April 2003

Cowling, R. M. and Richardson, D. M. (1995) *South Africa's Unique Floral Kingdom*, Cape Town, Fernwood Press

Crouch, N. R., Douwes, E., Wolfson, M. M., Smith, G. F. and Edwards, T. J. (2008) 'South Africa's bioprospecting, access and benefit-sharing legislation: Current realities, future complications, and a proposed alternative', *South African Journal of Science*, vol 104, pp355–366

Department of Environmental Affairs and Tourism (1997) 'White Paper on the conservation and sustainable use of South Africa's biological diversity', Government Gazette Notice 1095 of 1997, vol 385, no 18163

European Federation of Pharmaceutical Industries and Associations (2004) 'Inventive endeavors in the pharmaceutical sector and the potential impact of disclosure of geographic origin of genetic material', October 2004, www.efpia.org, accessed 25 May 2009

Glazewski, J., Meiring, A. and Fakir, S. (2001) Final Report. Report to the chairman of the board of the NBI on research and licence agreement between the National Botanical Institute and the Ball Horticultural Company

Gosling, M. (2001) 'South Africa's floral heritage sold to US company', *Cape Times*, 6 April

Henne, G. and Fakir, S. (1999) 'NBI-Ball agreement: A new phase in bioprospecting?', *Biotechnology and Development Monitor*, vol 39, p18

Institute of Biodiversity Conservation, Ethiopian Agricultural Research Organization, Health and Performance Food International bv. (HPFI), 'Agreement on access to, and benefit sharing from, teff genetic resources', Addis Ababa, December 2004

International Federation of Pharmaceutical Manufacturers and Associations (2006) 'Guidelines for IFPMA members on access to genetic resources and equitable sharing of benefits arising out of their utilization', 7 April, Geneva, Switzerland

Kidd, M. and Mayet, M. (2003) 'Access to genetic resources in South Africa', in Nnadozie, K. et al (eds) *African Perspectives on Genetic Resources. A Handbook on Laws, Policies and Institutions Governing Access and Benefit Sharing*, Chapter 16, Environmental Law Institute, Washington DC

Laird, S. (2008) 'Natura, Brazil: The use of traditional knowledge and community-based sourcing of "biological materials" in the personal care and cosmetics sector', in Laird, S. and Wynberg, R. *Access and Benefit Sharing in Practice: Trends in Partnerships Across Sectors*, vols I, II and III, CBD Technical Series 38, Montreal, Secretariat of the Convention on Biological Diversity, pp79–82

Laird, S. and Wynberg, R. (1996) *Biodiversity Prospecting in South Africa: Towards the Development of Equitable Partnerships*, Johannesburg, Land and Agriculture Policy Centre

Laird, S. and Wynberg, R. (2008) *Access and Benefit Sharing in Practice: Trends in Partnerships Across Sectors*, vols I, II and III. CBD Technical Series 38, Montreal, Secretariat of the Convention on Biological Diversity

PhytoTrade Africa (2007) 'PhytoTrade Africa's approaches, achievements and experiences of ABS', http://www.phytotradeafrica.com/downloads/PhytoTrade-and-ABS-June-07.pdf, accessed 25 May 2009

Rosenberg, D. (2006) 'Some business perspectives on the international regime', 7 November 2006, UK, GlaxoSmithKline

Stewart, K. and Cole, D. (2005) 'The commercial harvest of devil's claw (*Harpagophytum* spp.) in southern Africa: The devil's in the details', *Journal of Ethnopharmacology*, vol 100, pp225–236

Ten Kate, K. and Laird, S. (1999) *The Commercial Use of Biodiversity: Access to Genetic Resources and Benefit-Sharing*, London, Earthscan

Union for Ethical Biotrade (2007) 'Biotrade verification framework for native natural ingredients', http://www.uebt.ch/dl/Engl-UEBT-Nat-Ingredients-Ver-framework-2007-09-20(rev1)b.pdf, accessed 25 May 2009

Wang, J., Soisson, S. M., Young, K. et al (2006) 'Platensimycin is a selective FabF inhibitor with potent antibiotic properties', *Nature* 441, 18 May, pp358–361

World Conservation Monitoring Centre (1992) *Global Biodiversity: Status of the Earth's Resources*, London, Chapman & Hall

Wynberg, R. (2003) 'Biodiversity prospecting and benefit-sharing in South Africa', *Developing Access and Benefit-Sharing Legislation in South Africa*, Switzerland and Cambridge, UK, IUCN, Gland, pp56–81

Wynberg, R. (2004a) 'Bioprospecting delivers limited benefits in South Africa', *European Intellectual Property Review*, vol 26(6), pp239–243

Wynberg, R. (2004b) 'Rhetoric, realism and benefit-sharing – use of traditional knowledge of *Hoodia* species in the development of an appetite suppressant', *World Journal of Intellectual Property*, vol 6(7), pp851–76

Wynberg, R. (2005) 'Consolidated report of inputs arising from provincial consultation workshops on the access and benefit-sharing regulations', Prepared for the Department of Environmental Affairs and Tourism by the Environmental Evaluation Unit, University of Cape Town, 30 November

Wynberg, R. and Chennells, R. (2009) 'Green diamonds of the south. A review of the San-*Hoodia* case', in Wynberg, R., Schroeder, D. and Chennells, R. (eds) *Indigenous Peoples, Consent and Benefit-Sharing: Lessons from the San-Hoodia Case*, Springer, Berlin

Wynberg, R. and Swiderska, K. (2001) 'South Africa's experience in developing a policy on biodiversity and access to genetic resources: Participation in access and benefit-sharing policy', Case Study 1, London, International Institute for Environment and Development

Wynberg, R., Taylor, M. and Laird, S. (2007) 'Access and benefit sharing in South Africa: An analysis of legal frameworks and agreements', Environmental Evaluation Unit, University of Cape Town, in collaboration with Cheadle, Thomson & Haysom and People and Plants International, Prepared for the Department of Environmental Affairs and Tourism on behalf of the Southern African Development Community Biodiversity Support Programme, Report no 1/07/274

Chapter 12

The Process of Legislation on Access and Benefit Sharing in China – A New Long March[1]

Tianbao Qin

Introduction

As one of the countries with the richest biodiversity and TK in the world, China offers golden opportunities to bioprospectors, especially those from foreign countries who usually appropriate illegally GR and/or related TK.

China has realized the importance of legal control of foreigners' access and fair benefit sharing. Currently, there are a large number of laws and regulations related to GR which constitute a preliminary ABS framework, but they cannot meet the urgent demand for regulation, especially benefit-sharing requirements. This urges China to develop a special law on ABS. During the past 15 years, China's efforts have moved from general implementation of the Convention on Biological Diversity (CBD) obligations to a ministerial decree, then to a special legislation or regulation through three stages.

The game-playing among relevant competent authorities accounts for the following controversial issues during the drafting process: China's position on the exchange of GR; a future competent national authority; the form of the legislation; the scope of regulated GR; differentiated procedures for different types of bioprospecting and protection of TK.

Although there are still numerous controversies and difficulties, the author is convinced that the members of the drafting team and the Working Group on ABS Legislation (WGAL), will succeed in breaking through one obstacle after another by adopting flexible strategies and tactics, and adopting an ideal ABS legislation as soon as possible.

Background

With the third largest land area, China is also one of 12 countries with the richest biodiversity in the world. China has a vast territory with a complex climate, varied geomorphic types, a large river network, many lakes and a long coastline. Such complex natural conditions inevitably form diversified genes, species, habitats and ecosystems. China owns more than 30,000 species of higher plants and 6,347 species of vertebrate, accounting for 10 per cent and 14 per cent, respectively, of the world total.[2] China's biodiversity is rich not only in numbers but also in levels of endemism. Some groups have an ancient origin, complicated composition and a large number of rare species.

China consists of 56 different nationalities. Except for the Han nationality, almost all other minor nationalities live in either remote or mountainous areas. They have developed their own unique culture, religion and way of life in a specific environment where biological resources are usually abundant. Their TK and practices play an important role in nature and biodiversity conservation in China.

Given its rich biological diversity and TK, China attracts much interest from bioprospectors, especially from other countries. Most foreign bioprospectors have appropriated GR illegally and/or related TK in the form of patents without paying any compensation to China. A classic case is the patent application over wild soya, in China by Monsanto. In April 2000, Greenpeace accused Monsanto of trying to monopolize the world's soya. Monsanto had laid claim to patent rights over a naturally occurring gene-sequence discovered in wild soya, originating in China. The sequence connects to yield characteristics in the food plant and would have effectively given Monsanto exclusive rights to plants, seeds and progeny exhibiting high-yield traits. The patent application was filed simultaneously in over 100 countries apart from Europe and the USA. The patent would have particular ramifications for China, the source of 90 per cent of the world's strains of wild soya.[3] Finally, facing immense pressures from China and the international community, Monsanto withdrew most of its patent applications.

Unfortunately, China is not so lucky in most similar cases, especially in the field of plant GR for medical uses. For example, *Homalomena occulta* is one of China's medicinal plants. In 1998, the IP Australia granted an Australian organization, Redlands Nursery, the plant breeder rights even without determining what kind of plant variety was involved. In addition, the French pharmaceutical company Rhone-Poulenc and the British company GSK were granted a patent over a substance which is isolated from a native plant of China, *Artemisia annua,* used to treat cerebral malaria.[4]

For a variety of reasons, the exact number of incidents of such kinds of biopiracy is difficult to calculate. However, it is very likely to be a 'tip of the iceberg'. Under such circumstances, China has realized the importance of legal control of foreigners' ABS.

The existing legislation and institutions related to ABS in China

Current legal system[5]

China has no single legislation on GR. However, China's current environmental laws and regulations have formed a preliminary legal system. This system includes different levels such as Constitution, Laws, Administrative Regulations, Ministerial Decrees and Local Regulations.

The basic Constitutional framework related to ABS is found in several articles because the issue crosses sector boundaries. Especially, the *1982 Constitution* establishes the ownership on GR. As a socialist country, China defines public property over almost all natural resources in its Constitution.[6] In some cases, the Constitution also recognizes the collective ownership over some natural resources, such as certain land, wild plants and animals. It does not mention GR, but it can be implied that natural resources contain the biological, biochemical and genetic components. Therefore, the ownership of biological resources can be extended to cover their GR attached to or thereon.

Besides those constitutional provisions, China has its wild animal law and wild plant regulations covering ABS issues. According to the *1989 Wild Animal Protection Law*, the wild animal resources belong to the state and the state protects wild animals and their habitats, and any illegal hunting and destruction by any organization or individual is banned. Generally, hunting, killing, selling and buying of nationally protected animals (and their products) are strictly banned; however, it can be permitted when capture is needed for the demands of scientific research, breeding, exhibition and other special uses. Field surveys or taking photos, movies and videos by foreign people within the jurisdiction of China should be approved by the responsible ministry or its authorized institution.

The *1997 Wild Plants Protection Regulations* have similar provisions. The state will protect wild plants and their living environment, and prohibit illegal collection of wild plants or any destruction of their living environment. Collecting, selling and buying the nationally protected, first-grade plants are generally banned. When it is necessary to collect the first-grade protected plants for scientific research, artificial breeding and other special uses, the collector must apply for collection permission from the responsible ministry under the State Council or from the institution authorized by the ministry. Also, foreigners cannot be permitted to collect or buy the nationally protected wild plants distributed inside China. For field investigations of the habitat of nationally protected plants inside China, application should be made to the responsible local provincial department for its first examination and then submitted to the central responsible ministry for approval.

Most recently, China passed two very important pieces of legislation, the *2004 Seed Law* and the *2006 Animal Husbandry Law*, both of which stipulate the ABS framework for germ plasm and animal GR. First of all, the *Seed Law* declares the state has the sovereign right over germ plasm resources. Collecting and cutting natural germ plasm resources that are under special protection of the state are prohibited. Where such collecting or cutting is required for scientific research or other special purposes, the matter shall be subject to approval by the administrative department for agriculture or for forestry under the State Council, or under the government of a province, autonomous region or municipality directly under the central government. And any units or individuals that wish to provide germ plasm resources to people outside China shall apply to the administrative department for agriculture or for forestry under the State Council for approval.

Similarly, the *Animal Husbandry Law* stipulates that any entities or individuals that wish to cooperatively use nationally protected animal GR with entities or individuals outside China shall apply to the administrative department for agriculture at a provincial level, agree a benefit-sharing arrangement, and then get approval from the Ministry of Agriculture.

The legislation mentioned above has established a basic legal system to control access to wild animals and plants. However, the regulatory framework is not specially aiming at access to GR and benefit-sharing, and it has many serious flaws. First, the existing legal system is deficient in terms of applicability. China's existing laws are mainly applied to animals and plants and other biological resources, not to GR attached to those biological resources.

Second, there are serious loopholes in the existing legal framework in China. The current laws apply mainly to animal GR and plant GR, but another important part of GR – microbial GR – has no applicable rules in the existing legislations. Even as to plant and animal GR, the existing laws also apply solely to those GR of nationally protected wild animals and wild plants, excluding those GR that do not fall in the scope of nationally protected wild animals and wild plants.

Third, China's existing legal framework is not consistent with the integrated control requirements of GR. According to the basic principles of biological prospecting, the development and utilization of GR has no links with the distinction among animal, plant or microbial resources. The current system actually separated animal and plant GR, and promulgated a respective legislation for each of them, which cannot meet the requirements of integrated control.

Existing institutional arrangement[7]

As to the supervision and management of biological resources, China takes a sectoral approach. Currently, many departments have the power to regulate one or more aspects of biological resources, such as the environmental department, forestry department, agriculture department, urban construction department, traditional Chinese medicine department and marine department. This institutional arrangement can also be seen as the forerunner of the regulation of ABS.

In the area of biological resources, the State Environmental Protection Administration (SEPA, 2005) is mainly responsible for coordinating works of all other departments. Meanwhile, it also bears certain specific functions, such as coordination and supervision of environmental protection related to national wild plants and the integrated management of the national Nature Reserves.

The forestry department and the agriculture department are the two most competent departments for management of biological resources. The forestry department is mainly responsible for management of forest resources, wild animals and plants within forestry areas, and precious wild trees outside of forestry areas, forestry nature reserves and terrestrial wildlife. The agriculture department is mainly in charge of the supervision and management of other wild plants outside the jurisdiction of the forestry department, aquatic wildlife, fisheries and nature reserves.

Also, the urban construction department is mainly responsible for the supervision and management of city gardens, wild plants within scenic spots and wild animals in zoos. The traditional Chinese medicine department is mainly responsible for the supervision and management of resources of species of wild herbs. The marine department is mainly responsible for the supervision and management of marine nature reserves.

In addition to these major departments, there are also a number of other government departments and NGOs involved in the conservation and utilization of biological resources, such as the science and technology department, the Chinese Academy of Sciences, health departments and intellectual property departments.

Furthermore, in 1993, in order to implement actively the CBD, China established the 'Coordination Body of China's Implementation of the

Convention on Biological Diversity', which is led by SEPA and was composed of another 12 relevant, competent departments. Since then, in order to further strengthen protection of biological resources and to halt the loss of biological resources, China has established the 'Inter-ministerial Meeting of Biological Resources', which consisted of SEPA as its leader and 17 other departments at central governmental level in 2003. Its role is to provide advisory suggestions for central government's policy making, domestically and internationally.

We will find that the existing institutional framework in China cannot meet the need to regulate ABS. This deficiency is mainly expressed in the following. First, ABS involves a number of departments, and it requires a unified, integrated and coordinated regulation among the various involved departments. In China, the existing system divides GR into different groups according to their biological taxonomy and regulates them respectively. In the context that there is no specific ABS law designating the leading department and competent departments, it is difficult to build such a coordinated mechanism.

Second, there are serious problems of overlapping, duplication and omission of powers and mandates caused by an unclear division of powers among involved departments. For example, as to the collection of nationally protected wild aquatic species for medical uses within the marine nature reserve, at least three departments, namely, the State Oceanic Administration, Ministry of Agriculture and Traditional Chinese Medicine Authority, have the power to regulate this activity. By contrast, there is no one government department with the power to regulate collection of microbial GR according to existing laws.

Third, in many cases, China's GR are mainly managed by scientific research institutions, in particular those directly under relevant competent departments. Obviously, the exercising of administrative power by scientific research institutions is not effective, which results in the loss of GR in practice.

In summary, China has established a basic legal and institutional framework to regulate ABS, but it cannot meet the urgent demand for regulation, and especially lacks benefit-sharing requirements. This is motivation for China to develop a special law on ABS.

Main process of drafting legislation on ABS in China

As mentioned above, China has addressed legal and policy responses to ABS issues for more than 15 years, progressing through three stages.

First stage: National policy for general implementation of CBD obligations

China's realization of the importance of ABS issues can be traced back to the negotiation of CBD in the 1980s. To conserve global biodiversity and ensure its sustainable use, China actively participated in the negotiations of CBD. During the CBD's preparation, China expressed its concern on access to GR and contributed to CBD Article 15. At the United Nations Conference on Environment and Development (UNCED), Premier Li Peng, on behalf of the Chinese Government, signed the CBD on 11 June 1992, as the 64th signatory; and China ratified it on 5 January 1993, becoming one of the earliest countries to ratify the CBD.

The government of China is serious in implementing the CBD, which is indicated not only by the fact that China actively participates in various follow-up activities for CBD implementation organized by United Nations Environment Programme (UNEP) and all the meetings of the Conference of Parties to CBD, but also by the fact that China has initiated a series of actions and measures for implementing CBD in order to fulfil its obligations under CBD.

As early as November 1992, only several months after UNCED, China issued *A Ten-point Policy for Environment and Development*, which was formulated to materialize the spirit of UNCED. To implement Article 6 of the CBD, China began to formulate the *China Biodiversity Conservation Action Plan* in 1992 and officially launched it in June 1994. This action plan has identified the priority ecosystems and species for conservation, as well as the objectives in seven areas and 26 priority actions. Eighteen priority projects have been proposed according to the feasibility and urgency of conservation. Meanwhile, China also formulated and launched *China's Agenda 21*. This document put forward a realistic master strategy, policy and action plan for promoting the integrated development of society, economy, environment, resources and sustainable development and the relation between the population, environment and development. In particular, Chapter 15 deals with biodiversity conservation where the policy, goal, priority areas and projects for biodiversity conservation are put forward. In all these initiatives, the protection of GRs was placed on the list of priority objectives and areas.

In early 1995, China began to implement the *China Country Study for Biodiversity*. This project outlines and analyses the basic situation of biodiversity in China, evaluates the economic value of this biodiversity and estimates the full additional expenses and the benefits of biodiversity conservation for implementing CBD in China. Among them, the detailed overview of the GRs in danger and under protection was made and the policies and measures to protect them were also proposed.

Under such circumstances, *China's First National Report on Implementation of the CBD* states that a priority action for the country is to draft a GR policy or law that regulates principles, benefit-sharing issues and intellectual property rights (IPRs), among other issues. This is the first time that China proposed to draft an ABS law or policy. At this stage, China's main national policy on the ABS issues was aimed at implementation of CBD, which was a routine and somewhat passive reaction.

Second stage: Ministerial decree on ABS by SEPA

China's proactive and positive drafting of ABS law started at the beginning of the new century, especially influenced by the publication of the *Bonn Guidelines on Access to Genetic Resources and the Fair and Equitable Sharing of Benefits Arising out of Their Utilization* internationally, and the raising of awareness domestically.

In May 2000, COP 5 of the CBD established the Ad Hoc Open-Ended Working Group on Access and Benefit-Sharing with a mandate to develop guidelines and other approaches for submission to the Conference of the Parties at its sixth meeting.[8] After a year's discussion and negotiation, CBD adopted the Bonn Guidelines with some amendments at its COP 6.[9] In essence, they elaborate on the key provisions in the CBD on ABS, such as those on access to GR, benefit sharing, mutually agreed terms (MAT), prior informed consent (PIC), protecting TK, capacity building and enforcement. Although the Bonn Guidelines are not legally binding, they serve as a point of reference for policy, and legislative and contractual matters related to ABS for parties of the CBD. The adoption of the Bonn Guidelines drove China to reconsider the necessity and urgency of developing its national ABS law.

As early as the mid and late 1990s, some Chinese biologists had realized the significance of developing ABS law and published several academic articles to call for it. However, due to the limited coverage of academic journals, this issue did not arouse the attention of the general public. Fortunately, almost at the same time of drafting and adoption of the Bonn Guidelines, reports by some influential mass media exposed illegal appropriation of several precious GR by foreign companies and individuals, which had occurred recently or in past decades in China. For example, one report titled 'Planting Chinese Soya Bean, but Infringes American Patents?', by one of the most influential newspapers *Southern Weekend*, revealed Monsanto's attempt to apply more than 100 patents over Chinese wild soya bean in China and many other countries.[10] Since then, more media have joined the efforts in educating the public in general and scientific research, in particular the vital importance of benefit sharing. All these awareness-raising efforts provide further impetus to China.

Encouraged by the above international and domestic factors, in June 2002 SEPA commissioned Professor Wang Xi, from the Research Institute of Environmental Law, Wuhan University, to organize a research team to study existing foreign and international ABS laws and policies and to draft a ministerial decree, aiming at providing an ABS framework immediately for China to control the access of GR and to secure benefit sharing. Just after the team submitted a research report and a decree draft in October of the same year for SEPA's internal review, SEPA was granted a new mandate and the drafting of the legislation was given a new and better opportunity.

Third stage: Regulations on ABS by the State Council

In late 2002, the State Council of China authorized SEPA to coordinate all issues regarding ABS issues to ensure the implementation of the CBD. In 2003, China established the 'Inter-Ministerial Meeting of Biological Resources', consisting of 18 departments at central governmental level, which affirms again the leading role of SEPA in the field of ABS. The first session of the Inter-Ministerial Meeting was convened in August 2003, and it adopted the 'Action Plan of Joint Inspection of Enforcement of Biological Species Resources Conservation and Management Laws and Regulations', 'Action Plan of Nationwide Survey of Biological Species Resources' and other documents. Consequently, from 2003 to 2004, SEPA and another 12 ministries conducted a joint inspection for enforcement of current laws and regulations on biological species resources. Meanwhile, SEPA is leading a national project to make an inventory of all GR as well as TK in China, funded by the Ministry of Finance with US$5,000,000. This survey is expected to be continued in 2008 and over several years, investigating the current status of bioprospecting activities in China by overseas individuals and companies. These two actions laid a solid foundation for a further drafting in the near future.

In April 2004, the State Council issued the *Circular regarding Strengthening Conservation and Management of Biological Species Resources*, identifying seven priority areas of the Inter-ministerial Meeting, one of which is the establishment of an ABS framework as soon as possible. Subsequently, the State Council issued its comprehensive policy for environmental protection, the *Decision regarding Carrying out the Scientific Outlook of Development and Strengthening Environmental Protection*, which stresses markedly the necessity of drafting special Regulations on ABS. Also by this legal instrument, the drafting of the special ABS Regulations was listed formally into the legislative plan of the State Council.

Based on those authorizations, SEPA assembled the WGAL to develop the special ABS Regulations at the end of 2005.[11] Governmental officials from different ministries and experts designated by the relevant ministries have been participating in the ABS Regulations' development. This includes the participation of experts from many organizations and universities from the law, agriculture, forestry, fishery and medical sectors. SEPA also set up a drafting team mainly consisting of several environmental law professors, biologists and officials.[12] Until now, the drafting team has held eight formal meetings and more than ten informal meetings, and produced at least ten draft versions. During this course, the drafting team has reported regularly its work to WGAL and submitted its draft version for comment. Currently, the team and the WGAL have reached already a consensus on the following important issues.

Firstly, all members of WGAL recognized that China shall take some form of legal regulation of ABS in relation to the status quo of conservation of GR and bioprospecting, with public law as its main regulatory and contractual approach and bioprospectors' self-regulation as an important supplement.

Secondly, as to the scope of legal regulation of ABS, it is agreed that both GR and its related TK shall be included in future regulations because the WGAL believed the bioprospecting of China's GR will not be successful in most cases without contribution of related TK.

Thirdly, the legal regulation of ABS shall be based on PIC and MAT aimed at sharing the benefits arousing from the utilization of GR, accessed in China in a fair and equitable way. Both PIC and MAT, as two interrelated cornerstones of ABS, are major innovations of CBD for international environmental law. PIC is intended to reduce the imbalance in bargaining power between providers and users, while MAT aims at enhancing the bargaining power of providers of GR by stipulating statutory terms and conditions in ABS agreements. They hope to change the position of provider countries that are at a disadvantage when negotiating ABS with users that are usually transnational companies.

Lastly, the WGAL emphasized that the Bonn Guidelines and other important national legislations shall be models for China to learn from, especially their benefit-sharing provisions. Future benefit sharing will list all kinds of short-, medium- and long-term benefits, whether monetary or non-monetary. However, the general situation regarding legislation is not so optimistic. The drafting work moves very slowly, with many controversial issues unsolved.

Major controversial issues during the drafting process

On the one hand, the unsatisfied developing process lies in the complexity of GR and its regulation; on the other hand, it is also because of WGAL's failure to coordinate different interests in the developing process. To a certain extent, the game-playing among relevant stakeholders at central government level accounts for the following controversial issues.

China's position in the exchange of GR

The first question the WGAL must answer before its legislative process is: What is China's basic position in the international exchange of GR? More accurately, is China a provider or user country? The key point in this regard is that the Chinese position will determine several basic elements for the proposed legislation, such as whether it is necessary to regulate ABS and, if so, what kind of orientation the regulation shall take: encouragement or restriction, for example.

Historically and currently, China is both an exporter and importer of GR. The export of GR from China has made great contributions to agriculture in the world. The short-stalked rice gene originating in China has promoted the revolution of rice breeding in the world. The dissemination of soybean in the world has enriched plant protein for human beings. The import of overseas GR has also promoted agriculture production in China. The access to GR has played a significant role as to agriculture, medicine and biodiversity conservation throughout the whole world.

According to the preliminary results of a survey conducted by SEPA, GR exports exceed those imported into China, though there are no accurate statistics.[13] According to this situation, it is clear that China needs some kind of legal regulation of ABS and restriction of access to GR by other countries.

However, some scientists did not applaud this conclusion. Some agricultural scientists claim that more recently plant GR for agriculture and food introduced from abroad exceed those exported. For example, from 1971 to 1991, China introduced 110,000 genetic materials, but exported only 28,000 genetic materials.[14] And in the near future, the number of annuals introduced is likely to increase. Considering the importance of agricultural production and other economic activities, China should encourage, rather than regulate, access to GR; and even though some kind of regulation is needed, China should put facilitation of access as its focus in the proposed Regulations. This argument is supported strongly by the biotechnology sector, which believes the development of biotechnology and bio-industry in China will be hindered by the new legislation. According to them, bio-industry is a new, rising industry, and its development is rapid in China. Although China is leading in this area among developing countries, it falls behind the USA, EU, Japan and other, more developed, countries. In order to promote the development of China's bio-industry, China should encourage and support the access of GR and other bioprospecting activities.

The necessity of introducing some kind of legal control over ABS has now been recognized. However, as a compromise, the draft team has proposed that the orientation of the Regulations shall be 'reasonable restriction of access while appropriately encouraging legal bioprospecting'.

Future competent national authority

So far the main difficulty faced by the process has been the overlapping of functions and lack of coordination between the relevant ministries. Theoretically, the introduction of an integrated and coordinated management led by SEPA can correct some imperfect aspects of existing ABS administration, since SEPA is responsible for the implementation of the CBD, and also the focal point at the CBD for the ABS issue. This proposal seems to be supported by the State Council. According to the unwritten legislative practices in China, the leading ministry designated by the State Council to draft a new legislation usually will be the sole or the main competent authority of the subject matter of the legislation. As noted above, SEPA is currently leading in drafting the legislation on ABS, so the drafting team have proposed that SEPA shall be the main competent authority on ABS issues.

However, the largest obstacle in the designation of SEPA as the main competent authority comes from other ministries such as the Ministry of Agriculture (MOA). Currently, the MOA is in charge of ABS issues pertaining to crops, livestock and fishery production, according to the *Seed Law* and the *Animal Husbandry Law*. In such circumstances, it is not so difficult to understand that the MOA is opposed to the proposed competent authority.

To address this problem, SEPA has asserted repeatedly that it has no intention of overlapping the existing competence of the MOA and other ministries, and it will only be responsible for issues which are not covered currently by them. Even so, the officials and experts on behalf of these ministries have been suspicious of such institutional arrangements during the meetings of the WGAL.

Form of the legislation

Another issue related closely to the second one is the form of the proposed ABS legislation. Based on the study of the Secretariat of the CBD, forms of current ABS legislation in the world can be divided into the following five types: (1) authorization clauses in the existing framework of environmental law; (2) amendment or broad interpretation of existing legislation to cover ABS; (3) comprehensive legislation with broader goals including ABS provisions; (4) special ABS legislation; and (5) supra-national legislation of regional organizations.[15]

Among them, the comprehensive legislation such as the *1998 Biodiversity Law* of Costa Rica, and specific ABS legislation such as the *2002 Biodiversity Bill* of India, are two popular forms of ABS legislation. At the beginning, the drafting team expected to draft special ABS Regulations promulgated by the State Council, and then upgrade and incorporate them into a proposed comprehensive law for biodiversity conservation passed by China's legislature.

SEPA welcomed the proposal, while the MOA and other ministries disagreed with it. The MOA argued that China has just adopted the *2004 Seed Law* and the *2006 Animal Husbandry Law*, and both laws have clear ABS provisions so, although these two laws do not cover all kinds of GR, it is not necessary to draft new legislation. This problem can be solved by amending or interpreting broadly the existing clauses in the above two laws and other relevant legislation.

Scope of regulated GR

According to the CBD, 'genetic resources' means genetic material of actual or potential value, and 'genetic material' means any material of plant, ani-mal, microbial or other origin containing functional units of heredity; and 'biological resources' includes GR, organisms or parts thereof, populations or any other biotic component of ecosystems with actual or potential use or value for humanity.[16] From this definition it is therefore clear that GRs con-sist of any material of plant, animal, microbial or other origin, except human, possessing the mentioned qualities, and are themselves one component of biological resources.[17] However, from the scientific perspective, it seems that there is no substantial distinction between GR and biological resources. Therefore, some experts insisted that the proposed Regulations shall regulate all kinds of biological resources, not limit itself to so-called GR. By doing so, the provider country can share its benefits maximally that are produced by all accessed biological resources, without any loopholes. Criticism concerning this point came from the round-table discussions, for it is claimed that including all biological resources in the scope of control will definitely increase the workload and difficulty of competent authority, and further more, states would have to take the risk that those strict regulations would be virtually impossible to enforce.

Currently, this issue is still under discussion and a preliminary agreement has been achieved on a new proposal that any biological material from living beings, wild or domesticated, which may be utilized as such, as a whole or in its macroscopic parts, shall be excluded from the scope of control.

Differentiated procedures for different bioprospecting

During discussion, almost all agreed that the bioprospecting activities shall be divided depending on the nature of access, that is, academic research or commercial exploitation. However, as to respective procedures, opinions still vary.

One suggestion is that all kinds of bioprospecting activities shall be subject to certain procedures for approval. As to those purely for the purpose of academic research, it would be improper to hinder academic research and biological conservation efforts per se through the creation of an onerous, lengthy application and approval process. Therefore, the application and approval procedure for sampling for academic research should differ from that of commercial bioprospecting, with the one being much stricter than the other. Another suggestion goes further. Some members proposed that even a relatively easy and loose procedure for academic collection of GR is not enough, only a 'zero-tolerance regulation' is acceptable, and the proposed legislation shall take the commercial bioprospecting as its only target.

A similar situation occurs when involving foreign and domestic bioprospectors. One plan is that both shall be controlled by the Regulations, but the requirements for the foreigners shall be stricter. Another advises that all domestic bioprospectors shall be free of regulation, and only foreigners shall be controlled strictly by the proposed Regulations.

Protection of TK

Compared with protection of IPRs, the protection of TK of local communities in providing countries is insufficient, whether at international or national level. In most cases, TK forms the important basis of bioprospecting of GR. But due to its different nature from modern knowledge, the values of TK and its contribution in the ABS are seldom recognized and reasonably compensated. For the aims of preventing unauthorized appropriation of TK, fairly compensating the contribution of TK and promoting its sustainable utilization and development, China should take all necessary measures to protect its abundance of TK. Nobody has denied this proposal; however, controversy exists as to the form of protection of TK.

As to the forms of protection of TK, three options have been proposed. The first is to add comprehensive provisions to specific ABS legislation. The advantage of this option is that it can address all issues related to GR and its related TK in a single legal instrument, but some criticize it because this form may not fully take into account the special nature of TK and its protection.

The second is to provide special protection of TK through sui generis TK legislation. The merits of this kind of legislation are that its protection of TK will be more in line with the natural environment and local characteristics. Meanwhile, as pointed out by some experts, its possible defect is that this form may be detached from the ABS system.

The last option is the combination of ABS legislation and sui generis TK legislation. This form makes use of the advantages of the above two forms of individual legislation, maintaining the integrity of ABS while taking into account the specific protection of TK. Therefore, in theory, this form of legislation would be the best choice for China. Although the above-mentioned legislative form has an unparalleled advantage, it has an inherent deficiency; namely, it will definitely face more resistance and too much difficulty from all kinds of sectors.

Conclusion

Because of the tremendous changes in physics, chemistry and biology, and from industrial revolution to the biotechnology revolution around the world in the past decades, the importance of primitive resources of the industrial era such as fossil fuels, metals and minerals have been gradually replaced by GR. In the new era of genetics, genetic diversity is its primitive resource. At the same time, the importance of GR decides the importance of protection of resources and knowledge about using these resources.

A hopeful scenario for the next few years for China is that its legislation concerning the access of GR and related TK will ensure effective control over the GR within its territory, while sharing the benefits arising from their research and development in a fair and equitable way. Fortunately, China has already started its legislative process for establishing an ABS framework.

Due to the complexity of GR issues and the problematic relations among various stakeholders, China is unable to adopt and apply the crucial legal instrument currently. The legislative process is approaching a deadlock. Governments and bioprospecting groups will continue facing controversial issues such as access to TK.

More than 70 years ago, there was a world-shaking event in China; namely, the Long March by China's Red Army, which succeeded in breaking through one blockade after another by adopting flexible strategies and tactics. Seventy years later, we are facing a similar long march, not for a new war, but for new legislation on ABS. Although there are still numerous controversies and difficulties, the author is convinced that the members of the drafting team and WGAL, in the spirit of the Long March, will contribute their knowledge, wisdom and even courage; and succeed in breaking through one blockade after another by adopting flexible strategies and tactics, and adopting an ideal ABS legislation as soon as possible.

Notes

1 The research of this chapter was funded by the 'Legal Issues of Conservation of Biological Diversity' project (No 06BFX036) of the National Social Sciences Foundation (NSSF), China. The author is grateful to Ms Zhao Peipei and Mr Bryan Flannery for their kind support for this chapter. However, the views in the chapter are the author's own and do not represent those of NSSF in any way.
2 National Environmental Protection Agency (NEPA), China's National Report on Implementation of the Convention on Biological Diversity, Beijing, 1995, p1.
3 Greenpeace, 'Bio-pirate Monsanto at Large Again over Chinese Soya', available at www.i-sis.org.uk/isisnews/i-sisnews13-30.php, accessed 10 September 2008.
4 George, J. and Staden, J. van (2000) 'Intellectual Property Rights: Plants and Phytomedicinals – Past History, Present Scenario and Future Prospects in South Africa', *South African Journal of Science*, vol 96, p433.
5 See further, Qin T.B. (2006) *Legal Issues Concerning Access to Genetic Resources and Benefit-sharing*, Wuhan, Wuhan University Press, pp591–594.
6 Article 9, 1982 Constitution of the People's Republic of China.
7 See further, Qin, T. B. (2006) *Legal Issues Concerning Access to Genetic Resources and Benefit-sharing*, Wuhan, Wuhan University Press.
8 Secretariat for the Convention on Biological Diversity, International Regime on Access and Benefit-Sharing: Proposals for an International Regime on Access and Benefit-sharing, 7 January 2003, UNEP/CBD/MYPOW/6, p2.
9 For full text of the Guidelines, see Secretariat of the Convention on Biological Diversity, *Bonn Guidelines on Access to Genetic Resources and Fair and Equitable Sharing of the Benefits Arising Out of their Utilization*, UNEP/CBD/COP/6/20, VI/24, 27 May 2002.
10 'Planting Chinese Soya Bean, but Infringes American Patents', *Southern Weekend*, 25 October 2001.
11 According to the current legislative procedures for administrative regulations in China, there are four stages: (1) internal drafting by one leading competent

departments; (2) joint drafting by all relevant ministries; (3) comments, consultation and perfecting; and (4) approval by the State Council.

12 The author of this chapter is co-chair of this team.

13 Xue Dayuan (2005) *Current Status and Protection of Biological Genetic Resources*, Beijing, China Environmental Science Press, pp6–10.

14 Ibid, p16.

15 Secretariat of the Convention on Biological Diversity, International Regime on Access and Benefit-Sharing: Proposals for an International Regime on Access and Benefit-Sharing, UNEP/CBD/MYPOW/6, 7 January 2003, p9.

16 Article 2, CBD.

17 The only exception endorsed by CBD in the late 1990s.

Reference

State Environmental Protection Administration (2005) *China's Third National Report on Implementation of the Convention on Biological Diversity*, China Environmental Science Press

Chapter 13

The Role of the National Biodiversity Institute in the Use of Biodiversity for Sustainable Development – Forming Bioprospecting Partnerships

Jorge Cabrera Medaglia[1]

Introduction

The importance of technology transfer (including biotechnology) for food, agriculture, human health, environmental protection, among others, has been outlined by diverse studies and emphasized by entities such as the Food and Agriculture Organization of the United Nations and the United Nations Environment Programme (Krattiger, A., 2000) At the same time, the access and acquisition of these technologies is especially complex due to their proprietary character, basically because of the existence of intellectual property rights (IPRs) such as patents and plant breeder's rights.

In order to close this gap between those who have the control of these technologies and those who need them, especially developing countries, many different schemes have been essayed to facilitate the access and transfer of biotechnology, but mostly in the agricultural field. An interesting option on this subject has taken place in Costa Rica, via the negotiations undertaken by the National Biodiversity Institute (INBio). Through agreements on access and supply of biodiversity (samples and extracts), important technology has been acquired (not all cases involve biotechnology) that has helped to consolidate a minimum infrastructure which allows the adding of value and the discovery of new intelligent uses for GRs. As a private or public interest and non-profit institution, INBio has generated important experience on the subject of sharing the benefits derived from access to GRs since the signature of the Merck and Co. Agreement in 1991.

This experience is illustrative of the manner in which the objectives of the Convention on Biological Diversity (CBD) relative to the sharing of the benefits derived from access to GRs, including transference of technology, can truly be applied. In general, it shows the importance of collaborative agreements which allow our countries to access the technology and know-how necessary to add value to the elements of biodiversity and in this manner contribute to their conservation and sustainable use, thereby improving the quality of life of the inhabitants.

ABS: The legal regime in Costa Rica

The legislation that regulates access to genetic material, biochemical resources and TK is the Law of Biodiversity, No 7788, of 30 April 1998 (LB[2]). In relation to access policies, there is a National Biodiversity Strategy that contemplates a set of actions to be taken in the area of access to GRs.

Presently, there is also a 'General Access Procedure' (GAP) that functions as a by-law of the LB. This was approved by the Minister of Environment and Energy and the President through an executive decree (15 December 2003). The GAP was proposed by the National Commission for the Management of Biodiversity (NACOMB) in conformity with Article 62 of the above-mentioned law.

The LB was designed to implement the CBD in Costa Rica. The LB established that, without prejudice to the fulfilment of regulations relative to the trade of endangered species of flora and fauna, the application of sanitary and phytosanitary measures, and technical procedures and biosafety, the provisions on access to GRs will constitute neither a concealed restriction nor an obstacle to trade (Article 68 general rule of interpretation).

The general goal of the LB is to promote the conservation and sustainable use of biodiversity and to ensure the fair and equitable sharing of benefits derived from it (Article 1). The entire LB responds to this goal as put forth by the CBD. For example, it establishes:

- the environmental function of the land (Article 8)
- general principles of the law (Article 9); objectives (Article 10)
- criteria for applying the law (Article 11)
- the National System of Conservation Areas' (NASYCAs') administrative structure (including the administration of the national wild protected areas, Articles 22–43)
- the guarantee of environmental safety (biosafety and exotic organisms, Articles 44–48)
- the conservation and the sustainable use of the ecosystems and species (Articles 49–61)
- the regulations on access to GRs (Articles 62–76)
- intellectual property rights (Articles 77–85)

- education and public awareness and research and transfer of technology (Articles 86–91)
- environmental impact assessment (Articles 92–97)
- incentives (Articles 98–104) and procedures and sanctions (Articles 105–113).

All of these elements are in accordance with the three objectives of the CBD.

Scope of the law, exceptions and specific treatment for some sectors

The Legislation is applied 'on the elements of the biodiversity under the State's sovereignty, as well as on the processes and the activities carried out under its jurisdiction or control, independently of whether the effects of the actions are manifested inside or outside the national jurisdiction'. This Law will regulate specifically the use, management, associated knowledge, distribution of benefits and costs derived from the utilization of the elements of the biodiversity (Article 3). In the same way, Article 6 (public domain) establishes that 'the biochemical and genetic properties of the elements of the wild or domesticated biodiversity are of public domain. The State will authorize the exploration, research, bioprospecting, use and utilization of the elements of biodiversity that constitute goods of public domain, as well as the utilization of all genetic and biochemical resources, by means of the procedure of access established in Chapter V of this Law'. Also, in conformity with Articles 62 and 69, every research programme or bioprospecting on genetic material carried out in Costa Rican territory requires an access permit, unless covered by one of the exceptions foreseen by the law.

The exceptions of the Law (Article 4) refer fundamentally to access to human GRs and the exchange of genetic and biochemical resources that are part of traditional practices of indigenous peoples and local communities, and that have a non-commercial purpose. In addition, public universities were exempted from control for a term of one year (until 7 May 1999) in order for them to establish their own controls and regulations for non-commercial projects that require access. Apart from this, all the remaining sectors (pharmaceutical, agriculture, biotechnology, ornamental and medicinal herbs) are subject to the Law and must follow its access procedures. The GAP regulates access for commercial and non-commercial bioprospecting (including teaching), occasional economic utilization, constant use of genetic and biochemical resources and TK. The Law indicates that a concession will be required in the case of access to GRs for commercial use, without defining steps or requirements.

The Law is applied equally to genetic agricultural resources. The legislation foresees specifically that, in the case of duly registered ex situ collections, the regulation of the law will set the authorization procedure for access permits (Article 69). It would include any type of collection. The above-mentioned procedure was supposed to be determined by means of the already cited GAP; however, the Regulations do not have rules on this point. On the contrary, GAP establishes a moratorium on the access to GRs found in ex situ collections, unless the specific normative is approved. The GAP allocated six months for the drafting of these regulations; this period was later extended for one more year.[3] These regulations are especially complex due to the institutional structures that keep GRs in ex situ conditions. Furthermore, other applicable dispositions to ex situ collections can be found in different regulations, without direct relation to access, but in relation to conservation and maintenance.[4] There is no official record of the ex situ collections in the country.

As mentioned, the Law applies to all the elements of biodiversity found under the sovereignty of the State (Article 3) and to all basic research and commercial bioprospecting projects conducted in Costa Rican land (Article 69). In this respect, access regulations are applied to GRs in public or private land, terrestrial or marine environments, ex situ or in situ collection and indigenous territories. Nevertheless, there are some omissions relative to resources in marine areas. Hence, other legal rules can be applicable to obtain access to these biological resources. Specifically, the Costa Rican Institute of Fishing and Aquaculture (IFA) is the entity entrusted with granting fishing licenses, including research permits, but excluding permits for resources found in marine regions of wild protected areas (Law of Creation of IFA No 7384 of 29 March 1994, Article 5, and Attorney General's Opinion C-215-95 of 22 September 1995). In this case, access permits by the Technical Office of NACOMB (TO) are also required. Regarding access to indigenous land there are other applicable laws besides the LB, such as the Convention on Indigenous Peoples of the International Labor Organization (ILO 169 Convention) and the rules of the sui generis system of intellectual community rights that are being developed through a consultation process that is ongoing.

Institutional arrangements

The LB created the self-governed NACOMB (Article 14) as a separate legal entity, but belonging to the Ministry of Environment and Energy (MEE). NACOMB's duties include formulating the policies and responsibilities established in Chapters IV and V (access to genetic and biochemical elements and protection of associated knowledge) and VI of the LB.

Furthermore, it has to coordinate these policies with the relevant institutions. Additionally, it has to formulate and coordinate the policy for access to elements of biodiversity and associated knowledge, ensuring a suitable transference of technology, as well as the sharing of benefits, which are general procedures under Title V of the Law.

This entity has been formed by governmental bodies such as the MEE, which presides over it, the Ministry of Foreign Trade, the Ministry of Health, the Ministry of Agriculture, the IFA and National Commission of University Presidents; the Indigenous organization, the Farmers organization, the National Union of Chambers, the Costa Rican Federation for the Conservation of the Environment (FECON) and the Director of NASYCA (Article 15). NGOs are represented by FECON. It can also revoke the decisions of its TO regarding access matters (Article 14). In conformity with Article 62, NACOMB must propose policies on access to genetic and biochemical resources of ex situ and in situ biodiversity. It will also act as an obligatory consultant in procedures related to the protection of intellectual property rights on biodiversity. In addition, the Commission will execute its agreements and resolutions and will design its internal procedures by means of the TO's Executive Director (Article 16). To date, the TO has five full-time civil servants.

The TO will grant or deny access requests (Article 17, clause a); coordinate access issues with conservation areas, the private sector, indigenous peoples and rural communities (Article 17 para b); organize and keep an updated record of access requests and ex situ collections, as well as a record of the individuals and legal entities that devote themselves to genetic manipulation (para c); and compile and update regulations relative to the fulfilment of its agreements and directives (para d). Article 16 allows the NACOMB to name ad hoc expert committees in complex cases.

The Commission's activities are regulated by means of MEE's decree No 29680, published in *The Gazette* of 7 August 2001 and its modifications. Its members are designated for a two-year period. The Commission's responsibilities include the granting of access permits and the implementation of monitoring and evaluation procedures. To date, evaluation and monitoring procedures have not been implemented.

Evaluation of commercial and non-commercial bioprospecting initiatives

According to Article 71 (characteristics and conditions of access permits), the access requirements will be determined differently depending on whether the research has or does not have a commercial purpose. In the latter case, the non-commercial purpose will have to be verified. Nevertheless, GAP does not contemplate different requirements for bio-prospecting projects with commercial and non-commercial purposes in spite of the fact that Article 9 (permits for basic research) establishes that, if a project has commercial purposes, the interested party will have to fulfil additional requirements. In general, there is no clarity on the form this distinction would take.

Access procedures

The LB regulates the basic requirements for access, including the prior informed consent (PIC), transfer of technology, equitable sharing of benefits, the protection of associated knowledge, and the definition of the ways in which the above-mentioned activities will contribute to the conservation of species and ecosystems. It also mandates the designation of a legal representative in the country, when the person or organization requesting access is domiciled abroad (Article 63). The procedure to follow is clearly outlined in Article 64. It includes proof of the PIC of the owner of the property where the activity will be developed, whether it is an indigenous community, a private owner or public entity. Other interesting provisions incorporate the right of cultural objection (Article 66), the registry of access applications and the protection of confidential information, except in the case of biosafety concerns (Article 67).

The LB also regulates in detail commercial and non-commercial bio-prospecting permits (Article 69). These are valid for three years and can be renewed. They are given to specific persons or entities and are therefore not transferable. The permits are limited to the genetic and biochemical elements expressly authorized for specific areas or territories (Article 70). The permits will contain a certificate of origin, permission or prohibition to extract samples, periodic reporting obligation, monitoring and control, conditions relative to resulting property and any other applicable condition deemed relevant by the TO (Article 71).

The access request requirements are name and identification of the interested party, name and identification of the responsible researcher, exact location of the place and the elements of biodiversity that will be the subject of the investigation, indicating the owner and manager or holder of the premises. The applicant will also have to submit a descriptive chronology of activities, aims and purposes as well as a place for legal notifications. The application must be accompanied by the PIC (Article 72)[5] and a record of individuals or legal entities who are to conduct the bioprospecting (Article 73). The TO must also authorize those agreements contemplating access to genetic and biochemical elements (Article 74), signed between individuals, natives or foreigners, or between them and the institutions registered for such purposes. There is also a possibility to establish framework agreements with universities and other duly authorized centres (Article 74). It is established that up to 10 per cent of the research budget and 50 per cent of royalties will have to go to the conservation area, the private owner or indigenous community (Article 76). In cases in which the TO authorizes the continuing use of genetic material or of biochemical extracts for commercial purposes, applicants are required to obtain a separate concession from the interested party (Article 75).

First, in conformity with access procedure norms, interested parties must register with the TO using a specific form (Article 12). Later, the PIC must be negotiated in conformity with a guide, which stipulates the minimal points for discussion (Article 19) between the applicant and owner of the conservation area or indigenous land, resources or ex situ collections. This would include not only individuals, but other government entities such as municipal governments and the IFA.

The PIC is supposed to contain mutually agreed-upon terms that represent the fair and equitable distribution of benefits. Once obtained, this agreement must be endorsed by the TO. Even though the legislation is not clear, it is assumed that the PIC will be formalized in a private contract. The TO limits itself to endorsing the contract rather than negotiating it.

The TO's approval authorizes three fundamental aspects: the PIC's fulfilment of the requirements established in the Technical Guide; the number of samples to be taken; and the time frame for the reports to be presented (Article 13).

A request form and a completed Technical Guide (Article 9) must be submitted to the TO. In both cases there are requirements and documents that must be presented jointly. Additionally, the documents established in GAP's Article 9 must be attached. Additional requirements are established for those who request permits for basic research or bioprospecting (Article 9.4) and for those who need access permits for occasional or continuing economic utilization (Article 9.5).

The Law (Article 76) requires a determination of the administrative fee. The GAP also refers to this payment (Article 17 on administrative rates). After the TO extends a certificate of origin (Article 19 of GAP), it proceeds to publish the requests and final resolutions on its website within eight calendar days (Article 15).

GAP's Article 14 establishes some 'Criteria for the evaluation or approval of the request', based on the public environmental interest criteria embodied in the Law (Article 11.3).

Also, GAP's Article 24 allows the imposition of total or partial restrictions on access to the resources to ensure their conservation and sustainable use. These restrictions are issued by the TO in the resolution approving access. In this way, it can prohibit access, set limits and regulate the methods of collection, in application of the precautionary principle mentioned in LB's Article 11.2.

Once access is authorized, the monitoring and control phase begins (Article 20 of GAP) at the expense of the TO and in coordination with the authorized representatives of the place where access to the resources is taking place. Applicants will have to follow applicable sanitary and phytosanitary rules for the exportation of the materials.

Finally, an environmental impact assessment (EIA) can be requested by the TO based on some general provisions of the LB related to EIA, but not specific to bioprospecting activities (Article 92). The evaluation is the responsibility of the National Technical Secretariat (a body of MEE). To date no EIA has been requested of the National Biodiversity Institute (INBio) or any other bioprospector.

In any case, the current scheme would leave the negotiation of contracts (by means of the PIC) in the hands of the conservation areas and eventually of other public authorities, insofar as they are the owners of the lands or of the biological resources.

Characteristics of the access requirements

The procedures for access are not completely clear, especially under GAP. On the other hand, the requirements are established in Articles 63 and 72 of the LB, as well as in GAP's Articles 7–22. Only the TO and eventually the NACOMB shall grant access permits. A separate PIC should be obtained from other entities such as conservation areas, indigenous territories or public authorities who are owners of lands, or, in the case of marine resources, other authorities such as the IFA.

In this respect, access to flora and fauna found on private lands would eventually need other authorizations from state entities such as the MEE, particularly in cases of species in danger of extinction or with reduced populations. Access would be granted in conformity with the technical and scientific arrangements stated by the National System of Conservation Areas (NASYCA). Thus, even if the flora were in private lands (e.g. orchids), the NASYCA would give the permits for the manipulation of the resource (Wildlife Conservation Law, No 7317, Articles 14, 18 and 25, and its regulation No 26435, Article 20). In such a case it is not clear whether there should be a double authorization: from the TO for the GR and from the NASYCA for the biological one, as well as the landlord's consent regarding private property. In cases where collections are made in conservation areas, the PIC and the respective agreement are enough to obtain the access permit. The main difficulties arise when there is a question of privately owned wild, threatened flora.

It must be pointed out that there are no binding requirements that benefits must go towards the conservation of the resources. It is perfectly possible that a private owner, public institution or indigenous territory could grant the PIC without allocating benefits towards conservation since the legal authority of the TO is limited to endorsement. In these circumstances, it is valid to ask whether the TO would have the legal authority to revoke a previous consent because of a lack of benefits towards conservation derived from the access (Article 63). As one might expect, in those cases in which a conservation area grants the permits, it is assumed that the benefit will go in its entirety towards biodiversity conservation.

Analysis of monitoring issues of the permits

Two of the most relevant aspects of ABS that present larger difficulties are related to the monitoring of ABS conditions and the existing legal remedies against the non-compliance with the contract or permit (Cabrera Medaglia, 2004a). ABS legislation will always be difficult to enforce, due to the nature of GRs, particularly their wide availability and the ease of dissemination or replication.[6]

In general, most of the legislation lacks adequate monitoring systems (Normand, 2004; Ogolla, 2005). Monitoring and evaluation of the agreements is in most cases weak or absent.[7] Possibly this is one of the main difficulties of the ABS regimes. To this, the difficulties derived from the characteristic of GRs as information are added. This characteristic has brought to evidence the inconvenience of applying the traditional monitoring instruments. Probably, as the Expert Panel on ABS recommended (see First Report of the Expert Panel on ABS, para 88), monitoring could be more effective with the participation of an institution or local counterpart. This system has been considered by diverse countries (Bhutan, Bolivia, etc.). Regardless of its utility it must be acknowledged that research and development on their most advanced phases will be normally carried on outside the borders of the country of origin. For this reason, additional mechanisms to warrant the tracking of the materials, for example identification systems, must be explored (Cabrera Medaglia, 2004b). In the same way, mechanisms to oblige the users to present periodical reports, including reports on the applications for patents, together with the possibility of making audits to verify the compliance, are some of the indicated solutions (Cabrera Medaglia, 2004b). In general, countries do not have systems that allow them to practise audits to verify the compliance with the clauses stipulated on the contract itself.

Additionally, one more difficulty is the obstacle to effectively exercise legal remedies when contractual or legal violations occur, especially when it happens with companies located in other jurisdictions.

In Costa Rica, from the institutional perspective there is a duty (for the TO of the National Commission for the Biodiversity Management or CONAGEBIO) to enforce and inspect with the support of the representatives of the place where access will take place.

CONAGEBIO shall rely on a supporting TO for, amongst other functions, the transaction, approval, refusal and supervision of the applications for access to the elements or biochemical and GRs of biodiversity, as well as the associated TK in the terms of the present regulation (Article 5 of the Regulations).

The TO, in coordination with the authorized representatives of the site where the access to the elements or biochemical and GRs takes place, and in accordance with the established agreements and contracts in each phase of these rules, shall perform the pertinent tasks of verification and control.

To this effect, the officials will be able to do inspections in the property or place where the access is materialized, at any time that the respective permit remains valid or once the activities contemplated in the permit have been concluded. The officials shall act upon their control visits. The TO shall also attend any complaints and investigate the possible violation of the terms of the PIC, or the terms of the access permit.

Non-compliance with the agreements and commitments shall give rise to the permit's cancellation as stipulated in Article 27 of the present rules (Article 20 of GAP, 'verification and control').

The certificate of origin is also a form of monitoring the use of GRs (see Article 80 of the Biodiversity Law and 19 of the Regulations). To date one certificate has been issued as per the request of the applicant.

In addition, control and monitoring is carried out by the use of reports. For that reason, in the approval shall be established, among other conditions, 'the obligation of the interested party to present reports and of their periodicity' (Article 13 of the Regulations).

According to Article 24 of the Regulations, restrictions of the access permits of any kind, for basic research, bioprospecting or commercial use, shall be imposed on the applying natural person or legal entity. Access is personal, nontransferable and materially limited to the authorized elements or biochemical and GRs expressly indicated therein, and may only be undertaken within the areas or territory expressly indicated that the resolution issued by the Technical Office dictates. Finally, the contract (to obtain PIC) may establish rights to verify compliance, including audits.

In summary, the following are the main monitoring mechanisms in place in Costa Rica:

- Periodically reporting as mandated in the access permits (resolutions).
- The resolution granting access expressly indicates that the monitoring phase is open. There are no more details about how this monitoring phase will operate.
- There is no specific unit or procedure in the TO for monitoring. The TO lacks the expertise, human and technical resources for appropriate tracking of the use and transfer of the GRs.
- The TO does not provide specific guidance to the PIC provider for monitoring.
- Monitoring in the PIC contract is carried out through reporting.
- The TO has indicated its willingness to increase monitoring activities, especially with not well-known applicants.
- Monitoring mechanisms like those used by INBio, as appropriate, have not been developed.

Identification and analysis of the difficulties and successes in the implementation of the LB

As for the main difficulties in the practical application of the Law, we can mention:[8]

- The scope of access and the conceptualization of what is meant by accessing and using genetic and biochemical resources, in contrast to using organic or natural resources that do not involve access and are therefore not regulated by the applicable legislation on the matter. The specific sphere of access, especially as regards access to medicinal plants, nutraceutics and taxonomic research, is still giving cause for concern in various sectors. Likewise, the scope of the exceptions granted to public universities, what 'non-profit' stands for and so on are all conceptual barriers to the adequate functioning of the access system.
- The various interests permeating the issue of access and the opinions – often contradictory – about how and what for this topic should be regulated. Researchers and users are demanding clear, expedite and transparent rules that promote and encourage research into biodiversity, whereas other social groups are trying to restrict and control prospecting activities and the use of GRs for commercial purposes.
- The widely participative process that led to the writing of the Biodiversity Law affected, to some extent, its technical aspects, especially as regards some juridicially complex matters associated with administrative law and the jurisdiction of public authorities. Nowadays, after an unhurried process of reflection about the contents of the law, voids and contradictions have become evident, many of which should

require changes in the legislation that must be approved by the Legislative Assembly.
- Some critical aspects call for a more meticulous study. For instance, the access to the collections ex situ, the link between access and the conservation of biodiversity, the role of the state (through CONAGEBIO) in the negotiation/approval of access applications, sample exportation, how to establish efficient monitoring and control mechanisms, the content and the implications of framework agreements.
- The role of the State and procedures: absolute control or regulation and support. Historical inequities have probably led to perceiving the need for strict control to avoid the so-called biopiracy. The regulation mechanisms in some countries, for example the Philippines, have shown that, in spite of the proponents' good intentions, this kind of approach leads to disregard of CBD objectives and national laws. In this sense, some of the regulations promulgated so far have focused on controlling access rather than promoting it.

The INBio experience[9]

The National Biodiversity Institute (INBio) was created in 1989 as a non-governmental, non-profit association for private founding members and it has been declared to be of public interest. Its mission is to promote a new awareness of the value of biodiversity and thereby achieve its conservation and use for the sake of improving the quality of life.

In 1991, INBio developed the concept and practice of 'bioprospecting' as one of the answers to the need of using, in a sustainable way, Costa Rican biodiversity to benefit society. This concept continues gaining acceptance in government, scientific, academic and managerial circles, and it refers to the systematic search of new sources of chemical compounds, genes, proteins, microorganisms and other products that possess a current economic value or potential and can be found in our natural biological wealth. The use of the biodiversity presents opportunities and challenges to promote and to organize the infrastructure investments and human resources that add value and contribute to its conservation (Gámez and Sittenfeld, 1993).

INBio has a formal agreement with the MEE, which allows carrying out specific activities of the national inventory and of use of the biodiversity in the government's protected areas. INBio develops biodiversity prospecting actively in the protected wild areas of the country under that agreement, with the participation of the national and international academic and private sectors. Research is carried out in collaboration with investigation centres, universities and national and international private companies, by means of investigation agreements that include key elements such as access, research budget, benefit sharing, technology transfer, training, non-destructive activities and up-front payment for conservation.

The agreements between INBio and the partners specify that 10 per cent of the research budgets and 50 per cent of the future royalties are donated to the MEE to be reinvested in conservation. The research budget supports the scientific infrastructure in the country, as well as activities of added value aimed at conservation and sustainable use of the biodiversity.

INBio has had many agreements in both research and bioprospecting involving collaboration with industry, academia and others, many of which have exemplified the role of contract approach and its significance. In fact, studies carried out to date on benefit sharing for the use of the knowledge, and the different joint initiatives such as the Cooperative Biodiversity Groups, are based on contractual arrangements. As a result of these agreements, many benefits have been generated, including the following:

- monetary benefits through direct payments
- payment for supplied samples
- covering research budgets
- transfer of important technology which has enabled the development of the infrastructure at the Institute (biotechnology laboratory, etc.), which can be used for the investigation and generation of their own products
- training of the scientists and experts in state-of-the-art technology
- negotiation experience and knowledge of the market and the probabilities of searching for intellectual uses for biodiversity resources
- supporting of conservation through payments made to the Ministry of the Environment for the strengthening of the National System of Conservation Areas
- transfer of equipment to other institutions such as the University of Costa Rica
- future royalties and milestone payments to be shared 50:50 with the Ministry of the Environment
- establishment of national capabilities for assessing value of biodiversity resources
- royalties received from two products: a phytomedicine generated from the collaboration with Lisan (national company) and an industrial enzyme (Cottonase) for textile processing for cotton (an environmental friendly alternative for chemical scouring in cotton preparation) arising out of the Diversa (now Verinium) collaboration. The enzyme cleans better than chemical scouring and also reduces greatly the need for extensive waste, waste treatment and energy consumption. Another

product, a fluorescent protein, has also been developed (with Diversa), but no royalties have accrued yet to INBio.

The three tables below summarize the main collaborative agreements, benefits and research results.

Box 1 below presents in detail the results of the partnership between the INBio and the Costa Rican company Lisan, which has aimed to develop and market phytopharmaceutical and natural products.[10]

Box 13.1 *Joint partnership between INBio and Lisan Laboratories*[11]

Efforts carried out with funding from the Inter-American Development Bank and its Multilateral Investment Fund successfully ended with the collaboration of INBio and Lisan Laboratories – a generic pharmaceutical producer – in a 'collaborative research agreement' for the development of plant-derived pharmaceutical (phytopharmaceuticals) products. This has allowed the launching of a new line of products, 'Lisan Natura', with which Lisan Laboratories have acquired an advantage over local competitors that produce generic natural products without adequate quality control. Six products have been developed and registered as part of this collaboration.

In this case, INBio contributed its expertise and experience in plant extraction and chemistry, to a great extent derived from collaborations with international pharmaceutical companies. On the other hand, Lisan contributed its experience in quality control, product formulation and marketing. An initial confidentiality agreement was signed, which permitted the beginning of negotiations that resulted in the presentation of a research plan to the Executing Agency and its Advisory Committee, which in turn resulted in signature of the cited research agreement. The collaborative relationship covered four main phases: administrative, research, knowledge transfer and pre-commercial development.

Among the results to date, we can cite:

- A comprehensive laboratory procedure manual, including standardized extraction protocols.
- A business and research relationship between a research institution and a small enterprise.
- Material suppliers that comply with the Good Agricultural Practices standards.

- Six types of products, including a gel, tablets and creams with different therapeutic effects.
- Lisan Laboratories received an innovation award in 2003.

The experience showed that it is possible to generate alliances between the research and productive sectors that result in commercial products promoting biodiversity conservation and economic development. It has shown the feasibility of transforming knowledge into commercial products through alliances. Of course, investment in research and development in innovative products is needed for this to happen.

Among the main impacts and lessons learned:

- It was demonstrated that research and development can be led by developing country institutions.
- Phytopharmaceutical protocols were developed.
- New capacity-building and job creation opportunities through the introduction of non-traditional products were generated.
- Biodiversity was sustainably utilized.
- Benefits were generated along the whole production chain, from technicians to material-providing farmers.
- National technologies and knowledge were used.
- The benefits derived from profits generated by product marketing will be used to promote similar initiatives.
- Lisan Laboratories can offer high-quality phytopharmaceutical products distributed widely in the country.
- Under the agreement, INBio receives royalties derived from product marketing. These are divided 50:50 with the Ministry of the Environment to promote biodiversity conservation.
- The project has prevented illegal extraction by acquiring materials only from legal suppliers. These suppliers must grow the resources in a sustainable way and comply with food agriculture practices.
- Results and knowledge have been transferred to Lisan from INBio.
- It is possible to acquire patents for certain procedures and therapeutical applications.

Table 13.1 *Most significant research collaborative agreements with industry and academia. Period 1991–2002. These agreements involve a significant component of technical and scientific support from INBio†*

Industry or academic partner	Natural resources accessed or main goal	Application fields	Research activities in Costa Rica
Cornell University	INBio's capacity building	Chemical prospecting	1990–1992
Merck & Co.	Plants, insects, micro-organisms	Human health and veterinary	1991–1999
British Technology Group – ECOS	Lonchocarpus felipei, source of DMDP	Agriculture	1992–2005
Cornell University, Bristol Myers and NIH – International Cooperative Biodiversity Group (ICBG)	Insects	Human health	1993–1999
Givaudan Roure	Plants	Fragrances and essences	1995–1998
University of Massachusetts	Plants and insects	Agriculture	1995–1998
Diversa (now VERENIUM)	DNA from non-cultivable bacteria	Industrial applications	1995–present
INDENA SPA	Plants	Human health	1996–2005
Phytera Inc.	Plants	Human health	1998–2000
Strathclyde University	Plants	Human health	1997–2000
Eli Lilly	Plants	Human health and agriculture	1999–2000
Akkadix Corporation	Bacteria	Agriculture	1999–2001
Follajes Ticos	Palms	Ornamental applications	2000–2004
La Gavilana S.A.	Microorganisms	Agriculture	2000–present
Laboratorios Lisan S.A.	Plants	Human health	2000–2004

Industry or academic partner	Natural resources accessed or main goal	Application fields	Research activities in Costa Rica
Bouganvillea S.A.	Quassia amara	Agriculture	2000–2004
Agrobiot S.A.	Plants	Ornamental applications	2000–2004
Guelph University	Plants	Agriculture and conservation purposes	2000–2003
Chagas Space Program	Plants, fungi, marine organisms	Human health	2001–present
SACRO	Orchids	Conservation	2002–2008
Merck Sharp & Dohme	Training and education	IPR and bioprospecting	2002–2006
Industrias El Caraíto S.A.	Nutraceutics	Human health	2001–2004
Harvard Medical School – International Cooperative Biodiversity Group R21	Endophytic fungi	Human health	2003–2005
Universidad de Panamá – OEA	Plants	Human health	2003–2004
Harvard Medical School – National Cooperative Drugs Discovery Group (NCDDG)	Endophytic fungi	Human health	2005–2008
Ehime Women College	Plants	Human health	2005–2008
Laboratorios Vaco S.A.	Microorganisms	Industrial applications	2005–present
Harvard Medical School – International Cooperative Biodiversity Group (ICBG)	Endophytic fungi, microorganisms, lichens and marine organisms	Human health	2005–present
Instituto Pfizer	Microorganisms	Human health	2005–2006

Industry or academic partner	Natural resources accessed or main goal	Application fields	Research activities in Costa Rica
PNUD-BIOTRADE-UNCTAD-CAF	Implementation of the National Program of Biotrade	Biotrade	2005–2006
CONICIT	Spiders (DNA)	Molecular taxonomy	2004–2005
CONICIT	Plants	Human health	2005–2006
Korean Research Institute of Bioscience and Biotechnology (KRIBB)	Plants	Human health	2008
Harvard Medical School – Medicine for Malaria Venture (MMV)	Endophytic fungi	Human health	2007–present
CONICIT	Microorganisms	Industrial applications	2008
CONICIT	Establishment of Aedes aegypti bioassay	Human health	2007–present
Consejo Superior de Investigaciones Científicas de España (CSIC) Fundación CR USA	Microorganism	Enzymes of industrial applications	2008
Consejo Superior de Investigaciones Científicas de España (CSIC) Fundación CR USA	Microorganism	Human health	2008
BID-Fondo Chileno Universidad Adolfo Ibañez/ Octantis	INBio's capacity building	Entrepreneurialism	2008

† See Cabrera Medaglia, 2004c for further details on the legal system.

Note: These agreements involve a significant component of technical and scientific support from INBio

Source: Tamayo et al, 2004; Guevara, L. and Martin, N., personal communication

Table 13.2 *Monetary and non-monetary benefits of bioprospecting*

Monetary benefits	Non-monetary benefits
100% of research budgets	Trained human resources
Technology transfer and infrastructure	Empowerment of human resources
Up-front payments for conservation	Negotiations, expertise developed
Significant contribution for GCA and Universities	Market information
Milestone and royalty payments to be shared with MINAE	Improvement of local legislation on conservation issues

Table 13.3 *Outputs generated since 1992 as a result of RCA with INBio*

Project	Initiated	Output
Merck & Co.	1992	27 patents
BTG/ECOS	1992	DMDP on its way to commercialization
NCI	1999	Secondary screening for anti-cancer compounds
Givaudan Roure	1995	None
INDENA	1996	None
Diversa	1998	2 products/Publication
Phytera Inc.	1998	None
Eli Lilly & Co.	1999	None
Akkadix	1999	52 bacterial strains with nematocidal activity
CR-USA	1999	1 compound with significant anti-malarial activity
LISAN	2000	2 phytopharmaceuticals in the market
Caraito	2000	Industrial protocol for a nutraceutical production
Follajes ticos	2000	4 novelties of ornamental plants
Bougainvillea	2001	Biopesticide in the process of commercialization
La Gavilana	2001	Biopesticide in the process of commercialization
Agrobiot	2001	Kit Eco educational
SACRO	2002	In vitro plants for conservation

Source: Tamayo et al, 2004; Guevara, L. and Martin, N., personal communication

Negotiating strategies and contracts' content: Some key issues for consideration [see Cabrera Medaglia (1997) and Laird (1994)]

Objectives
Negotiations concerning access to GRs should aim to identify and promote the mutual interests of the two parties to the agreement – provider and recipient – so that the agreement captures and expresses an understanding of shared interests and objectives. In some negotiations involving parties with diverse backgrounds, this can entail building respect and understanding for the values and cultural backgrounds.

Working plan
This plan should be clear and within the scope of the technical and legal possibilities of the parties. The working plan could evolve and be more suitable to the research status. So a constant communication is required to the extent that, in some cases, it is even indicated in the same contract as telephone conferences carried out on specific dates. To a certain extent, main obligations could be established in the plan and only refer to it in the articles related to the agreement.

Definitions
A key aspect, from different points of view, are definitions. First, definitions allow the understanding of the scope of obligations. For this reason, it is crucial that parties get into an agreement on the main terms. For example, definitions of 'samples', 'materials', 'extracts', 'by-products', 'analogous', 'fragments', 'chemical entities' and 'affiliates' are especially relevant because juridical aspects such as conditions for third-party transfer could depend on each of them. Second, some definitions such as the one for products could have implications in relation to the benefits to be distributed, so, the wider the definition, the more possibilities exist to demand for benefits of a result of the investigation.

Exclusiveness and use
Exclusiveness should address diverse situations: exclusiveness in samples collection and their delivery; exclusiveness in obtaining extracts and fractions; and, in case that activity is proven in them, exclusiveness given to the partner for its later research, future extension of the period of exclusiveness under agreed conditions, and so on. In addition, those allowed uses could be restricted only for pharmaceutical, agricultural or industrial research, in which case the provider could not use materials for other activities, without the corresponding authorization. It is also possible that, within the

agreement's frame, technology transfer has occurred to the resource-provider party, such as research protocols. Whether this technology can be used freely by this party (meaning for purposes other than those indicated in the collaboration agreement) should also be stipulated.

Property of the research results

Ownership of the research findings needs to be considered from the beginning. Joint research will produce results and generate different types of information. It must be clarified who has ownership and what are the rights over such information. Nonetheless, it is important to determine whether the attained information could be used by each partner for different activities, for new research agreements with other institutions or for private use in their researches.

Materials transfer

The possibility of transferring materials to third parties should be carefully analysed, taking two factors into account. On the one hand, it might be required that materials are sent to third parties in order to perform screening and sequencing activities, when there are no installed capacities or when it is more cost-efficient. Also, it is possible that third parties may want to have access to material that present, for example, an activity. In some cases, it might be necessary to deposit materials (chemical or genetic) in material banks on which access by third parties depends upon specific regulations and contractual models. When facing these scenarios, even later access and transference could have a price, over which it might or might not have been a royalty agreed upon. This access and transference could also be restricted to some activities. On the other hand, third-party transference faces the inconvenience of hindering the tracking and verifying of results. This transference normally takes place through an agreement in which the provider is not involved and thus holds no right to claim non-compliance by the third party.

Report, verification and tracking

The nature of GRs hinders their proper tracking. Nonetheless, there are some interesting mechanisms. The first consists of activity reports, including patent request reports. It is important that such reports become an instrument for usage control and not merely another requirement. Secondly, contractual mechanisms of verification that allow third parties to perform audits (and visits to the company laboratories), specially foreseen when there are royalties, or gaining access to research logbooks and laboratory notes. In the case of audits it is necessary to clearly define who pays for them, how often they can take place, under which conditions and what consequences would bring any detection of irregularities. Thirdly, it can be requested that each item or sample be identified with a tracking number, which will also identify the use of the provided resource in the research activity chain.

Intellectual property rights

IPRs are a key negotiating factor. Regarding intellectual property, it is important to regulate the following:

- invention property
- collaboration and communication mechanisms between both parties
- applicable legal frame to determine who is the inventor
- obligation to state the country where the sample comes from (country of origin)
- whether IP depends upon the inventive contribution
- who faces the expenditure of requesting, maintaining and litigating it. If the user does, but the patent is registered under the name of the provider partner:
  - whether there is a right to first refusal option
  - whether there is the possibility for one part to take over the costs of applying for and maintaining the patent, should the other part decline to do so
  - whether the other part gets any licence, and if so, under which terms (free of charge, restricted to some countries, restricted to some uses, only for certain purposes, etc.).

Benefits

Non-monetary and longer-term benefits may be preferred over short-term or monetary benefits (see the illustrative list of benefits presented in the Annex to the Bonn Guidelines).

Confidentiality/exceptions

Prospecting contracts are particularly confidential, especially about issues dealing with payment, royalty, samples, technologies and research results. This includes three aspects:

- confidentiality of the contractual terms
- confidentiality of information provided by each party
- confidentiality of research reports.

In addition to confidentiality, there is a non-use clause on the provided information, except for collaboration purposes, valid for five and up to ten years. Upon termination of the agreement, parties must give back all materials and information, except for possibly keeping a copy in the partner's file.

Warranties

It is important that parties declare themselves able to meet all contractual obligations. It is advisable to make sure materials delivered are 'as is', meaning they are experimental in nature and therefore no accountability is claimed. This limitation is essentially aimed at issues with the other party, but can extend to the research of commercialized by-products.

Contract termination

Termination causes are included under the terms of contract termination. For example, whether a contract can be terminated by one of the parties after previous notice or it needs to be a consensual termination, specifics of noncompliance and its effects (including the possibility of repairing the situation within a given time-frame), end of contract time and extensions.

Dispute resolution and applicable law

It is important to consider legal mechanisms for solving eventual disputes, following a hierarchy of options that begins with conversation between the parties and that ends with arbitration or judicial processes. Also, it is necessary to define which law will be applicable in the event of disputes because in some cases the parties operate in different territories and are regulated by different laws (see Chaytor et al, 2000).

Lessons learned

The most important inferences that can be summarized from the above are as follows:

*   There must be a clear institutional policy for the criteria demanded in prospecting contract negotiations. In INBio's case, they are transfer of technology, royalties, limited quantity and time access, limited exclusiveness, not causing a negative impact on the biodiversity and direct payment for conservation (Mateo, 1996).
*   Existence of national scientific capabilities and, consequently, the possibilities of adding value to biodiversity elements increases the negotiating strengths and benefit sharing which are to be stipulated in contract agreements (Sittenfeld et al, 2003).

- Knowledge of operational norms, as well as of changes and transformations taking place in the business sector, and of the scientific and technological progress that underlie these transformations helps in defining ABS mechanisms. It is essential to possess knowledge of how different markets operate and of the ABS practices that already exist in these markets (Mateo, 1996). For information on how ABS operates in the different sectors, see Ten Kate and Laird, 1999 and Laird, 2002.
- There must be internal capacity for negotiations, which includes adequate legal and counselling skills relating to the main commercial and environmental law aspects. Possibly one of the key facts understood by the Institute is to acknowlege that negotiations involve a scientific aspect (of crucial importance to define key areas of interest such as a product, for example), a commercial aspect, a negotiation aspect and the respective legal aspects.
- There must be innovation and creativity capabilities for obtaining compensation. An ample spectrum of potential benefits exist. In the past, interesting benefit-sharing formulas, other than the traditional ones, were developed through the appropriate use of negotiations and included, for example, fees for visiting gene banks having collected material. The contractual path fortunately permits parties to adapt themselves to the situation in each concrete case, and from there proceed to stipulate new clauses and dispositions.
- There must be understanding in key subjects such as: IPR; restrictions on third-party transference of the material (including subsidiaries) and the obligations of such parties; precision of the key definitions provided they condition and outline other important obligations (such as products, extracts, material, chemical entity); precision of the property and ownership (IPR and others) of the research findings; and joint relationships.
- There must be proactive focus according to institutional policies. An active approach on negotiations, according even to the institution's own outlined policy that permits an understanding of national and local requirements, has resulted in important benefits. The existence of a Business Development Office at INBio, with highly qualified expert staff, attending seminars and activities within the industry, the distribution or sharing of information and material, and direct contacts, all enable an answer to be given, to a larger or smaller extent, to institutional challenges.
- There must be macro policies and legal, institutional and political support. It has been pointed out that, confronted with prospecting, the so-called macro policies have to exist (Sittenfeld and Lovejoy, 1998);

that is to say that clear rules on aspects related to what has been called the bioprospecting framework, which imply biodiversity inventories, information systems, business development and access to technology, have to exist.

To this must be added other elements, such as the existence of trustworthy partners, one of the most relevant aspects in joint undertakings (see Sittenfeld and Lovejoy, 1998).

Conclusion

The Costa Rican case has shown interesting individual features that make it worthy of mention, although it does not necessarily constitute an example to be followed in other nations. Peculiar circumstances of the national reality (see Mateo, 1996, for these special situations), such as the size of the country, the structure of the central government, its political, educational and social situation, have led to the establishment of important conditions of its own. It is an example of a nation that decided to take a road instead of continuing to discuss the difficulties that exist to travel on it.

Notes

1 The opinions here expressed are of a personal nature.
2 A series of topics were considered for the formulation of the dispositions relative to access, distribution of benefits and protection of TK. These included basic definitions, scope of activities to be covered by the regulations, the procedure for PIC, mutually agreed terms, competent authority, distribution of benefits and sanctions. Some relevant topics such as the need to distinguish between access for agricultural or pharmaceutical purposes, or between research for commercial or academic purposes and the need for prompt and special mechanisms for ex situ collections, were scarcely considered. These areas constitute some of the deficiencies of the legislation that must be corrected with an appropriate regulation.
3 The regulations for access to GRs found in ex situ conditions were approved by decree No 33677 of 27 April 2007.
4 For example, the decree of creation of the National Commission of Plant Genetic Resources No 18661-MAG of 9 September 1988, and the Law of Seeds No 6289 of 4 December 1978 and its by-law.
5 However, according to the regulations the application must be presented as the first step in the process, and later on the PIC negotiated with the provider must be submitted before the access permit is granted.
6 Barber et al (2002).
7 University of Columbia (1999).

8   The author is grateful to Ms Marta Liliana Jiménez, CONAGEBIO Executive
    Director, for her comments and observations. A further difficulty is the action of
    unconstitutionality presented against several articles of the Law.
9   Based on information provided by the Bioprospecting Unit of INBio.
10 The author wishes to thank Ana Sylvia Huertas and Ana Lorena Guevara (INBio)
    for the information supplied.
11 UNDP (2005) 'Roadmap to commercialisation: Costa Rica, sharing innovative
    experiences', vol 10, *Examples of the Development of Pharmaceutical Products from
    Medicinal Plants*, New York.

References

Barber, C., Glowka, L. and La Vina, A. (2002) 'Developing and implementing national
    measures for genetic resources, access regulation and benefit sharing', in Laird, S.
    (ed) *Biodiversity and Traditional Knowledge. Equitable Partnerships in Practice*, Reino
    Unido, Earthscan
Cabrera Medaglia, J. (1997) Contratos Internacionales de Uso de Diversidad Biológica.
    Una nueva forma de cooperación Norte-Sur, Revista de Relaciones Internacionales
    56–57. Escuela de Relaciones Internacionales de la Universidad Nacional, Primer y
    Segundo Semestre de 1997, Heredia
Cabrera Medaglia, J. (2004a) *A Comparative Analysis of the Implementation of Access and
    Benefit-sharing Regulations in Selected Countries*, Proyecto ABS, Bonn, IUCN-ELP
    Disponible en el Centro de Derecho Ambiental de la UICN a través del sitio web:
    www.iucn.org/themes/law, last date visited 15 May 2009
Cabrera Medaglia, J. (2004b) *Access and Benefit Sharing in Costa Rica: Lessons Learned
    from the Monitoring and Tracking of Genetic Resources in Access Contracts*. Documento
    de investigación preparado para el Centro de Derecho Internacional del Desarrollo
    Sostenible, Montreal, Canada
Cabrera Medaglia, J. (2004c) 'Costa rica: Legal framework and public policy', in
    Carrizosa, S., Brush, S. B., Wright, B. D. and McGuire, P. E. (eds) *Accessing
    Biodiversity and Sharing the Benefits: Lessons from Implementing the Convention on
    Biological Diversity*, IUCN Environmental Policy and Law Paper No 54, Cambridge,
    UK, IUCN-ELP, pp101–122
Chaytor, B., Gerster, R. and Herzog, T. (2000) 'Exploring the creation of a mediation
    mechanism', May 2000, available at http://www.gersterconsulting.ch/docs/
    JWIP_Mediation_Mechanism.pdf, accessed 15 March 2009
Gámez, R. and Sittenfeld, A. (1993) 'Biodiversity prospecting in INBio', in Reid, W. et
    al (eds) *Biodiversity Prospecting*, World Resources Institute
Krattiger, A. (2000) 'An Overview of ISAAA from 1992 to 2000', ISAAA Briefs No 19,
    New York, Ithaca
Laird, S. (1994) 'Biodiversity prospecting contracts', in Reid, W. et al (eds) *Biodiversity
    Prospecting. Sustainable Use of Genetic Resources*, Estados Unidos, World Resources
    Institute
Laird, S. (ed) (2002) *Biodiversity and Traditional Knowledge. Equitable Partnerships in
    Practice*, London, Earthscan

Mateo, N. (1996) 'Wild biodiversity: The last frontier? The case of Costa Rica', in Bonte-Friedheim, C. and Sheridan, K. (eds) *The Globalization of Science: The Place of Agricultural Research*, The Netherlands, ISNAR

Mateo, N. (2000) 'Bioprospecting and conservation in Costa Rica', in Svarstad, H. and Dhillion, S. (eds) *Responding to Bioprospecting. From Biodiversity in the South to Medicines in the North*, Oslo, Spartacus

Normand, V. (2004) 'Level of national implementation of ABS', Documento presentado en el Taller de Expertos Internacionales sobre Acceso a Recursos Genéticos y Distribución de Beneficios, Cuernavaca, Mexico, October 2004

Ogolla, D. (2005) 'Legislative regimes on access and benefit sharing: Issues in national implementation', in *Report International Expert Workshop on Access to Genetic Resources and Benefit Sharing*, Ciudad del Cabo, Noruega y Suráfrica, September 2005

Sittenfeld, A. and Lovejoy, A. (1998) 'Biodiversity prospecting frameworks: The INBio experience in Costa Rica', in Guruswamy, L. D. and McNeely, J. A. (eds) *Protection of Global Biodiversity, Coverging Strategies*, Durham and London, Duke Press University

Sittenfeld, A., Cabrera, J. and Marielos, M. (2003) 'Bioprospecting frameworks: Policy issues for island countries', *Insula International Journal of Island Affairs*, February, Year 12 no 1

Tamayo, G., Guevara, L. and Gámez, R. (2004) 'Biodiversity prospecting: The INBio experience', in Bull, A. T. (ed) *Microbial Diversity and Bioprospecting*, Washington DC, ASM Press, Ch 41, pp445–449

Ten Kate, T. and Laird, S. (1999) *The Commercial Use of Biodiversity. Access to Genetic Resources and Benefit-Sharing*, London, Earthscan

Universidad de Columbia, Escuela de Asuntos Internacionales, Taller sobre Estudios de Política Ambiental (1999) *Access to genetic resources: an evaluation of the development and implementation of recent regulation on access agreements*, New York

Chapter 14

Australian ABS Law and Administration – A Model Law and Approach?

Geoff Burton

Introduction

This chapter examines Australia's implementation of the *Bonn Guidelines on Access to Genetic Resources and Fair and Equitable Sharing of the Benefits Arising Out of Their Utilization* (the Bonn Guidelines) in order to see to what extent its national policy, its ABS law and its ABS administrative arrangements represent the development of a model ABS law and implementation. The chapter will also analyse the Australian law to identify legal solutions it has developed to address policy issues going beyond the Bonn Guidelines. It will consider these solutions to see what contribution they may make to the current debates underway in the development of the International Regime on Access and Benefit Sharing.

Background

Convention on Biological Diversity

Australia ratified the CBD on 18 June 1993. In 1996 it released its *National Strategy for the Conservation of Australia's Biological Diversity* (National Biodiversity Strategy). This strategy identifies Australia's policy goal for access to GR in the following terms:

> *ensure that the social and economic benefits of the use of genetic material and products derived from Australia's biological diversity accrue to Australia.*[1]

A national inquiry into access to GR in Commonwealth (i.e. federal) areas was undertaken in 1999–2000 (the Voumard Inquiry). This inquiry involved comprehensive public consultation including industry, the scientific community and environment interests and indigenous communities. It made over 70 recommendations for the establishment of a practical ABS regime for Australia.[2] In a related development, the issue of enhanced access to biological resources was integrated into Australia's National Biotechnology Strategy in 2000.[3] The development of draft legislation in the form of amendments to the *Environment Protection Biodiversity Conservation Regulations 2000* then followed, with final legislation coming into force in December 2005.

During this policy development period, Australia was active in its support for the evolution and later adoption of the Bonn Guidelines by the CBD in 2002. Australia's international support for the Guidelines was later reflected domestically by the adoption of a common framework to implement the Bonn Guidelines by all nine Australian governments.[4] This intergovernmental agreement is the *Nationally Consistent Approach for Access to and the Utilisation of Australia's Native Genetic and Biochemical Resources* (the *Nationally Consistent Approach*). Unusually in the Australian federal experience, this agreement was drafted, negotiated and agreed within several months of the CBD's adoption of the Guidelines. It is a policy framework consisting of 14 binding general policy principles and a further 11 agreed common elements to be considered by all Australian governments in taking action to implement the Guidelines.

Australia

Australia is a constitutional federation of six sovereign states, two self-governing territories and a national government. It has a 'common law' legal system derived from Britain and shared with Canada, the United States, New Zealand, India, and other countries with historic ties to the former British Empire. Australia is an economically developed country with an annual per capita income of US$50,150 (2008).[5] It has a well-supported, sound scientific and research community that includes 470 biotechnology companies with 73 publicly listed companies, worth AU$22.7 billion in 2008.[6]

Physically, Australia is large, with a continental landmass of 7.7 million square kilometres and an administered marine jurisdiction of some 10 million square kilometres.[7] Its size and geographic isolation has resulted in rich biodiversity, with 10 per cent of the world's species found within its borders and high levels of endemism.[8] It is a mega-diverse country. As a developed biodiverse country, Australia is both a user and a provider of GR. This has informed its thinking about ABS issues and is reflected in its policy development and policy outcomes.

National ABS law by regulation

Australia's federal ABS law is found at *Part 8A – Access to biological resources in Commonwealth areas* of the *Environment Protection Biodiversity Conservation Regulations 2000*.[9] Authority for making federal ABS regulations is found in section 301 of the supervening *Environment Protection Biodiversity Conservation Act 1999*.[10] This section is broad in its scope and allows for the ABS system to be established by regulation.

While regulations are considered as subordinate law in Australia, it should be noted that under the terms of Australia's Constitution, regulations made by the national government override state and territory law to the extent of any conflict.[11] In this case the regulations operate to avoid any such inconsistency by applying only to biological resources owned or managed by the national government. This federal jurisdiction includes defence lands, certain national parks, Australia's external territories and Australia's 10 million square kilometres of ocean resources. In all, it includes about 5 per cent of the world's biodiversity.

ABS objectives

The (six) objectives of the federal access to biological resources law are set out at regulation 8A.01 of *Part 8A of the Environment Protection Biodiversity Conservation Regulations 2000*. This states:

> *For section 301 of the Act, the purpose of this Part is to provide for the control of access to biological resources in Commonwealth areas to which this Part applies by:*
>
> *(a) promoting the conservation of biological resources in those Commonwealth areas, including the ecologically sustainable use of those biological resources*
>
> *(b) ensuring the equitable sharing of the benefits arising from the use of biological resources in those Commonwealth areas*
>
> *(c) recognizing the special knowledge held by Indigenous persons about biological resources*
>
> *(d) establishing an access regime designed to provide certainty, and minimise administrative cost, for people seeking access to biological resources*

> *(e) seeking to ensure that the social, economic and environmental benefits arising from the use of biological resources in those Commonwealth areas accrue to Australia*
> *(f) contributing to a nationally consistent approach to access to Australia's biological resources.*

These objectives are self-evidently consistent with both the CBD and the Bonn Guidelines. For example, purposes (a) and (b) reflect the three objectives in Article 1 of the Convention, while (c) foreshadows responsibilities to indigenous and local communities under Articles 8(j) and 10(c), and paragraphs (d) and (e) address Article 15. Paragraph (f) itself relates to the overarching agreement made by all Australian governments to implement the Bonn Guidelines, that is, the Nationally Consistent Approach.

The Australian system

Anyone wishing to access native biological resources for the purpose of research and development on its genetic or biochemical makeup, and to be taken from lands or waters administered by the Australian federal government, must apply for a permit. This may be done online or in writing. If the access sought is for a commercial purpose, then the permit fee is a nominal AU$50. Access for non-commercial purposes such as taxonomy is free.

The national competent authority (delegate of the Minister) will approve the permit for a commercial purpose if the collection causes no environmental harm and the applicant has entered into a benefit-sharing agreement.

Access for non-commercial purposes does not require a benefit-sharing agreement – only that no environmental harm is done and that the permission of the manager of the area where the collection is made has been given. The applicant provides information in the form of a Statutory Declaration.[12] In the Declaration, the applicant also undertakes to negotiate a benefit-sharing agreement if he later wishes to commercialize, offer to provide a taxonomic copy of any species collected, provide a copy of his research outcomes and to seek permission before transferring the material to any third parties.

The requirement for a non-commercial permit holder to return to negotiate a benefit-sharing agreement in the event that he or she wishes to commercialize their research is an important provision in two ways. Firstly it takes into account that accidental or serendipitous discovery is a continuous feature of science, and secondly it removes any temptation to choose a non-commercial permit for the sake of convenience or to avoid entering into a benefit-sharing agreement. Permits may be issued in as little as two working days.

In the event that the applicant wishes to obtain material from indigenously owned land or use their traditional indigenous knowledge, then the benefit-sharing agreement must be between the applicant and the landowners. In such cases the permit will be issued if the national competent authority is satisfied that the specified conditions for prior informed consent (PIC) and mutually agreed terms (MAT) have been met. Benefits accrued under such agreements go to the landowners, not to the government. This respects the private property rights of indigenous landowners. The disposition of benefits is solely a matter for the landowners.

To assist applicants, two model benefit-sharing agreements have been published by the Australian Government. One is for consideration when access to publicly owned areas is sought and the other one is for consideration when seeking GR from indigenous peoples' privately owned lands; however, they have broader application as examples of contemporary private international law. Applicants are not required to use these agreements. They were introduced to facilitate parties to reach an agreement by reducing costs and uncertainty and the need for legal research. They are detailed and conform to the Australian law and the CBD. A copy is annexed at page 455 of this book. The model agreements crystallize Australia's understanding of the CBD Article 15 and the MAT elements of the Bonn Guidelines (see Table 14.1). These model agreements have been designed to be easily downloaded.[13]

All permits are entered into a public register, which is viewable online.[14] This creates a fully transparent system of virtual certificates of origin, source and legal provenance. It allows instant electronic verification of evidence of PIC and MAT at no cost. Commercial, cultural or environmentally sensitive information may not be included on the viewable register. This transparency voids any accusation or suspicion of misappropriation of resources or of biopiracy. It allows a user of resources to meet any disclosure requirement in foreign intellectual property systems and reduces any possible legal uncertainty over the origins and circumstances of the material collected. Moreover, it safeguards the user's interests by ensuring that the permit records are always available in the event that their records are lost or become misplaced over time.

In 2008, applications for access to biological resources in federal areas were made and subsequently granted at the rate of more than one a week. This reflects a steady growth in applications since the scheme's commencement in 2006.[15]

Comparison with the Bonn Guidelines

To test whether the Australian ABS law and administrative structure implements the Bonn Guidelines, the table below sets out relevant Guideline provisions and the Australian legal and administrative response.

Table 14.1 *Relevant Bonn Guidelines provision and their implementation in Australia*

Relevant Bonn Guidelines provision	Australian law, policy & administration response to Bonn Guidelines provisions
Scope	
9. 'excluding human genetic resources'	Paragraph 9: Regulation 8A.03 excludes access to human remains.
Other relevant regimes	
10. 'The guidelines should be applied in a manner that is coherent and mutually supportive of the work of relevant international agreements and institutions'	Paragraph 10: Principle 2 of the Nationally Consistent Approach requires government action to be consistent with the country's other international obligations.
10. 'The guidelines are without prejudice to the access and benefit-sharing provisions of the FAO International Treaty for Plant Genetic Resources for Food and Agriculture'	Paragraph 10: Regulation 8A.05(1)(c) excludes resources controlled under the International Treaty on Plant Genetic Resources for Food and Agriculture.
Objectives	
11. (a) 'To contribute to the conservation and sustainable use of biological diversity'	Paragraph 11: Reflected at regulation 8A.01 as quoted above.
(b) 'To provide Parties and stakeholders with a transparent framework to facilitate access to genetic resources and ensure fair and equitable sharing of benefits'	Paragraph 11(b): Part 8A sets out a transparent and detailed system with the aim of facilitating access and ensuring fair and equitable sharing of benefits.

Relevant Bonn Guidelines provision	Australian law, policy & administration response to Bonn Guidelines provision
(d) 'To inform the practices and approaches of stakeholders (users and providers) in access and benefit-sharing arrangements'	Paragraph 11(d): The level of detail provided at Part 8A sets out the rights and obligations of all parties under the Australian system and all considerations to be taken into account in reaching decisions. In addition, the Minister has published a model ABS Agreement under regulation 8A.07 to assist stakeholders.
(l) 'Taxonomic research, as specified in the Global Taxonomy Initiative, should not be prevented, and providers should facilitate acquisition of material for systematic use and users should make available all information associated with the specimens thus obtained'	Paragraph 11(l): Special, simplified, requirements exist for the collection of resources for non-commercial purposes. Permits for such collections are free. This is set out at regulations 8A.12–14.

A. National focal point

13. 'Each Party should designate one national focal point for access and benefit sharing and make such information available through the clearing-house mechanism. The national focal point should inform applicants for access to genetic resources on procedures for acquiring prior informed consent and mutually agreed terms, including benefit-sharing, and on competent national authorities, relevant indigenous and local communities and relevant stakeholders, through the clearing-house mechanism'	Paragraph 13: A national focal point has been established for each Australian jurisdiction: these are fully contactable in writing, telephone or by interactive website and the clearing-house mechanism (CHM).[16] It is the responsibility of the national focal point to inform applicants for access to GR on procedures for acquiring PIC and MAT, including benefit-sharing. This information is included in publications and online.

Relevant Bonn Guidelines provision	*Australian law, policy & administration response to Bonn Guidelines provision*

B. Competent national authority(ies)

14. 'Competent national authorities, where they are established, may, in accordance with applicable national legislative, administrative or policy measures, be responsible for granting access and be responsible for advising on:	Paragraph 14: Australia has registered its National Competent Authority with the CBD. This is Mr Ben Phillips, who heads the team responsible for all the actions listed from (a) to (h) in respect of federal areas.
(a) the negotiating process	
(b) requirements for obtaining PIC and entering into MAT	
(c) monitoring and evaluation of access and benefit-sharing agreements	
(d) implementation/enforcement of access and benefit-sharing agreements	
(e) processing of applications and approval of agreements	
(f) the conservation and sustainable use of the genetic resources accessed	
(g) mechanisms for the effective participation of different stakeholders, as appropriate for the different steps in the process of access and benefit sharing, in particular, indigenous and local communities	

Relevant Bonn Guidelines provision	Australian law, policy & administration response to Bonn Guidelines provision
(h) mechanisms for the effective participation of indigenous and local communities while promoting the objective of having decisions and processes available in a language understandable to relevant indigenous and local communities'	

C. Responsibilities

16. 'Recognizing that Parties and stakeholders may be both users and providers, the following balanced list of roles and responsibilities provides key elements to be acted upon:	
(a) Contracting Parties which are countries of origin of genetic resources, or other Parties which have acquired the genetic resources in accordance with the Convention, should:	Australian Response:
(i) be encouraged to review their policy, administrative and legislative measures to ensure they are fully complying with Article 15 of the Convention	Paragraph 16(a)(i): The access regulations comply with Article 15 and it is Australian government policy that the operation of the system will be subject to annual and continuous review.[17]
(ii) be encouraged to report on access applications through the clearing-house mechanism and other reporting channels of the Convention	(ii) All access permits are available for viewing online at https://apps5a.ris.environment.gov.au/grid/public/perrep.jsp

Relevant Bonn Guidelines provision	*Australian law, policy & administration response to Bonn Guidelines provision*
(iii) seek to ensure that the commercialization and any other use of genetic resources should not prevent traditional use of genetic resources	(iii) Section 8A.03 (3)(a) exempts indigenous person's use of biological resources.
(iv) ensure that they fulfil their roles and responsibilities in a clear, objective and transparent manner	(iv) Legal requirements under other legislation exists for the regulation of public service conduct and mechanisms exist to provide for administrative and legal review of decisions taken, e.g. the Office of Commonwealth Ombudsman and the Administrative Appeals Tribunal.
(v) ensure that all stakeholders take into consideration the environmental consequences of the access activities	(v) Regulation 8A.16 provides an environmental impact assessment process for any action likely to have more than a negligible impact on the environment.
(vi) establish mechanisms to ensure that their decisions are made available to relevant indigenous and local communities and relevant stakeholders, particularly indigenous and local communities (vii) support measures, as appropriate, to enhance indigenous and local communities' capacity to represent their interests fully at negotiations'	(vi) & (vii) Regulation 8A.10 provides that access to resources on Indigenous owned lands can only occur where the Indigenous owners have given their informed consent to a benefit sharing agreement, and sets out the criteria to confirm that such consent was freely given and fully informed. The Indigenous landowners determine the nature and subsequent distribution of benefits. Decisions are published and are available to indigenous and local communities. Indigenously owned land leased to the federal government is jointly managed.
(b) 'In the implementation of mutually agreed terms, users should:	Australian Response:
(i) seek informed consent prior to access to genetic	(i) Part 8A of the regulations sets out extensive requirements for the provision of PIC to access.

Relevant Bonn Guidelines provision	Australian law, policy & administration response to Bonn Guidelines provision
resources, in conformity with Article 15, paragraph 5, of the Convention	
(ii) respect customs, traditions, values and customary practices of indigenous and local communities	(ii) Respect for Indigenous people's rights and customs underlies the provisions at regulations 8A.03 (3)(a), 8A.04 (1)(c) & (i). In particular, 8A.08 requires '…protection for, recognition of and valuing of any indigenous people's knowledge to be used…' and for transparency for any dealing in Indigenous knowledge together with evidence of consent and benefit sharing.
(iii) respond to requests for information from indigenous and local communities	(iii) Indigenous owned land in federal areas is jointly managed with the owners. Consultative mechanisms are in place, as determined by the joint management boards and the plan of management for each national park.
(iv) only use genetic resources for purposes consistent with the terms and conditions under which they were acquired	(iv) This is a permit condition and penalties for its breach are found at regulation 17.08. Regulation 17.09 details the manner and circumstances under which permits may be varied. Review of the benefit-sharing agreement is also provided for at clause 2 of the Model Agreement.
(v) ensure that uses of genetic resources for purposes other than those for which they were acquired, only take place after new prior informed consent and mutually agreed terms are given	(v) Variations to permits are required for any new or additional uses. This is set out at regulation 17.09. The Model Agreement also provides for review of the Agreement at any time at the request of one of the parties.
(vi) maintain all relevant data regarding the genetic resources, especially documentary evidence of the	(vi) Regulation 8A.18 requires a public register of permits, while regulation 8A.19 requires the user to keep and share records of all collections made with the provider and the government.

Relevant Bonn Guidelines provision	Australian law, policy & administration response to Bonn Guidelines provision
prior informed consent and information concerning the origin and the use of GR and the benefits arising from such use	
(vii) as much as possible endeavour to carry out their use of the genetic resources in, and with the participation of, the providing country	(vii) This decision is left for the parties to a benefit-sharing agreement to determine according to their mutual interests.
(viii) when supplying genetic resources to third parties, honour any terms and conditions regarding the acquired material. They should provide this third party with relevant data on their acquisition, including prior informed consent and conditions of use and record and maintain data on their supply to third parties. Special terms and conditions should be established under mutually agreed terms to facilitate taxonomic research for non-commercial purposes	(viii) This is a standard benefit-sharing agreement requirement at Model Agreement clause 5.3.[18] In addition, regulation 8A.13 provides for special conditions for taxonomic research as non-commercial research.
(ix) ensure the fair and equitable sharing of benefits, including technology transfer to providing countries, pursuant to Article 16 of the Convention arising from the commercialization or other use of genetic resources, in conformity with the mutually	(ix) Technology transfer, as part of a benefit-sharing agreement, is negotiated by the parties to the agreement. Item (d) Schedule 4 of the Model Agreement illustrates this.[19] Breaches of the agreement are dealt with as contract breaches unless offences under the legislation have also occurred, in which case criminal penalties may apply.

Relevant Bonn Guidelines provision	Australian law, policy & administration response to Bonn Guidelines provision
agreed terms they established with the indigenous and local communities or stakeholders involved'	
(c) Providers should:	Australian Response:
(i) only supply genetic resources and/or traditional knowledge when they are entitled to do so	(i) Regulation 8A.02 applies the regime to federally owned, managed or leased areas and to native species. Parties using TK must disclose this and the terms and circumstances on which it was obtained.
(ii) strive to avoid imposition of arbitrary restrictions on access to genetic resources.	(ii) Imposition of arbitrary restrictions is contrary to Australian administrative and criminal law.
(d) 'Contracting Parties with users of genetic resources under their jurisdiction should take appropriate legal, administrative, or policy measures, as appropriate, to support compliance with prior informed consent of the Contracting Party providing such resources and mutually agreed terms on which access was granted. These countries could consider, inter alia, the following measures:	
(i) mechanisms to provide information to potential users on their obligations regarding access to genetic resources	(i) Australia's national access laws do not address 'user measures' except by way of example of transparency in origin and terms on which material is obtained.
(ii) measures to encourage the disclosure of the country of origin of the genetic	(ii) The establishment of the Genetic Resources Database (GRID) creates a system of virtual certificates of origin and provenance, and

Relevant Bonn Guidelines provision	Australian law, policy & administration response to Bonn Guidelines provision
resources and of the origin of traditional knowledge, innovations and practices of indigenous and local communities in applications for intellectual property rights	supports users to meet any form of disclosure requirement introduced by other nations.
(iii) measures aimed at preventing the use of genetic resources obtained without the prior informed consent of the Contracting Party providing such resources	(iii) Australian international cooperation in civil and criminal matters is dealt with in other legislation.
(iv) cooperation between Contracting Parties to address alleged infringements of access and benefit-sharing agreements	(iv) International cooperation in civil and criminal matters is dealt with in legislation administered by the Criminal Justice Division of the Commonwealth Attorney General's Department.
(v) voluntary certification schemes for institutions abiding by rules on access and benefit-sharing	(v) Australian national institutions such as the Australian National Botanic Gardens participate in such schemes as the Common Policy Guidelines for Participating Botanic Gardens.
(vi) measures discouraging unfair trade practices	(vi) Australian law through the Trade Practices Act 1974 and related legislation discourages unfair trade.
(vii) other measures that encourage users to comply with provisions under subparagraph 16(b) above.	(vii) Regulation 8A.05 provides for the recognition and exemption of collections or specified resources where they are already being managed consistent with the purpose of the regulations.
III. Participation of stakeholders	Australian Response: Stakeholders were extensively canvassed during the National Inquiry into Access to Genetic Resources and during the lengthy legislative development stage.[20]

Relevant Bonn Guidelines provision	Australian law, policy & administration response to Bonn Guidelines provision
	Consultation with Indigenous and environment groups, industry and the scientific community is ongoing.[21]

IV. Steps in the access and benefit-sharing process

A. Overall strategy	Australian Response:
22. Access and benefit-sharing systems should be based on an overall access and benefit-sharing strategy at the country or regional level. This access and benefit-sharing strategy should aim at the conservation and sustainable use of biological diversity, and may be part of a national biodiversity strategy and action plan and promote the equitable sharing of benefits	Paragraph 22: The Australian ABS system is part of Australia's National Biodiversity Strategy, its National Biotechnology Strategy and its Nationally Consistent Approach for Access to and the Utilisation of Australia's Native Genetic and Biochemical Resources.

C. Prior informed consent	Australian Response:
24. As provided for in Article 15 of the Convention on Biological Diversity, which recognizes the sovereign rights of States over their natural resources, each Contracting Party to the Convention shall endeavour to create conditions to facilitate access to genetic resources for environmentally sound uses by other Contracting Parties and fair and equitable sharing of benefits arising from such uses. In accordance with	Paragraph 24: The federal ABS System implements Principle 5 of the Nationally Consistent Approach for Access to and the Utilisation of Australia's Native Genetic and Biochemical Resources to: '...facilitate the ecologically sustainable access and use of Australia's genetic and biochemical resources'. Australian ABS laws do not discriminate between domestic or foreign users. All are subject to the same rules. Discrimination is illegal under Australian law. Thus access and use by other Contacting Parties is protected and facilitated.

Relevant Bonn Guidelines provision	Australian law, policy & administration response to Bonn Guidelines provision
Article 15, paragraph 5, of the Convention on Biological Diversity, access to genetic resources shall be subject to prior informed consent of the contracting Party providing such resources, unless otherwise determined by that Party	

1. Basic principles of a prior informed consent system

26. The basic principles of a prior informed consent system should include:	Australian Response:
(a) Legal certainty and clarity	Paragraph 26(1)(a): Part 8A and Part 17 of the Environment Protection Biodiversity Conservation Regulations 2000 is framed in 'plain words drafting' for clarity and legal certainty.
(b) access to genetic resources should be facilitated at minimum cost	(b) The schedule of Fees at Schedule 11 of the regulations provides for only a nominal access permit fee of AUD$50. Non-commercial access permits are free.
(c) restrictions on access to genetic resources should be transparent, based on legal grounds, and not run counter to the objectives of the Convention	(c) Part 8A seeks to meet this requirement by establishing the legal grounds of access and use in detail.
(d) Consent of the relevant competent national authority(ies) in the provider country. The consent of relevant stakeholders, such as indigenous and local	(d) Access approval is granted by the National Competent Authority. The property rights of Indigenous landowners are recognized and their informed consent to a benefit-sharing agreement based on MAT between themselves and the user is required for approval by the Authority.

Relevant Bonn Guidelines provision	Australian law, policy & administration response to Bonn Guidelines provision
communities, as appropriate to the circumstances and subject to domestic law, should also be obtained	

2. Elements of a prior informed consent system   Australian response:

27. Elements of a prior informed consent system may include: (a) competent authority(ies) granting or providing for evidence of prior informed consent (b) timing and deadlines (c) specification of use (d) procedures for obtaining prior informed consent (e) mechanism for consultation of relevant stakeholders (f) process	Paragraph 27(a) to (f): These elements are present in the requirement for a permit and are set out at Regulation 17.02 (2)(ga) and at Regulation 8A.10. The model Agreement also sets these elements out.

Competent authority(ies) granting prior informed consent   Australian response:

28. Prior informed consent for access to in situ genetic resources shall be obtained from the Contracting Party providing such resources, through its competent national authority(ies), unless otherwise determined by that Party	Paragraph 28: The machinery for prior informed consent for access to in situ GR sits within Part 8A and Part 17 of *Environment Protection Biodiversity Conservation Regulations 2000.*

Relevant Bonn Guidelines provision	*Australian law, policy & administration response to Bonn Guidelines provision*
29. In accordance with national legislation, prior informed consent may be required from different levels of Government. Requirements for obtaining prior informed consent (national/provincial/local) in the provider country should therefore be specified	Paragraph 29: Details of the requirements for obtaining consent from the different levels of government are set at the Australia government access to GR website. A consistent approach has been agreed in the *Nationally Consistent Approach for Access to and the Utilisation of Australia's Native Genetic and Biochemical Resources* and, in the case of the federal jurisdiction, national legislation introduced.
30. National procedures should facilitate the involvement of all relevant stakeholders from the community to the government level, aiming at simplicity and clarity	Paragraph 30: This was done at the policy development stage and is reflected in the detailed legislation at Parts 8A and 17, and in its administrative transparency via the use of an online public register.
31. Respecting established legal rights of indigenous and local communities associated with the genetic resources being accessed or where traditional knowledge associated with these genetic resources is being accessed, the prior informed consent of indigenous and local communities and the approval and involvement of the holders of traditional knowledge, innovations and practices should be obtained, in accordance with their traditional practices, national access policies and subject to domestic laws	Paragraph 31: Respecting established legal rights of indigenous and local communities associated with the GR being accessed or where TK associated with these GR has been comprehensively undertaken. This is reflected in provisions at Regulations 8A.03, 8A.04, 8A.07, 8A.08, 8A.10, 17.02, 17.03A and 17.03B.

Relevant Bonn Guidelines provision	Australian law, policy & administration response to Bonn Guidelines provision
32. For ex situ collections, prior informed consent should be obtained from the competent national authority(ies) and/or the body governing the ex situ collection concerned as appropriate	Paragraph 32: This has been done, and with simplified conditions and additionally with provision at Regulation 8A.06, whereby ex situ collections are administered consistently with the objectives of the regulations, they may be exempted from the operation of the regulations. This measure avoids unnecessary duplication and reduces costs.
Timing and deadlines	Australian response:
33. Prior informed consent is to be sought adequately in advance to be meaningful both for those seeking and for those granting access	Paragraph 33: This is a requirement of the Australian law and is set out at Regulations 8A.10, 8A.12 and 8A.13.
Decisions on applications for access to genetic resources should also be taken within a reasonable period of time	Applications as can be granted in as little as two days.
Specification of use	Australian response:
34. Prior informed consent should be based on the specific uses for which consent has been granted. While prior informed consent may be granted initially for specific use(s), any change of use including transfer to third parties may require a new application for prior informed consent. Permitted uses should be clearly stipulated and further prior informed consent for changes or unforeseen uses should be required.	Paragraph 34: Regulation 17.02 requires applicants to specify use. Regulation 17.09 deals with variations while 17.11 deals with transfers of permits. Permitted uses are clearly set out and may be electronically verified by reference to permit conditions published on the GRID.[22] The concept of special simplified conditions for non-commercial use as found at regulations 8A.12 & 8A.14 was introduced with the needs of taxonomy in mind.

Relevant Bonn Guidelines provision	Australian law, policy & administration response to Bonn Guidelines provision
Specific needs of taxonomic and systematic research as specified by the Global Taxonomy Initiative should be taken into consideration	
35. Prior informed consent is linked to the requirement of mutually agreed terms	Paragraph 35: A permit cannot be granted until a benefit-sharing agreement has been reached (Regulation 8A.11).

Procedures for obtaining PIC

36. An application for access could require the following information to be provided, in order for the competent authority to determine whether or not access to a genetic resource should be granted. This list is indicative and should be adapted to national circumstances:	Paragraph 36: This has been done. The considerations listed 36(a) to (o) are set out at Regulations 17.02(ga), 17.03, 17.03A and 17.03B.
(a) legal entity and affiliation of the applicant and/or collector and contact person when the applicant is an institution	
(b) type and quantity of genetic resources to which access is sought	
(c) starting date and duration of the activity	
(d) geographical prospecting area	
(e) evaluation of how the access activity may impact on conservation and sustainable	

Relevant Bonn Guidelines provision	*Australian law, policy & administration response to Bonn Guidelines provision*
use of biodiversity, to determine the relative costs and benefits of granting access	
(f) accurate information regarding intended use (e.g.: taxonomy, collection, research, commercialization)	
(g) identification of where the research and development will take place	
(h) information on how the research and development is to be carried out	
(i) identification of local bodies for collaboration in research and development	
(j) possible third party involvement	
(k) purpose of the collection, research and expected results	
(l) kinds/types of benefits that could come from obtaining access to the resource, including benefits from derivatives and products arising from the commercial and other utilization of the genetic resource	
(m) indication of benefit-sharing arrangements	
(n) budget	
(o) treatment of confidential information	

Relevant Bonn Guidelines provision	Australian law, policy & administration response to Bonn Guidelines provision
Process	Australian response:
38. Applications for access to genetic resources through prior informed consent and decisions by the competent authority(ies) to grant access to genetic resources or not shall be documented in written form	Paragraph 38: All decisions are documented and are published on the GRID.[23]
39. The competent authority could grant access by issuing a permit or license or following other appropriate procedures. A national registration system could be used to record the issuance of all permits or licences, on the basis of duly completed application forms	Paragraph 39: Regulation 8A.05 establishes a public register of permits. This is the GRID.
40. The procedures for obtaining an access permit/licence should be transparent and accessible by any interested party	Paragraph 40: All procedures are publicly available from the website.[24]
D. Mutually agreed terms	
41. In accordance with Article 15, paragraph 7, of the Convention on Biological Diversity, each Contracting Party shall 'take legislative, administrative or policy measures, as appropriate ... with the aim of sharing in a fair and equitable way the results of research and development and the benefits arising from the commercial and other utilization of	

Relevant Bonn Guidelines provision	Australian law, policy & administration response to Bonn Guidelines provision
genetic resources with the Contracting Party providing such resources. Such sharing shall be upon mutually agreed terms.' Thus, guidelines should assist Parties and stakeholders in the development of mutually agreed terms to ensure the fair and equitable sharing of benefits	Paragraph 41: Mutually agreed terms are required and their terms are set out at Regulation 8A.08.
1. Basic requirements for mutually agreed terms	
42. The following principles or basic requirements could be considered for the development of mutually agreed terms:	
(a) legal certainty and clarity	Paragraphs 42(a) to (g): All the considerations listed in the Bonn Guidelines are set out in the ABS access law.
(b) minimization of transaction costs, by, for example:	
(i) establishing and promoting awareness of the Government's and relevant stakeholders' requirements for prior informed consent and contractual arrangements	
(ii) ensuring awareness of existing mechanisms for applying for access, entering into arrangements and ensuring the sharing of benefits	In particular: Paragraphs 42(b)(i) & (ii) – The Genetic Resources Management Team of the Department of the Environment, Water, Heritage and the Arts conducts public awareness activities.[25]

Relevant Bonn Guidelines provision	Australian law, policy & administration response to Bonn Guidelines provision
(iii) developing framework agreements under which repeat access under expedited arrangements can be made	Paragraph (iii): Regulation 17.00 allows multiple or repeat collections.
(iv) developing standardized material transfer agreements and benefit sharing arrangements for similar resources and similar uses (see appendix I for suggested elements of such an agreement) (c) Inclusion of provisions on user and provider obligations (d) Development of different contractual arrangements for different resources and for different uses and development of model agreements (e) Different uses may include, inter alia, taxonomy, collection, research, commercialization (f) Mutually agreed terms should be negotiated efficiently and within a reasonable period of time (g) Mutually agreed terms should be set out in a written agreement	Paragraphs (iv)(c) – (g) & 43 & 44 – Regulation 8A.08 establishes the elements for a reasonable benefit-sharing agreement. Regulation 8A.07 provides for the Minister to publish model benefit-sharing agreements as a guide for applicants. Two such model agreements have been published: one for publicly owned areas and one for Indigenous privately owned lands. They are available on line at http://www.environment.gov.au/biodiversity/science/access/model-agreements/index.html
43. The following elements could be considered as guiding parameters in contractual agreements.	Paragraph 43: The Model Agreements take into account all the elements outlined in the Guidelines at 42(c) to (g) and those at 43 and . . .

Relevant Bonn Guidelines provision	Australian law, policy & administration response to Bonn Guidelines provision
These elements could also be considered as basic requirements for mutually agreed terms:	
(a) Regulating the use of resources in order to take into account ethical concerns of the particular Parties and stakeholders, in particular indigenous and local communities concerned	
(b) Making provision to ensure the continued customary use of genetic resources and related knowledge	
(c) Provision for the use of intellectual property rights include joint research, obligation to implement rights on inventions obtained and to provide licences by common consent	
(d) The possibility of joint ownership of intellectual property rights according to the degree of contribution	
2. Indicative list of typical mutually agreed terms	
44. The following provides an indicative list of typical mutually agreed terms:	44. In addition, special facilitating access arrangements for non-commercial uses are introduced at regulations 8A.12 and 8A.13.
(a) Type and quantity of genetic resources, and the geographical/ecological area of activity	

Relevant Bonn Guidelines provision	Australian law, policy & administration response to Bonn Guidelines provision
(b) Any limitations on the possible use of the material	
(c) Recognition of the sovereign rights of the country of origin	
(d) Capacity-building in various areas to be identified in the agreement	
(e) A clause on whether the terms of the agreement in certain circumstances (e.g. change of use) can be renegotiated	
(f) Whether the genetic resources can be transferred to third parties and conditions to be imposed in such cases, e.g. whether or not to pass genetic resources to third parties without ensuring that the third parties enter into similar agreements except for taxonomic and systematic research that is not related to commercialization	
(g) Whether the knowledge, innovations and practices of indigenous and local communities have been respected, preserved and maintained, and whether the customary use of biological resources in accordance with traditional practices has been protected and encouraged	
(h) Treatment of confidential information	

Relevant Bonn Guidelines provision	Australian law, policy & administration response to Bonn Guidelines provision
(i) Provisions regarding the sharing of benefits arising from the commercial and other utilization of genetic resources and their derivatives	
3. Benefit-sharing	Australian response:
45. Mutually agreed terms could cover the conditions, obligations, procedures, types, timing, distribution and mechanisms of benefits to be shared. These will vary depending on what is regarded as fair and equitable in light of the circumstances	Paragraph 45: The model benefit-sharing agreements have these attributes – as required by the Nationally Consistent Approach for Access to and the Utilisation of Australia's Native Genetic and Biochemical Resources.
Types of benefits	Australian response:
46. Examples of monetary and non-monetary benefits are provided in appendix II of the Guidelines	Paragraph 46: The *Nationally Consistent Approach for Access to and the Utilisation of Australia's Native Genetic and Biochemical Resources* requires at *Common Element 9* that all Australian governments consider the range of monetary and non-monetary benefits set out in Appendix II to the Bonn Guidelines.
Timing of benefits	
47. Near-term, medium-term and long-term benefits should be considered, including up-front payments, milestone payments and royalties. The time-frame of benefit-sharing should be definitely stipulated. Furthermore, the balance among near-term, medium-term and long-term benefit should be considered on a case-by-case basis.	Paragraph 47: Timing of benefits is considered on a case-by-case basis.

Relevant Bonn Guidelines provision	Australian law, policy & administration response to Bonn Guidelines provision
Mechanisms for benefit sharing	Australian response:
49. Mechanisms for benefit-sharing may vary depending upon the type of benefits, the specific conditions in the country and the stakeholders involved. The benefit-sharing mechanism should be flexible as it should be determined by the partners involved in benefit-sharing and will vary on a case-by-case basis.	Paragraphs 49–50: The mix of benefits and their distribution are determined on a case-by-case basis by the resource provider. The publication of two forms of model contracts is to demonstrate the range of considerations that may be taken into account in a negotiation and to provide some assistance with the form and structure of an acceptable agreement.[26]
50. Mechanisms for sharing benefits should include full cooperation in scientific research and technology development, as well as those that derive from commercial products including trust funds, joint ventures and licences with preferential terms	In the case of indigenously owned land, the determination of benefits and their distribution are matters for the owners and their statutory advisors.
A. Incentives	Australian Response:
51. The following incentive measures exemplify measures which could be used in the implementation of the guidelines: (a) The identification and mitigation or removal of perverse incentives, that may act as obstacles for conservation and sustainable use of biological diversity through access and benefit-sharing, should be considered	51: Removal of obstacles to access to GR has been undertaken through providing clear terms of access, legal certainty, the publication of model contracts, the creation of national focal points and national competent authorities, an online application facility and the timely processing of applications.

Relevant Bonn Guidelines provision	Australian law, policy & administration response to Bonn Guidelines provision
(b) The use of well-designed economic and regulatory instruments, directly or indirectly related to access and benefit-sharing, should be considered to foster equitable and efficient allocation of benefits	
(c) The use of valuation methods should be considered as a tool to inform users and providers involved in access and benefit-sharing	
(d) The creation and use of markets should be considered as a way of efficiently achieving conservation and sustainable use of biological diversity	
B. Accountability in implementing access and benefit-sharing arrangements	Australian response:
52. Parties should endeavour to establish mechanisms to promote accountability by all stakeholders involved in access and benefit-sharing arrangements 53. To promote accountability, Parties may consider establishing requirements regarding: (a) reporting; and (b) disclosure of information.	52: Reporting provisions are included in permits and benefit-sharing agreements. In addition to a transparent process, the creation of an online publicly accessible register of permits (GRID) assists in promoting accountability and compliance. External auditing of agreements and reporting responsibilities are included in the model benefit-sharing agreements. Paragraphs 53 & 54: Under permit Regulations 8A.13 dealing with the Statutory Declaration made by an applicant of non-commercial use, the user undertakes to report to the access

Relevant Bonn Guidelines provision	*Australian law, policy & administration response to Bonn Guidelines provision*
54. The individual collector or institution on whose behalf the collector is operating should, where appropriate, be responsible and accountable for the compliance of the collector	provider the results of his or her research. Where an agent is used this is to be disclosed.

C. National monitoring and reporting

Australian Response:

55. Depending on the terms of access and benefit-sharing, national monitoring may include:

Paragraphs 55–57: The operation of the ABS system is subject to internal and external auditing and annual reporting. In addition, the operation of GRID allows external, public verification of permits issued, and the material collected together with the terms on which permits are granted.

(a) Whether the use of genetic resources is in compliance with the terms of access and benefit-sharing

(b) Research and development process

(c) Applications for intellectual property rights relating to the material supplied

56. The involvement of relevant stakeholders, in particular, indigenous and local communities, in the various stages of development and implementation of access and benefit-sharing arrangements can play an important role in facilitating the monitoring of compliance

D. Means for verification

57. Voluntary verification mechanisms could be developed at the national level

Relevant Bonn Guidelines provision	*Australian law, policy & administration response to Bonn Guidelines provision*
to ensure compliance with the access and benefit-sharing provisions of the Convention on Biological Diversity and national legal instruments of the country of origin providing the genetic resources	
58. A system of voluntary certification could serve as a means to verify the transparency of the process of access and benefit-sharing. Such a system could certify that the access and benefit-sharing provisions of the Convention on Biological Diversity have been complied with	Paragraph 58: The establishment of the GRID system for transparency is the primary tool for ensuring transparency of the process of ABS. The system demonstrates that the ABS provisions of the CBD have been complied with.

E. Settlement of disputes

59. As most obligations arising under mutually agreed arrangements will be between providers and users, disputes arising in these arrangements should be solved in accordance with the relevant contractual arrangements on access and benefit sharing and the applicable law and practices	Paragraph 59: Dispute settling provisions are contained in benefit-sharing agreements. In addition, reviews of permit decisions are available through the Office of the Commonwealth Ombudsman, litigation under the Administrative Decisions Judicial Review Act and by direct request to the decision maker or to the Minister.

F. Remedies

61. Parties may take appropriate effective and proportionate measures for violations of national legislative, administrative or policy measures implementing the access and benefit-sharing	Paragraph 61: The *Environment Protection Biodiversity Conservation Regulations 2000* set the penalty for offences under the regulations at 50 Penalty Units. In 2008 this amounts to AUD$5,500. However, it should be noted that breaches of the regulations may also involve conduct that involve offences under

Relevant Bonn Guidelines provision	Australian law, policy & administration response to Bonn Guidelines provision
provisions of the Convention on Biological Diversity, including requirements related to prior informed consent and mutually agreed terms	the Commonwealth Crimes Act 1914. Where this is the case, the levels of penalty may be much greater.

Source: Burton, G., January 2009

Comparison conclusion

It is clear from the review of policy, administrative and legislative actions of the Australian government that the ABS system introduced by Parts 8A and 17 of the *Environment Protection Biodiversity Conservation Regulations 2000* follows closely the advice and instruction set out on the Bonn Guidelines for providing countries. Its relative clarity and comprehensive coverage warrant it being regarded as model ABS legislation.

Beyond the Bonn Guidelines: Influences for current debates about the international regime

The Australian regulations also seek to resolve some of the policy difficulties identified in current debates about the nature and scope of an international regime as they impinge on its domestic system.

Derivatives

This concern arises out of the perception that the CBD definition of 'genetic resources' does not allow for control of extracts or components of organisms of value, but which do not have elements of heredity. The Australian solution is to provide a definition for the otherwise undefined term 'access'. Access to biological resources is defined at regulation 8A.03 as:

> *taking of biological resources of native species for research and development on any genetic resources, or biochemical compounds, comprising or contained in the biological resources*

By linking the biological object to the intended purpose of its collection and use, this definition avoids any possible confusion with wild harvest, forestry, commodity trade or other more conventional uses. A common criticism of attempts to expand the ambit of GR to cover derivatives of organisms is that it will have unintended consequences, such as affecting the ordinary trade in products made from natural materials such as wood, honey, essential oils or commodity trade such as fish or livestock. Defining 'access' in the above terms avoids this problem and avoids having to attempt to alter the meaning of the CBD definition of GR.

Moreover, it is within the spirit of the CBD and the intent of Article 15 and within the scope of Article 3, which affirms countries' national sovereignty over their resources.

Finally, this approach has another advantage particularly in their marine jurisdiction. Countries control the use of living resources in their territorial waters and exclusive economic zones, but do not, generally, claim ownership over those resources. A focus on regulating access avoids any need to assert ownership and is therefore consistent with existing bodies of national law regulating the use of natural resources. Such an approach may also have positive implications for the eventual management of living resources in waters beyond national control.

Respect for national sovereignty

The Australian system is only applied to species native to Australia.[27] Species from other countries are not covered. Australia does not seek to take advantage of its possession of foreign species, accidental or otherwise. This accords with the spirit of Article 3 of the CBD. With its high degree of endemism, yet high reliance on imported GR for agriculture, Australia seeks to set a model example in showing respect for the national interests of other countries.

Accreditation of ex situ collections

Objective (l) of the Bonn Guidelines identifies the importance of taxonomy and avoiding action that would damage its conduct. Sub-paragraph 16(a)(viii) provides that special terms and conditions should be established under MAT to facilitate taxonomic research for non-commercial purposes. The Australian federal regulations provide for this responsibility at Division 8A.3. The regulations go beyond protecting the conduct of taxonomy to cover all non-commercial scientific research – subject to certain safeguards. Taxonomic ex situ collections and living collections created to support taxonomy and conservation was also thought to warrant special treatment.

These concerns led to the development of provisions at regulation 8A.05 of the regulations to protect the interests and conduct of those ex situ collections. This development arose during the national inquiry, or

Voumard Inquiry. It had received submissions from a variety of scientific ex situ collections on this issue and recommended creating a special exemption for ex situ collections. The basis on which this is done is innovative. Regulation 8A.05 sets a test for the grant of an exemption from the operation of the regulations: this is whether or not the operation of the collection is administered in a manner consistent with the stated purposes of the regulations. If so, they may be exempted.

Thus, where ex situ collections operate in accordance with existing international and sectoral voluntary schemes for CBD compliance or otherwise meet the terms of the test, then they may be accredited as compliant and made exempt. In this way, collections do not have to deal with the regulatory and procedural burden of two CBD compliance schemes or systems. Moreover, they are able to maintain their existing collaborative systems with similar institutions, while able to demonstrate to third parties that they meet the accreditation requirements of the national law of Australia: a double benefit.

The adoption of accreditation of institutions to an international standard is a concept that ought to be further considered in light of the Australian practical experience.

Legal certainty, compliance and verification

Providing legal certainty for any party considering investing in research and development of GR is important in maximizing the amount of research undertaken and in maximizing the economic value of GR as a vital ecosystem service. A low-cost system of 'virtual' certificates of origin and evidence of legal provenance is one way to do this. The Genetic Resources Information Data Base (GRID) was established under section 515A of the Environment Protection Biodiversity Conservation Act 1999.[28] This database sets out the full particulars of each permit, including its unique identifier, the date of issue, to whom and for what purpose the permit is issued, the species to be sampled and in what amounts. It is accessible online[29] to anyone undertaking legal 'due diligence' testing before investing in research. It accordingly demonstrates, at no cost, where the source material was obtained, from whom and upon what terms. Moreover, GRID also progressively lists the identity of the resulting samples collected and gives each sample its own unique identity. GRID now contains details of thousands of samples.[30]

By verifying compliance with Australian ABS law, the value of any biological discovery is increased compared with any similar discovery based on unverifiable sources with its attendant risks of litigation, damage to shareholder value or even criminal association. Open verification means that no accusations of misappropriation or biopiracy can be made. The reduction in perceived risk facilitates the commercialization of scientific discovery based on natural resources.

It should also be noted that the GRID system implemented by Australia is a low-cost database and one based on open-source software. It was designed at the outset to be shared with other interested jurisdictions after an initial settling-in period.

Its growing use by applicants demonstrates its acceptance and functionality. In 2006, the first year of operation, there was one application; in 2007, 14 applications; while that doubled last year to 28. Currently, applications are being received at the rate of one a week.[31]

Disclosure of source, origin and provenance

A number of countries have introduced, or are in the process of introducing, disclosure requirements in intellectual property applications. These vary in complexity and in their mandatory application. By providing transparency about what material has been collected, by whom, when and where and on what terms, the GRID system supports domestic and foreign researchers to protect the intellectual property in their discoveries in all potential markets. Accurate research record-keeping is an important consideration in patent examiners' intellectual property application assessments.

Given the long research and development time-frames involved in the creation of new products (up to 15 years), the early and permanent records of what was collected for research will assist the researcher or patent attorney in later preparing patent applications. Moreover, in the event of any patent dispute over the identity of the biological resource used to create the invention, the independent existence on GRID of the permit record and of the sample collected is a deterrent to misplaced patent challenges, and a source of confidence to the owner of the IP and indeed to a patent examiner or patent review body.

Confidence about the origins of source material used has the additional value of encouraging innovation by being a disincentive to adopting the alternative route of trade secrets.

Electronic verification facilitates the commercialization process by informing the market about the value of the intellectual property concerned. In addition, transparency provides an important innovation signal to governments responsible for protecting ecosystem services. If governments are clear about which areas are of scientific interest, or are giving rise to the development of new and valuable bio-derived products, then they have a better basis for allocating scarce conservation dollars. This is especially important in the field of micro-organisms that have no iconic status and where community awareness of their contribution to society is limited.

Conclusion

It is clear that the design and operation of the Australian federal ABS system demonstrates, at its highest conceptual level, support for conservation, innovation and economic development by sustaining and nurturing the use of GR as a vital ecosystem service while protecting non-commercial academic science. Its growing success is demonstrated by the steady increase of applications for access to its federally managed GR.

Notes

1 Objective 2.8 of the *National Strategy for the Conservation of Australia's Biological Diversity*; see http://www.environment.gov.au/biodiversity/index.html, accessed 29 May 2009.

2 *Commonwealth Public Inquiry into Access to Biological Resources in Commonwealth Areas*; see http://www.environment.gov.au/biodiversity/publications/index.html, accessed 29 May 2009.

3 See http://www.biotechnology.gov.au/index.cfm?event=object.showContent& objectID=0B674DD3-BCD6-81AC-1871247366BECE18, accessed 24 January 2009.

4 The *Nationally Consistent Approach for Access to and the Utilisation of Australia's Native Genetic and Biochemical Resources* was adopted by the Australian Council of Governments in October 2002. See: http://www.environment.gov.au/biodiversity/ publications/access/nca/index.html, accessed 29 May 2009.

5 Australian Department of Foreign Affairs and Trade. See http://www.dfat.gov.au/ geo/fs/aust.pdf, accessed 29 May 2009.

6 Australian Department of Innovation, Industry Science and Research 2009. See http://www.innovation.gov.au/Section/AboutDIISR/FactSheets/Pages/AustralianBio technologySectorFactSheet.aspx, accessed 29 May 2009.

7 Australian Government, *Geoscience Australia*; see http://www.ga.gov.au/oceans/mc_ LawSea.jsp, accessed 29 May 2009.

8 *Number of Living Species in Australia and the World*, Chapman, A. D. (2005), Australian Biological Resources Survey.

9 This can be downloaded at http://www.comlaw.gov.au/ComLaw/Legislation/ LegislativeInstrumentCompilation1.nsf/0/FAA515B854C46E02CA2570C900200F 31?OpenDocument, accessed 29 May 2009. The regulations deal with many other aspects of environmental management, so care should be taken to download only those parts of interest.

10 301 Control of access to biological resources:

(1) The regulations may provide for the control of access to biological resources in Commonwealth areas.

(2) Without limiting subsection (1), the regulations may contain provisions about all or any of the following:

(a) the equitable sharing of the benefits arising from the use of biological resources in Commonwealth areas

(b) the facilitation of access to such resources

(c) the right to deny access to such resources

(d) the granting of access to such resources and the terms and conditions of such access.

11 This is set out at Section 9 of the Constitution as follows: 'When a law of a State is inconsistent with a law of the Commonwealth, the latter shall prevail, and the former shall, to the extent of the inconsistency, be invalid'. See http://www.comlaw.gov.au/ comlaw/comlaw.nsf/440c19285821b109ca256f3a001d59b7/57dea3835d797364ca 256f9d0078c087/$file/constitutionact.pdf, accessed 29 May 2009.

12 This is a legal document with penalties for dishonesty.

13 Copies available online at http://www.environment.gov.au/biodiversity/ science/ access/model-agreements/index.html, accessed 29 May 2009.

14 This is the Genetic Resources Information Data Base or GRID. See https://apps 5a.ris.environment.gov.au/grid/public/perrep.jsp, accessed 29 May 2009.

15 Verified from discussion with Mr Ben Phillips, the Australian National Competent Authority, 2009.

16 See http://www.environment.gov.au/biodiversity/science/access/index.html, accessed 29 May 2009.

17 This is stated on p9 of the document, *Genetic Resources Management in Commonwealth Areas – Sustainable Access Shared-Benefits*, Australian Government Department of the Environment and Heritage, 2005.

18 See http://www.environment.gov.au/biodiversity/science/access/model-agreements/ index.html, accessed 29 May 2009.

19 Ibid.

20 Chapter 1, paragraphs 1.30–1.38, Inquiry into Access to Biological Resources in Commonwealth areas, available at http://www.environment.gov.au/biodiversity/pub- lications/inquiry/index.html, accessed 29 May 2009.

21 Author's discussion with the national competent authority, 2009.

22 Available at http://www.environment.gov.au/biodiversity/science/access/permits/ apply.html, accessed 29 May 2009.

23 Available at http://www.environment.gov.au/biodiversity/science/access/permits/ apply.html, accessed 29 May 2009.

24 Ibid.

25 Author's discussion with the national competent authority, 2009.

26 Author's discussion with the national competent authority, 2009.

27 Regulation 8A.03(1).

28 Section 515A: Publication of information on the Internet.
Without limiting the operation of section 170A, the Secretary must publish on the Internet each week a list of:

(a) all permits issued or granted under this Act in the immediately preceding week

(b) all matters required by this Act to be made available to the public in the immedi- ately preceding week.

29 See https://apps5a.ris.environment.gov.au/grid/public/perrep.jsp?resetNav=Y, accessed 29 May 2009, and then choose the period of permits to be viewed.

30 See https://apps5a.ris.environment.gov.au/grid/public/cerrep.jsp, accessed 29 May 2009, and follow the same procedure as for viewing permits.
31 Author's discussion with the national competent authority, 2009.

PART FOUR

CORE PROBLEMS OF PROVIDER COUNTRY MEASURES

Chapter 15

PIC in Access to TK in Brazil

Sandra A. S. Kishi

Introduction

Many scholars agree that prior informed consent (PIC) is the key issue in Article 15 of the CBD (Firestone, 2003, p25; Hendrickx et al, 1993, p252). In the Brazilian legal system PIC is required for the authorization of access to associated TK. According to Article 7, II, of Medida Provisória (MP) 2186-16/2001, TK associated to GRs is information or practice – individual or collective – of the indigenous or local community, with real or potential value, associated to genetic heritage. New laws and recent case law in Brazil have turned PIC into a binding regulatory requirement. PIC is not a summary declaration, but a whole procedure with several meetings and discussions. Only then will it be possible to guarantee the self-determination of traditional peoples and the identification in advance of the holders of knowledge and the representatives of a traditional community, according to its own form of social organization, as well as the determination of the geographic origin of the knowledge and the state of the knowledge at the moment of access. Therefore, the PIC procedure consists of a legal link or a bridge between indigenous law and the law of the surrounding society, holding together the overall legal system in a sufficiently flexible and effective manner.

This chapter discusses problems of conceiving PIC of traditional communities on the background of Brazilian law and practice. It advocates for a comprehensive notion of PIC, that ensures the broad participation of the affected communities, rather than a simple declaration by a representative of the community. For this purpose an anthropological study can be crucial.

Prior informed consent

The CBD was ratified by the Brazilian National Congress and incorporated into the country's constitutional legal order as a constitutionally guaranteed fundamental law. It became part of Brazil's legal system following its ratification by the National Congress on 2 February 1994, by Legislative Decree 2, and came into force for Brazil on 29 May 1994. It can therefore be immediately enforced, rather than merely guiding the introduction of implementing legislation. In its Article 15.5, the Convention refers to PIC (a procedure) in contrast to previous acquiescence (an act), which is the term used in MP 2186-16/2001. The use of the word 'approval' of holders of TK (Article 8(j), CBD) is the fundamental reference for PIC.

Brazil's current implementing legislation on this matter (Provisional Measure or MP 2.186-16-2001), in the light of the CBD, requires the PIC of the concerned indigenous community, after consulting the official Indian Affairs agency, when access takes place on indigenous land. In addition, the MP 2186-16/01 requires PIC (prior acquiescence) from the environmental authority for access to a component of the genetic patrimony belonging to a strictly endemic or threatened species; from a responsible agency, when access takes place in a protected area; from the owner of private land, when the access takes place on it; from the National Defense Council, when access takes place in a national-security area; and from the Navy, when access takes place in Brazil's jurisdictional waters, on the continental platform or in the exclusive economic zone.

PIC of traditional communities is needed both for access to community-held associated TK on the genetic patrimony and for access to biological material on the lands they traditionally occupy, considering Article 11, IV (b) of the MP 2186-16/2001. In Brazil, whenever research involves access to associated TK, PIC will be required.

Representation in PIC

In practice, one of the main problems is how to identify the traditional community in the PIC procedure. Brazil has no specific law that regulates the legal capacity of Indians. Brazil's new Civil Code (Article 4 – sole paragraph) did well in recognizing that the state's guardianship over Indians under the Indian Statute does not provide for legitimate representation of their interests and stipulates that 'the capacity of Indians shall be regulated by special legislation'.

Representation is a major legal issue because traditional communities do essentially possess a legal personality, although not precisely as either public law or private law legal persons. It is certain, though, that their right to self-determination is consolidated in the Federal Constitution (Article 4/III).

The legal capacity of traditional communities in Brazil

Indigenous peoples have a constitutional right of self-representation, which should not be exercised by public authorities, particularly by those in the Executive Branch, due to frequently contradictory interests between local communities and the state. They therefore have legal personality to be parties to PIC and to access and benefit-sharing contracts, with no need to be represented by any guardian or caretaker agency and with all due respect for their own traditional methods of choosing their representatives.

Representation of communities in PIC and the practical case of the Krahô tribe

It is interesting to analyse how the non-participation of all representative organizations of ethnic groups holding TK can prejudice the formalization of the benefit-sharing (BS) contract. This was very clear in the actual case of access by the Federal University of São Paulo (UNIFESP) to plants used in medicinal rituals and traditional practices by several ethnic groups of the Krahô peoples in the State of Tocantins. Out of 400 species collected, 138 were scientifically identified as having potential neurological function, and 11 of them have already been targeted for pharmacological and phytochemical studies. The lack of representation of knowledge holders caused a breakdown in the negotiations aimed at drafting the contract. During the access process, two associations said they represented the Krahô peoples: the Vyty-Cati (an association made up of three Krahô villages and other 'Timbiras' peoples) and the Kapéy (an association including several other Krahô villages). Only one of them – the Vyty-Cati – however was initially consulted. The Kapéy did not participate from the beginning in the PIC procedure nor agree to the use of genetic material collected by the UNIFESP based on their uses and customs. As a result, in 2002 the Kapéy decided not to authorize the continuation of the research project and conditioned any further discussions on the prior payment of an indemnification of 5 million reals for moral damages (collection of genetic material in a manner that violated their uses and customs), plus an up-front prospecting fee of 20 million reals (Castilho, 2003, p467). Following new negotiations, an agreement to replace the 25 million reals with a health clinic and a vehicle to be used on the Krahô peoples' territory was reached.

Access to TK that had already occurred was validated by the prior consent of the villages represented by the Kapéy, who also agreed to the continuation of the research with inclusion of ethnic Krahôs using other samples of collected genetic material. However, the procedure was suspended, and the contract on use and BS was not concluded.

This experience demonstrates that, in addition to the lack of debate (Article 11, VI, 6-16/2001) and intense exchanges of information about access in the PIC procedure and the inability of the national legislation to protect this traditional Krahô knowledge, there were no instruments to protect the knowledge, such as inventories (Article 215 para 1 and Article 216 para 1, Federal Constitution/88 and Articles 11/II(d), 14/III(b) and 15/IX(b) of MP 2186-16/2001). Likewise, there was no independent anthropological study, which could have identified all the peoples holding the TK at the beginning of the PIC procedure, prior to access, so as to avoid the exclusion of any village from the exchange of information and mutual agreements. Such a study could also have helped to protect the shamanic practices in the collection and use of the biological resources.

An independent anthropological study is required by Article 4/II of the Genetic Patrimony Management Council (CGEN) Resolution 6/2003 and Article 6 para 2 of the CGEN Resolution 12/2004, and must provide at least the following information: (1) indication of the community's forms of social organization and political representation; (2) assessment of the extent to which the community has been informed about the content of the proposal and its consequences; (3) assessment of the social and cultural impacts arising from the project; (4) detailed description of the procedure used to obtain acquiescence; (5) assessment of the degree of respect for the process in which PIC is to be obtained.

The anthropological study should be based on contributions from various fields of science: sociology, ethnology, ethnobotany, biology, parataxonomy, genetic engineering, ethnopharmacology, biochemistry, law, environmental economics and others that interface with the subject matter. It also includes the participation of the traditional communities that are involved. An anthropological study will determine whether it is the shaman, the chief, the entire community or only part of it that holds the TK, and explain how the knowledge has been held over time.

It is possible that the outcome of the Krahô case prompted CGEN to condition PIC on prior independent anthropological studies in order to ensure due representation of knowledge holders identified in the studies, as provided in Resolution 6/2003 as well as Resolution 12/2004 (Brazil, CGEN).

Restoration in case of procedural failure

If any knowledge-holding traditional people fails to participate in PIC, the omission can be made good by a revision of the consent and the authorization as well as contract add-ons, in addition to possible administrative fines, as provided in MP 2186-16/01 (Article 30). For example, if a traditional people did not participate in a PIC procedure for lack of an independent anthropological study, when there was a resolvable doubt about the true knowledge holders, the case may lead to payment of damages, in addition to re-writing the PIC, the authorization and the contract, if the identification of knowledge holders would have been possible through an anthropological study.

This study can count on the participation of the traditional community. But it is not the anthropological study that will define the group position to the access. This expert opinion does not substitute the process of consent and formation of representation, but it eases the indemnification of the peoples who share the same TK and enhances the knowledge level of the related provider communities about the project content and its consequences. In addition, it helps to identify which common systems of social and political organization are recognizable and applicable in PIC.

Representation and disagreement on consent among providing communities

Once the communities holding the same associated TK have been identified, what happens if one of them consents and the other or others do not? Whose case would succeed? The independent anthropological study and other types of evidence (such as detailed reports) appended to the PIC procedure would be useful to structure an appropriate PIC process. The anthropological study is done in advance and must be presented together with the PIC because its purpose is to prove that the rights to otherness and to self-determination have been observed. It also assures that the rights of traditional peoples are recognized in the PIC process and that their forms of social organization are respected. In addition, it approaches the communities to be informed in a language they can understand about the social and cultural impacts caused by access. It is these rights that form the grounds for refusal of access in case any of the community holding the TK does not consent.

Dissemination of TK

In the example of access to the TK of some medicinal-herb vendors ('*erveiras*') at the Ver-o-Peso Market – who had provided information on the manipulation of the breu branco herb to the Natura cosmetics company, and in which the issue of who the real knowledge holders was raised – we see the purpose of the independent anthropological study as a tool not only for identifying the holders, but also to indicate the origin of the knowledge as well as the nexus between the source and the holders, the *erveiras* vendors, in that particular case. On that subject too, Kleba argues that an anthropological expert opinion is capable of determining whether the knowledge is disseminated or still related to specific communities (Chapter 7 – A Socio-legal Inquiry into the Protection of Disseminated Traditional Knowledge – Learning from Brazilian Cases, in this book). The study also shows the course of transmission of the TK. Successive transmissions of TK cannot erase the line of succession of ancestral knowledge from its source up to the immediate provider, even when the remote origins are not specifically investigated. If the course of transmission of such knowledge from its origin is neglected, the danger arises that in the future there will be no associated TK left. In addition, there will be no chance at all for PIC because with ongoing processes of development and transmission of knowledge, and a succession of academic-scientific publications about it, there will only be ex situ TK.

The 'Oriximiná' case of access: Bioprospecting or access for research in PIC procedure

The current case in Brazil of access to TK of the Quilombola community of Oriximiná, State of Pará, by the Universidade Federal Fluminense – RJ – is being considered as a reference because there was an appropriate anthropological study, an appropriate PIC with information exchange, and an appropriate contract. In this case, the PIC was given as a summary declaration, but embraced the exchange and explanation of information on the project through meetings and discussions.

The anthropological study in the Oriximiná case provided ethnographic data and the anthropological view from field researches, and identified ways of social organization and political representation, as well as the level of the knowledge of the communities' members about the project content and its consequences. For four months there were telephone calls, emails, meetings and conversations until the collective decision of the Oriximiná community was taken to approve the access. A researcher was reported to have said that medicines may be created, of which the Quilombolas would be co-inventors, and someone from the Oriximiná communities answered: 'We think that it's gonna be good ... because it's trying to help to rescue our culture, things that we learned, but were being forgotten'. In this case, the PIC and the agreement did not address the BS methods; they were left for later consideration in an addendum to the agreement once the potential of commercial use would become clear. Nonetheless, in the initial agreement the commercial use on the access had to be specified by the researcher–user. The PIC was limited to stipulate the publishing of the researcher's

PhD thesis and the registration in a film of all the research phases. The PIC could have stipulated the prepaying of benefit shares, a bioprospection tax or the facilitation on the access to biotechnology, but nothing was foreseen in that regard.

Ownership regime over TK

Regime of customary law in the Brazilian juridicial system

In Brazil, associated TK includes individual or collective information or practices (Article 216 of the 1988 Federal Constitution) of an indigenous or local community, with real or potential value, associated with genetic patrimony (Article7/II of MP 2186-16/2001), integrated into Brazil's cultural heritage (MP 2186-16/2001 and Article 215 para 1 and Article 216/I and II of the 1988 Federal Constitution), and with recognition for the indigenous and local community's right to decide upon its use (Article 8 para 1 of MP 2186-16/2001).

The Preamble to Brazil's 1988 Constitution affirms that the country works for 'a fraternal, pluralistic ... society', while its Article 3 sets forth the objective of a free and just society in solidarity, and the quest to reduce social inequities. Article 215 para 1 of the Constitution holds that: 'The State shall protect the manifestations of people's, indigenous and Afro-Brazilian cultures and of other groups that share in the nation's process of civilization'. The nation's cultural diversity has been recognized, including local and indigenous peoples' cultures under the protection of the Constitution. Article 216 of the Constitution lays out the content of Brazil's cultural heritage, which covers forms of expression and the ways of life, creation and thinking of the different groups that make up Brazilian society and the indigenous peoples (Article 231 of the Federal Constitution). We can also mention Article 129/III and V of the Federal Constitution regarding the constitutional functions of the Public Prosecutor's Office to institute civil investigations and public civil suits to protected public and social property, the environment and other diffuse and collective interests, and to defend in court the rights and interests of the Indian populations. A systematic interpretation of these constitutional provisions leads us to conclude that rights regarding associated TK can be legally classified as collective rights or diffuse interests. In a discussion on Article 232 of the Federal Constitution, Luciano Mariz Maia notes that 'Indians individually, or their communities and organizations, have standing to file suit in defence to individual or collective rights, that refer to them, their communities or organizations' (Maia, 1993, p290). In Brazil, diffuse and collective interests are defended in court through class action suits in which the Public Prosecutor is a qualified plaintiff (Law 7347/1985, Article 5).

According to Mazzilli, the publication of studies by Mauro Cappelletti (Mazzilli, 1992, p20) beginning in 1974 gave rise to a critical discussion of the traditional dichotomy between the public interest (individuals versus the state) and private interests (amongst individuals), and to the emergence of an intermediate category of collective interests, the meta-individuals, affecting groups of individuals who have something in common. These trans-individual interests 'refer to holders dispersed within the collectivity' and 'affect a determined (or difficult-to-determine) group of individuals'. This discussion of the content of public interest leads us to Renato Alessi's conception of 'primary public interest (interest of the public good) as opposed to secondary public interest, that is, the way in which agencies of the Administration see public interest' (see Mazzilli, 1992, p20). Diffuse or collective interests are primary public interests. This is why leading jurists in Brazil (Leite, 2003, p242; Mirra, 2002, p12) consider that the environment, for example, is both 'a good for the common use of the people' (Article 225, chapeau, CF88) and also 'diffuse' (Article 129/III, CF88). For Mazzilli, these are 'the most authentic of diffuse interests (the par excellence example being the environment)' (Mazzilli, 1992, p19). Analogous to that logic applied to the environment, TK associated with the genetic heritage is also a diffuse-interest good involving either: (a) the collective interests of a group of persons linked together or to the opposing party by a basic legal relationship (Article 81/II, of Law 8078/92) 'that grants them unity of action and a differentiated legal situation' (Krieger et al, 1998, p207); or else (b) the legally recognized diffuse interests of an indeterminate plurality of subjects who potentially might include all participants in the general community whose rules protect this kind of interest; or, alternatively, (c) homogeneous individual interests held 'by an identified or identifiable person, whose homogeneity with the interests held by other rights holders, considering a common origin, generates a numerous and uniform series of interests of the same nature, allowing for their collective defence' (Mazzilli, 1992, p19). Whatever the practical situation of access to associated TK, there are means available to defend it under one of these three categories of trans-individual interests (Article 129/III, 1988 Federal Constitution and Article 1/IV, Article 5 and 8, para 1 of Law 7347/1985). The possession of these meta-individual interests is legally recognized, allowing for their collective defence and differentiating them entirely from the notions of *res nullius* (with no owner), free access and *res publica* (owned by the state). For these reasons, statutory law in Brazil regarding the legal nature of associated TK is absolutely incompatible with its protection through patents, which are restricted to the individual appropriation of knowledge. That's because, according to the Brazilian laws about intellectual property, the invention patent has three requirements: novelty, inventive activity and industrial application, being the system ruled by the principle of absolute novelty. Article 10 of Law 9279/96 considers that an invention is not discovery, providing that the whole or parts of living beings, biological matter found in nature, genomes or germ plasm of any natural living being and natural biological processes are not inventive activities. The principle of absolute novelty of patented inventions in Brazil precludes any written or oral disclosure. This is not adaptable and does not fit with the nature of TK. The Brazilian law defines TK as the 'information about knowledge or individual or collective practice, associated to GRs, of a native Brazilian or local community' (Article 7, IV, MP 2186-16/2001). This set of knowledge, practices and processes is qualified as 'traditional' due to the way it is transmitted or constructed in each act of delivery to every new generation of these indigenous or local communities.

Disseminated TK in Brazil's draft bill on access

Brazil's new draft bill on access (Article 7/XIX) introduces the concept of TK that is widely disseminated in Brazilian society or, more specifically, 'not recognized as being directly associated with the culture of identified indigenous, Quilombola or traditional communities'. For such TK, the bill provides for 'free use by all'. This rather vague provision gives rise to legal insecurity in practice as to what such TK entails. What would be 'widespread' or 'disseminated' in Brazilian society that is not 'directly' associated with traditional 'culture'? Disseminated in all Brazilian society, considering the whole Brazilian territory? These seem too vague and impracticable.

It would appear that the new draft law refers to another category of knowledge that is part of Brazilian culture shared by the entire society, such as expressions of folklore, which are manifestations of traditional and popular culture. Folklore is defined in the Recommendation on the Safeguarding of Traditional and Popular Culture approved by the 1989 UNESCO General Conference as the totality of tradition-based creations of a cultural community, expressed by a group or individuals and recognized as reflecting the expectations of the community.

Kleba (Chapter 7, in this book) also observes that the draft bill does not provide an answer to the question on how DTK must be for it to be no longer directly associated. In this situation, that juridicial instrument of independent anthropological study could be useful to clarify the link to the primary source of knowledge, or even to clarify that there had indeed been a 'shortcut', that is, discovery of 'a short and rapid way' of arriving at an industrially useful product. Considering this matter, see also Kamau, (Protecting TK Amid Disseminated Knowledge – A New Task for ABS Regimes?, item 4.4, in this book).

The legal nature of PIC

In Brazil, PIC is a rule and not a moral or political proposition. It is required by Article 15 of the CBD, which has been incorporated as a legal rule into Brazilian law. The MP 2186-16/2001 (Article 16/8 and 16/9) established PIC as a prerequisite to authorize access, although under the mistaken nomenclature of prior acquiescence (*anuência* or 'nod of approval'), in disagreement with the terms of the CBD. Consent must be informed. The mere prior acquiescence (a single act in an overall PIC procedure) of associated TK holders, unaccompanied by intense and ongoing exchange of information and an independent anthropological study (see above), is likely to hinder the authorization of access or may cause the cancellation even of authorizations issued by the CGEN itself.

According to the classic doctrine of civil law, consent of the parties is one of the subjective requirements of any contract. Today, under Brazil's new Civil Code (Article 421), the legal system for contracts includes a social function and is guided by the principle of objective good faith (Nery and Nery, 2002, p181). This rule on their social function (Brazil, new Civil Code, Article 421, Law 10406/2002) thus permeates contracts with social purpose, by placing value on probity and good faith as essential elements to convey security in legal relations.

PIC creates a formal mechanism to facilitate the exchange of information amongst all players in the process (traditional community, user, owners of land upon which the GR to be accessed are found, the government, public prosecutors, FUNAI [Indigenous People National Foundation], for example) and to aid in the due process of formalizing the access contract.

There are two legal features which could be used to understand PIC: (1) an offer for a civil law contract, allowing access and ensuring BS (relation between user and community); and (2) a pre-condition for the authorization and for the terms of the Material Transfer Agreement between user and governmental body. PIC can be both these characteristics together. It consists of a type of previous contract, which is a regulatory basis to control access and to put the provider in a better position to negotiate the final terms of the contract itself, as the prevision of CDB/COP/4/23, on 19 February 1998, and as explained by Márcia Bertoldi (2005, p137). In

Brazilian legislation, the MP 2186-16/2001 (Article 11/IV/b) does make the PIC a pre-condition for authorization of access.

If well administered, PIC as a procedure for the exchange of information, data and studies under the aegis of the national development policy for traditional peoples and independent anthropological studies will assure the right to self-determination, the right to deny access, the right to have the communities' traditions and cultural values recognized, the right of peoples to development, the right to be represented in accordance with their own will and forms of organization, and the right to allow their differences to prevail as a fundamental human right – as provided by Article 1/III of the 1988 Federal Constitution. In Brazil, Decree 6040/2007 creates the National Sustainable Development Policy for Traditional Peoples and Communities. PIC, therefore, cannot be dressed as mere acquiescence in which an ungrounded 'yes' or 'no' would be nothing more than a procedural step, forgetting the word 'informed', which is part of the legal order. Otherwise there can neither be legitimacy in the access procedure nor an equitable distribution of benefits. The implementation of the PIC procedure, in the light of existing instruments for protection of TK and its above-mentioned basic principles, makes PIC a tool that catalyses the different interests involved and constitutes the true regulatory link of indigenous or minority law with statutory law.

PIC procedure

The formal features of PIC must ensure that traditional communities have access to information and must facilitate their effective participation, which will of course presume long and intense discussions in the indigenous language or dialect, with the cooperation of professionals qualified in the various interfacing sciences, including anthropologists, ethnobiologists, ethnobotanists, ethnopharmacologists, biologists and also members of the Federal Prosecutor's Office, and of indigenous-support associations or NGOs active in the defence of indigenous peoples' interests. In practice, this entire process will take time and go through iterations at its own pace, as these peoples' traditional cultural values must be respected, including their frequent practice of not naming or electing representatives to speak in their name, and with all decisions taken by consensus (Kishi, 2004, p334). There is no standard form of procedure. There is a minimum of legal and material content, but it is neither standard nor uniform. Each PIC, based on the participating traditional communities, will adopt its own procedure. The CGEN will continue to be responsible for adapting regulations to concrete cases; for example, where the communities for whatever reasons do not want to sign the PIC in spite of their forms of organization, habits and customs having been respected, and them having agreed to participate in the project. At such a time, the CGEN will accept other forms of evidence, such as the independent anthropological study for example, together with a detailed report on the form in which consent is obtained, a term of responsibility of the applicant and the formal opinion of the official indigenous agency (FUNAI), as stipulated in CGEN Resolutions 5/2003 (Article 4) and 9/2003 (Article 6), as amended by Article 1 of Resolution 19/2005.

The costs of the PIC process must be borne by the institution applying for access to the genetic patrimony or to the TK associated with it. This is why PIC must become a due process of participation and feedback, rather than a mere term of acquiescence. It must produce detailed and well-discussed clauses that are the fruit of mutual agreement, assuring the right to refuse access with no onus upon the TK holder.

The legal and material content of PIC

Material content

Laurel Firestone (2003, pp28–48) suggests a structural model for PIC that must contain basic information from the access seeker, who must disclose the nature and the target of the activity, as well as explaining all potential risks that may ensue. These requirements include: (1) disclosure of the project methodology, foreseeable consequences, full identification of the access-seeking legal entity or individual, of the applicant's sponsors and of public or private entities, NGOs and civil associations that are partners or collaborators in the research, as well as possible development of the goods to be accessed; (2) identification of benefits to be shared with the people or person whose consent is being solicited, demonstrating the mechanisms and agreements proposed to share the benefits of access, including an indication of the royalties that the provider will receive for the use of the accessed good; (3) indication of possible alternative activities and procedures; (4) assurance that the provider of the GR or TK will be informed and participate, particularly sharing in discoveries that occur during the course of the access activities, considering the community's willingness to continue its collaboration or not during any further stage of use of the accessed good, in which case an additional PIC will be required; (5) precise information on the intended use and any commercial interest arising from the collection of the desired material, as well as the purpose of the gathering of biodiversity resources and the possible, current and potential uses of the good to be accessed; (6) guidelines the researcher is following and previous practices used in similar projects; (7) assessment of access-related risks and possible environmental and socio-economic impacts, also considering future generations.

Formalities and the legal content of PIC

I would add to those requirements: (1) widespread publicity of the PIC procedure throughout the entire traditional community; (2) the existence of a formal, written instrument in the two languages, if necessary, of the parties involved; (3) during the process of informing the traditional peoples involved in the PIC, organization of discussions and debates in the form of public hearings, with the seeker of access to GRs or to associated TK taking responsibility for organizing and holding these forums or decision-making spaces, so as to assure the full participation of all interested communities that hold TK to be accessed; (4) the presence of interlocutors and of qualified, as well as the Federal Prosecutor's Office and the FUNAI; (5) the PIC agreement must be filed in a public office responsible for protecting traditional community rights and in the public office responsible for authorizing access; (6) recognition, through some form of public record or certification, of the source and the specifications of the GR or of the TK to be accessed, in order to safeguard the analysis and the evolution of the state of the art in later patent requests or for other forms of IPRs, while assuring secrecy regarding the content of the TK; (7) failure to observe any of these prerequisites renders the PIC null when so requested in court either by the indigenous community or by the Public Prosecutor who, when not a party, should act in all stages of the suit as *custos legis*, as provided in Article 232 of the Federal Constitution.

Conclusion

Constitutional rules, the rules of the CBD and of the national law discussed in this chapter are grounds for the immediate adoption of a sui generis system using the legal tools of today's laws in Brazil, which recognize the collective character of IPRs, even when only one individual holds the TK. The features of traditional IPR protection – such as the principle of absolute novelty of patented inventions, which precludes any written or oral disclosure – are not adaptable to and compatible with the nature of TK, which is a public interest good that is inalienable, imprescriptible, unrenounceable and indivisible.

The sui generis system for protecting TK means that PIC can be used to include the entire range of instruments for the protection of TK and of the indigenous or local communities that hold it. In this case, the public authority in Brazil should act in its administrative rather than proprietary capacity, through the CGEN, with the collaboration of stakeholders in the access process – traditional communities, the Public Prosecutors, NGOs and society – to ensure the collective character of intellectual property rights with indications of the source and geographic origin. PIC should preclude any possibility that TK falls into the public domain after a term of IPR protection as patents do, as well as ensure representation of traditional communities through their own forms of organization and following their habits and customs. Under this protection system, the independent anthropological study will effectively ensure the appropriate representation of affected traditional communities and the non-static nature of TK, while it brings to the procedure relevant ethnic, social and anthropological information to identify and protect the specific dynamics of this living knowledge. All this would help to facilitate access to TK and BS through efficient PIC procedures as a pre-contract basis for fair and equitable BS, considering concrete and symbolic cases in Brazil.

References

Bertoldi, M. R. (2005) 'Regulação internacional do acesso aos recursos genéticos que integram a biodiversidade', in Benjamin, A. H. and Milaré, E. (eds) *Revista de Direito Ambiental*, no 39, São Paulo, Editora Revista dos Tribunais, pp127–146

Castilho, E. W. V. (2003) 'Parâmetros para o regime jurídico sui generis de proteção ao conhecimento tradicional associado a recursos biológicos e genéticos', in Mezzaroba, O. (ed) *Humanismo latino e estado no Brasil*, Florianópolis, Fundação Boiteux, Treviso, Fondazione Cassamarca, pp453–473

Firestone, L. (2003) 'Consentimento prévio informado, princípios orientadores e modelos concretos', in Lima, A. and Bensusan, N. (eds) *Quem Cala Consente? Subsídios para a proteção aos conhecimentos tradicionais*, documentos ISA 8, São Paulo, Instituto Socioambiental

Hendrickx, F., Koester, V. and Prip, C. (1993) 'Convention on Biological Diversity – Access to genetic resources: A legal analysis', *Environmental Policy and Law*, 23/6, pp250–258

Kishi, S. A. S. (2004) 'Principiologia do acesso ao patrimônio genético e ao conhecimento tradicional associado', in Barros-Platiau, A. F. and Varella, M. D. (eds) *Diversidade Biológica e Conhecimentos Tradicionais*, vol 2, Belo Horizonte, Editora Del Rey

Krieger, M. G., Maciel, A. M. B. and Carvalho Rocha, J. C. (eds) (1998) *Dicionário de Direito Ambiental, Terminologia das Leis do Meio Ambiente*, Porto Alegre/Brasília, Universidade/UFRGS/Procuradoria Geral da República

Leite, J. R. M. (2003) *Dano Ambiental: Do Individual ao Coletivo Extrapatrimonial*, 2nd edn, São Paulo, Editora RT

Maia, L. M. (1993) 'Comunidades e Organizações Indígenas. Natureza Jurídica, legitimidade processual e outros aspectos jurídicos', in Fabris, S. A. (ed) *Direitos Indígenas e a Constituição*, Porto Alegre, pp251–294

Mazzilli, H. N. (1992) *A Defesa dos Interesses Difusos em Juízo,* 4th edn, São Paulo, Editora RT

Mirra, Á. L. V. (2002) *Ação Civil Pública e Reparação do dano ao Meio Ambiente,* São Paulo, Editora Juarez de Oliveira

Nery, N. Jr. and Nery, R. M. A. (2002) *Novo Código Civil e Legislação Extravagante Anotados,* São Paulo, Editora Revista dos Tribunais

Chapter 16

Design and Functions of Databases on TK – The Case of Venezuela

María Julia Ochoa

Introduction

For centuries, indigenous and local communities have created and developed technologies, uses and practices related to the natural resources existing on their lands, which are usually regions with high biological diversity, such as, the neotropical forests, which are located in South and Central America and the Caribbean and include the Amazon Rainforest. These indigenous communities' technologies, uses and practices are known as traditional knowledge (TK).[1] This knowledge has been useful in developing products and processes mainly by pharmaceutical and agricultural industries, which hold intellectual property rights in order to hinder the production and putting on the market of unauthorized copies. Technologies, uses and practices traditionally developed by indigenous and local communities, on the contrary, do not enjoy such protection.

This chapter will deal with one of the existing options to regulate the access to TK related to GRs: the construction of databases. In several countries, different kinds of databases have been developed and their use has been regulated. Some of these experiences will be briefly described here; so they can be compared with the experience in Venezuela, a neotropical country with large regions rich in biodiversity and significant indigenous and local populations. Some aspects of the regulation of the access to indigenous and local communities' TK related to GRs will be exposed, laying emphasis on the Venezuelan perspective. Specifically, the construction of databases is discussed in the first section. Databases offer the possibility to use the stored information as part of the prior art in patent registration procedures. The stored information can also be treated as a prior requirement for enjoying specific positive rights, so they could be a useful mechanism to protect TK. In the second part of the chapter, the experiences of some countries are described and a remarkable case taken from the Venezuelan experience is explained and analysed. This allows us to make an assessment from a practical, normative point of view.

Databases: Functions and forms

Functions

Databases can offer two kinds of protection: preventive and positive. The preventive protection is related to the use of stored information in patent examination procedures as a source of prior art; that is, as information freely accessible to the public before the patent application has been submitted to the patent office. The positive protection, on the contrary, connects information contained in TK databases to some positive rights granted in favour of the indigenous or local communities that have developed the respective TK.

The preventive protection offered by databases can hinder people from obtaining intellectual property rights on products based on unauthorized use of TK. If the stored TK is treated as prior art, it can be determined whether an invention is really new and involves an inventive step (Leistner, 2004, p59). However, it is important to keep in mind that this is only possible if the information can be accessible to the patent offices during the examination procedure, and the included technical information regarding characteristics and applications of the biological material to which the stored TK relates is sufficient.

The Intergovernmental Committee on Intellectual Property and Genetic Resources, Traditional Knowledge and Folklore of the World Intellectual Property Organization (WIPO CGRTKF) has collected information on worldwide existing databases of TK. The WIPO CGRTKF states that, though many databases do exist and their number is increasing, they are not sufficiently specific and do not contain technical information enough to be used as a source for prior art.[2] If databases cannot offer technical security in the manipulation of collected information, they could lead to the publication of TK that has not been available before, thus destroying the possibility of its protection as a trade secret[3] or by other intellectual property mechanisms.

The positive protection is offered by some legislation in relation to TK contained in databases. Thus, the recognition of positive rights in favour of indigenous or local communities, for example the right to restrict the use of their TK or to share the benefits obtained from the use of TK, has been released to the registration or inclusion of their TK into databases.

These two forms of protection are, in principle, plausible. However, important matters are usually put aside in the discussions held at international level. For example, how can TK, on the basis of its value for developing other products or processes, be separated from knowledge, practices and contexts that are essential for its existence, but which are considered irrelevant and do not deserve legal protection (Agrawal, 2005, p374)? What consequences could this separation have?

In spite of the fact that in discussions towards protecting TK the referred matters are not considered, it is worth describing some national experiences related to TK databases so as to see how they have been operationalized. Some legal systems do not mention using TK databases during the patent examination procedures, although they could be consulted, even when there is no legal obligation for the patent office. However, in some countries this is not possible. In the USA, for instance, TK, which is not known or used in the USA, may be considered prior art only if it has been patented or included in 'a printed publication' (Leistner, 2004, p76). According to US patent law: 'A person shall be entitled to a patent unless -: the invention was known or used by others in this country, or patented or described in a printed publication in this or a foreign country, before the invention thereof by the applicant for patent'.[4]

Forms

Databases of TK existing in a few countries are discussed below. It is interesting to observe which forms this kind of database can adopt and how they can be used in particular cases to implement either a preventive (India, China and the USA) or positive protection (Peru).

India

As stated by the Indian Biological Diversity Act (2002), the respect and protection of the knowledge of local people relating to biological diversity include the 'registration of such knowledge at local, state or national levels, and other measures for protection, including sui generis regimes'.[5] Furthermore, one function of the Biodiversity Management Committee is to record knowledge relating to biological diversity.[6] The Indian Patents Amendment Act (2005) contains a provision similar to the Andean Decision No 486 on intellectual property rights,[7] according to which: 'where an application for a patent has been published but a patent has not been granted, any person may, in writing, represent by way of opposition to the Controller against the grant of patent on the grounds ... that the invention so far as claimed in any claim of the complete specification is anticipated having regard to the knowledge, oral or otherwise, available within any local or indigenous community in India or elsewhere'.[8] However, a relationship between both of these provisions does not exist.

In practice, the Indian government has designed a protection for TK whose main goal is to facilitate its use as prior art in the patent examination procedures, and also to use it in order to create a sui generis protection.[9] The National Institute of Science Communication of the Indian Council of Scientific and Industrial Research has created the Traditional Knowledge Digital Library (TKDL),[10] and it has planned to sign agreements with international patent offices for the use of this database confidentially, so as to hinder unauthorized uses of the collected TK.[11]

China

In Chinese law there are no provisions on the use of databases of TK as prior art. There is, however, a preventive protection through the use of the Traditional Chinese Medicine Patent Database.[12] Due to the fact that the preventive protection of TK is very important in China, where more than 90 per cent of the patent applications are related to traditional medicine, the State Intellectual Property Office has created the Traditional Chinese Medicine Patent Database,[13] which contains literature on traditional Chinese medicine.[14] Furthermore, in the context of special protection of traditional medicine products, the National Committee on the Assessment of the Protected Traditional Chinese Medicine Products has created the Protected Traditional Chinese Medicinal Products Database.[15]

The USA

The creation of databases in the USA is related to trademark regulations. The creation of the Database of Official Insignia of Native American Tribes was recommended in a report required by the Trademark Law Treaty Implementation Act and established by the US Patent and Trademark Office (USPTO).[16] Its main goal is to assist the agency in reviewing trademark applications in order to protect federal- and state-recognized insignia of Native American tribes. It is included within the USPTO's database of material that is not registered, but is searched to make decisions regarding the ability of trademarks to be registered. Thus, if in the examination procedure a sign is found that is confusingly similar to a Native American tribe's official insignia, then the official insignia will be considered before making a determination of its ability to be registered.[17]

In connection with TK-related biological resources, there are in practice some interesting initiatives, such as the Traditional Ecological Knowledge Prior Art Database (TEK*-PAD) of the American Association for the Advancement of Science, which is an index and search engine of an existing internet-based public domain documentation concerning the use of indigenous knowledge and plant species.[18] Nevertheless, the use of such databases is not included in US patent law.

Peru

From a legal perspective, the Peruvian regulations are of interest because they contain a specific positive, as well as preventive protection related to the existence of databases (registers) of TK. The Peruvian Law No 27811 contains three kinds of registers: (1) the public national register of indigenous peoples' collective knowledge,[19] which contains knowledge in the public domain;[20] (2) the secret national register of indigenous peoples' collective knowledge that cannot be accessed by the general public;[21] and (3) the local register of indigenous peoples' collective knowledge that is administered by the indigenous peoples according to their customary norms.[22] The objectives of these registers are to protect the indigenous peoples' collective knowledge and to protect their rights recognized in the law.

On the one hand, indigenous peoples can claim for damages when their collective knowledge in the public domain is published or used without authorization.[23] On the other hand, the patent office is obligated to defend the indigenous peoples when the use of their knowledge is discussed.[24] Additionally, the patent office shall send the information contained within the national public register to the more important patent offices abroad[25] so as to make possible its use as prior art. Although these provisions could be a useful instrument for the protection of TK, their practical application seems to be more difficult than expected. So far the registers established by the Peruvian law have not been satisfactorily created.[26]

The Venezuelan experience

A significant attempt to create a database of TK has occurred in Venezuela. Its results have been, however, not very encouraging. The feeding of this database necessitated access to biological resources of indigenous areas and related TK (it included at the beginning only indigenous Piaroa's lands in the Amazonas state). Concrete reasons for its failure and some guidance to prevent similar situations from happening will be revealed below in the analysis of this case.

Background

Before describing the facts relating to the Venezualan case, it is worth pointing out some legal aspects. Within the regional legal framework applicable in Venezuela – the Andean Decisions No 391 containing the common regime on access to GRs and No 346 containing the common regime on intellectual property – there are no provisions on the creation of databases of TK in order to offer either preventive or positive protection. The Andean Decision No 391 mentions only a register of GRs and their by-products[27] and establishes that any act connected to the access to GRs shall be contained in a register.[28] According to the Venezuelan Law on Biological Diversity, the biodiversity office shall create a database of TK diversity.[29] However, neither regional nor national norms contain specific provisions for the creation of such databases of TK.

In 2000 a project to create a TK database in Venezuela received particular attention. That year the biodiversity office signed an access contract with the Foundation for the Development of Physical, Mathematical and Natural Science (FUDECI),[30] whose main goal was the creation of a TK database so as to hinder the loss of indigenous peoples' information about the use of plants and animals in the production of food and medicaments (Febres, 2002, p108). The database, called Biozulua, was to contain information on taxonomic identification of each collected species as well as their general and scientific names, their active biological components, and their local, national and international uses. The register was meant to include digital data, such as photos and videos of each plant and animal with descriptions of their uses, and was to be made according to the model of the International Committee for Documentation of the International Council of Museums.[31] The implementation of this project began with the inclusion within the designed software of information existing in publications. When the discussions with the indigenous communities took place in 2002, there was a fifth version of the software with 556 registers, 640 photos and seven videos.[32]

Even though the FUDECI researchers showed willingness to work, it was not clear from the beginning of the project how the collected information would be used and there was no authorization of the indigenous groups for access to the TK included in the database. When the communities were consulted in order to obtain their authorization to continue with the collection, there were disagreements between the FUDECI researchers and the indigenous representatives (Castillo, date unknown, p11). The indigenous representatives eventually demanded to have the collection of biological resources on their lands stopped and the database given back. During the discussions between the FUDECI and indigenous representatives, the lack of clear regulations was mentioned as a decisive cause of this situation, which is still unresolved.[33]

Limits and lessons

Obtaining the consent of the communities to access. One of the main problems related to the request for communities' authorization in cases like that described seems to be the time when the authorization should be required. It would appear that in the described case the discussions with the indigenous groups began too late. Everything indicates that if FUDECI had informed the indigenous representatives from the very beginning of the project, they would have been more open or at least more informed during the discussions.

Aspects that could be revised in this case are related to the form and time of getting the authorization to access. According to the mentioned legal framework applicable in Venezuela, each access to biological resources, especially if it involves access to TK, requires the authorization of the involved indigenous or local community. Thus, the communities' authorization is in every case required and, consequently, the absence of such consent would lead to the failure of the access project. Hence, without consent, legally valid access cannot exist.

A fictitious wall. It is clear that, linked to historical reasons and to a growing awareness of their rights, there is distrust among the indigenous and local communities of researchers, scientists and government workers. This lack of confidence leads to a reluctance to give any kind of consent or to enter into any contract involving their knowledge. In some cases, it could erect a fictitious wall between the interests of GRs or TK users in accessing and obtaining their approval, and the desire of the indigenous peoples to prevent any access in a way they consider inappropriate.

Assessment

Capacity building of indigenous and local communities could contribute to preventing similar situations. This should be a process that involves, among other aspects, information about the use of the material or knowledge to be accessed. In this sense, it is a tool that enables the communities' consent, which is highly informed. The consent requirement established in laws like the Venezuelan one constitutes prior informed consent, which is a legal figure introduced by the CBD. This can also be considered, together with the sharing of benefits and the conservation of biological diversity, as an objective of the Convention, since it is the basis of materializing both of these main objectives. However, in relation to access processes, these objectives have not only a teleological character, that is, they should not be considered only as goals. They also have a normative character because they constitute a fundamental normative guidance in the implementation of each access activity. Due to the basic importance of capacity building as the basis of the prior informed consent, a question can be asked at this point: Could capacity building be also considered an objective with a normative character within these processes, in other words, not just a mere requirement? The answer seems to be affirmative.

Prospect

Identifying a priority order of objectives in access processes. Assuming that capacity building is, in addition to prior informed consent, benefit sharing and conservation of biological diversity, an objective with a normative character in any access process, could a priority order of these objectives be identified?

Such a priority order would not be referred to priority in terms of time, but in terms of importance, which would be based on an essentially practical point, that is, the feasibility of access processes. That means, which objective should be achieved before the next objective shall be tackled? So, in answering the question as to which objective should be at the beginning of this priority order, a very simple reasoning could be useful: without capacity building and the information that it implies, there would not be prior informed consent and, without this, there would neither be access nor benefits sharing. Consequently, capacity building should be at the beginning in this priority order. But it does not mean that capacity building and discussions to obtain the authorization should be the first phase of each project. If that is so, it could be difficult because in discussions on authorization there are naturally specific aspects that can be discussed only when the access project does exist and the work has begun, with financial investment, personal efforts, and so on. The fact that capacity building is at the beginning of this priority order means that it should have the consideration of first importance, because if it is not taken into account, it is difficult and in many cases impossible to achieve the rest of the mentioned objectives in a manner compatible with the CBD.

But which particular features should this capacity building have? The capacity building process should be a comprehensive process, which is not limited to teaching activities about the specific characteristics of a particular access process, or about the legal and institutional framework on access to GRs and TK. Even though in access processes we find the same problems existing in the use of current intellectual property mechanisms (i.e. geographical indications) to protect expressions of TK, since the ignorance of relevant laws and the lack of commercial knowledge, as well as of the infrastructure, also become important here, other aspects such as the lack of infrastructure should be taken into consideration. The goal of these capacity-building processes would be to help these communities to use modern instruments for managing the ownership of knowledge in a manner compatible with their values, in order to take benefit of the commercial value of that knowledge or to prevent its use in a way they consider inappropriate (Sunder, 2007, p116). This could only be achieved if capacity building is understood as a permanent process between communities, users and competent institutions, governmental or non-governmental. Moreover, such capacity-building processes would contribute to the strengthening of the relationship between biotechnology and biological resources and related TK as a cooperative relationship, rather than an essential conflict.

Increasing a respectful integration of communities and their knowledge into the state. The comprehensive approach of capacity building exposed here would contribute to emphasize the fundamental role the indigenous and local communities play as citizens in the life of the state they are part of. The approach of capacity building that we attempted to outline is not only comprehensive in terms of content, but also in terms of time. If capacity building is carried out on a sporadic basis, no stable linkage can arise from it. It seems clear that, for example, confidence – whose lack is one problem identified here – can really exist by creating stable linkages. From this perspective, capacity-building processes may also involve an integration process of these communities in the state as a whole. But such an integration is not a simple one: it is necessary to be developed on the basis of respect of cultural diversity and considering the essential value of the knowledge of local and indigenous communities not only for the whole population of their countries, but also for all humankind.

Challenges of this approach. Some tasks related to the implementation of this approach remain naturally thus far as challenges; for instance, questions about who should be responsible for such processes: Should governments, NGOs or each indigenous or local group carry out these processes? Also, there are questions about the required financial support: Should there be a public fund? Is international cooperation necessary? Further, there are questions about which communities will be in the processes: Should every indigenous and local community or only the communities that own TK take part in these processes?

Conclusion

Adjustments in the legal framework, as well as more clarity in current regulations, are necessary. But so far, it is extremely important to think about ways to make the existing principles and legal provisions effective. By considering the capacity building of local and indigenous communities as more than a mere prerequisite for access to biological resources and TK, some problems related to this access could be alleviated; for example, lack of clear rules or communities' reluctance to give authorization for access. Moreover, a comprehensive capacity-building cannot only bring more confidence into the discussions on access processes, but can also help to develop the communities' capacity to exploit their own resources and abilities, and increase a respectful integration of these communities and their knowledge into the state.

Notes

1 By delimiting TK as a subject to be legally protected, several definitions have been proposed. Consensus exists in relation to some aspects, which have been included in a working definition contained in the Revised Draft Provisions for the Protection of Traditional Knowledge, elaborated by the World Intellectual Property Organization. So, TK 'refers to the content or substance of knowledge resulting from intellectual activity in a traditional context, and includes the know-how, skills, innovations, practices and learning that form part of TK systems, and knowledge embodying traditional lifestyles of indigenous and local communities, or contained in codified knowledge systems passed between generations. It is not limited to any specific technical field, and may include agricultural, environmental and medicinal knowledge, and knowledge associated with genetic resources'. See document WIPO/GRTKF/IC/10/5.

2 This study can be found in the document 'Inventory of Existing Online Databases Containing Traditional Knowledge Documentation Data', prepared by the WIPO CGRTKF Secretariat, Geneva, 2002 (WIPO/GRTKF/IC/3/6).

3 This problem can be hindered through databases with limited access. For instance, in the USA, the Tulalip people from Washington created the database 'Story Base' with two different kinds of information. Information 'A' can be consulted only by members of the Tulalip people; information 'B', on the contrary, is public. More information on that database is included in the document WIPO/GRTKF/IC/3/6, pp23–24. In the Peruvian Law No 27811 there are provisions about a public and a secret register (Articles 17 and 18).

4 Title 35 United States Code, Section 102(a).

5 Section 36(5), Indian Biological Diversity Act.

6 Section 41(1), Indian Biological Diversity Act. Additionally, the National Biodiversity Authority and the State Biodiversity Boards shall consult the Biodiversity Management Committees, while making any decision relating to the

use of biological resources and knowledge associated with such resources (Article 41(2)).

7 Andean Decision No 486, Article 75: 'The competent national authority may, either ex officio or at the request of a party, and at any time, declare a patent null and void, where: ... (h) when pertinent, the products or processes whose protection is being requested have been obtained or developed on the basis of traditional knowledge belonging to indigenous, African American, or local communities in the Member Countries, if the applicant has failed to submit a copy of the document certifying the existence of a license or authorization for use of that knowledge originating in any one of the Member Countries'.

8 Section 23, Indian Patents Amendment Act (2005).

9 More detailed information is included in the document 'Safeguarding and Inventory-Making Methodologies', Sub-Regional Experts Meeting in Asia on Intangible Cultural Heritage, Bangkok, 2005, p3, www.accu.or.jp/ich/en/pdf/c2005subreg _Ind1.pdf, accessed 21 January 2008.

10 The origin of the TKDL was the claim of the Council of Scientific and Industrial Research from India for the re-examination of the patent No US 5 401 504, which was granted for the wound healing properties of turmeric. In a landmark decision, the US Patent and Trademark Office revoked this patent after ascertaining that there was no novelty because the innovation had been used in India for centuries. See document 'Safeguarding and Inventory-Making Methodologies', Sub-Regional Experts Meeting in Asia on Intangible Cultural Heritage, Bangkok, 2005, p3, www.accu.or.jp/ich/en/pdf/c2005subreg_Ind1.pdf, accessed 21 January 2008; and Leistner, M. (2004) 'Analysis of different areas of indigenous resources', in Von Lewinsky, S. (ed) *Indigenous Heritage and Intellectual Property. Genetic Resources, Traditional Knowledge and Folklore*, The Hague, Kluwer Law International, p77.

11 Other initiatives in India have been the Health Heritage database of the Unit for Research and Development of Information Products, a member of the Indian Council of Scientific and Industrial Research, the Inmedplan database of the Foundation for Revitalization of Local Health Traditions and the Plants of Ayurveda and Siddha database (see document WIPO/GRTKF/IC/3/6, p17 and annex 2).

12 See document WIPO/GRTKF/IC/5/INF/4, annex p40.

13 The English version of this database is available at http://211.157.104.69/englishver-sion/help/help.html, accessed 24 January 2008.

14 Other records have been made by the Chinese Academy of Science; for example, the Commonly-Used Traditional Chinese Medicines and the Effective Composition of Traditional Chinese Medicine. The Institute of Botany of the Chinese Academy of Science has made a database of plant species of China.

15 More information available (only in Chinese) at www.zybh.gov.cn, accessed 24 January 2008.

16 Section 302(a), Trademark Law Treaty Implementation Act as referred in the doc-ument WIPO/GRTKF/IC/5 /INF/4, p5.

17 More information at http://www.uspto.gov/web/offices/com/speeches/01-37.htm, accessed 2 February 2008.

18 See the document WIPO/GRTKF/IC/3/6, annex 2. In the USA there have also been created databases of TK related to agricultural and rural development – for example, the International Documentation Abstracts of the Centre for Indigenous Knowledge

for Agriculture and Rural Development (see http://www.ciesin.org/IC/cikard/docunit.html, accessed 19 January 2008); as well as databases on biological and GRs – for example, the National Plant Germplasm System of the US Department of Agriculture's Agricultural Research Service (http://www.ars-grin.gov/cgi-bin/npgs/html/index.pl, accessed 19 January 2008).
19 Article 17, Peruvian Law No 27811.
20 Article 15, Peruvian Law No 27811.
21 Article 18, Peruvian Law No 27811.
22 Article 24, Peruvian Law No 27811. The first two registers are administered by the Institute for Protection of Competition and Intellectual Property (INDECOPI).
23 Articles 42–43, Peruvian Law No 27811.
24 Article 16, Peruvian Law No 27811.
25 Article 23, Peruvian Law No 27811.
26 So far, only a few biological resources' names have been listed in INDECOPI's website.
27 Article 8, Andean Decision No 391. In Venezuela, as a former member state of the Andean Community, the Andean legislation on access to GRs (Andean Decision No 391), as well as on intellectual property (Andean Decision No 486), was applicable. However, after the retreat of Venezuela from the Andean Community in April 2006, the regulation of these matters became unclear, although regional norms are still applied in practice.
28 Articles 16 and 21, Andean Decision No 391.
29 Article 54(5), Venezuelan Law on Biological Diversity.
30 FUDECI is a non-governmental organization established in 1973 by the Venezuelan Academy of Science. Its activities have been focused on the conservation of natural resources of the Amazonas state in Venezuela.
31 See document WIPO/GRTKF/IC/3/6, p21.
32 According to Ramiro Royero, who was director of FUDECI and was in charge of running the project, in 'Seminario sobre el desarrollo de un sistema sui generis de propiedad intelectual para la protección de los conocimientos tradicionales indígenas', Tobogán de La Selva, Amazonas, Venezuela (undated manuscript), pp14, 27.
33 See 'Seminario sobre el desarrollo de un sistema sui generis de propiedad intelectual para la protección de los conocimientos tradicionales indígenas', Tobogán de La Selva, Amazonas, Venezuela (undated manuscript), p2.

References

Agrawal, A. (2005) 'Indigenous knowledge and the politics of classification', in Stehr, N. and Grundmann, R. (eds) *Knowledge. Critical Concepts*, London, Routledge
Castillo, O. *Conocimientos tradicionales colectivos de los pueblos indígenas de la Amazonia venezolana: Una evaluación de la implementación nacional de normas y compromisos internacionales sobre conocimiento tradicional relacionado con los bosques y asuntos conexos* (undated manuscript)
Febres, M. (2002) *La Regulación del Acceso a los Recursos Genéticos en Venezuela*, Caracas, Universidad Central de Venezuela

Leistner, M. (2004) 'Analysis of different areas of indigenous resources', in Lewinsky, S. von (ed) *Indigenous Heritage and Intellectual Property. Genetic Resources, Traditional Knowledge and Folklore*, The Hague, New York, Kluwer Law International

Sunder, M. (2007) 'The invention of traditional knowledge', *Law & Contemporary Problems*, Duke Law School, vol 70, pp97–124

Chapter 17

Biopiracy or Fallacy? Identifying Genuine Biopiracy Cases in Ecuador

Monica Ribadeneira Sarmiento

Introduction

With increasing frequency, it is possible to see multiple cases of access to GRs that are qualified as 'biopiracy'. This tendency is increasing so fast that it seems there are no legitimate cases of valid access to GRs. Indeed, many potential access applications are under a shadow of suspicion and doubt.

On more than one occasion, existing reports give the feeling or impression that biopiracy is a political denomination, as well as being vindictive, but not a national or international legal entity that could be presented to courts in order to get reparation or compensation for the country of origin of the GRs. In the available information about access to GRs and benefit sharing, there is a complex mixture of scientific information, political vindications, international law conflicts, national legal-framework gaps, criticisms and different adopted positions, but not serious proposals.

The author believes it is necessary to have a debate and information free of fundamentalism and based on an objective approach, which helps to distinguish the successful and legitimate cases from the irregular and illegal cases.

These and other reasons were the origin of the Ecuadorian Working Group on Prevention of Biopiracy (EWGPB) to which the author belonged between April 2005 and April 2008. This chapter contains the author's independent comments and personal opinions on ways of identifying genuine biopiracy cases and suggests a new approach of doing so, drawing from lessons learned during her work.

Working group on biopiracy prevention

In 2004, the Andean Amazon Biopiracy Prevention Initiative,[1] supported by the International Development Research Centre of Canada (IDRC), was founded. Its aim was to prevent and confront biopiracy through the implementation of the national and international legal framework.[2]

Tasks of the EWGPB

The EWGPB started its work in April 2005 with the followings tasks:

- to analyse cases in order to determine whether they are biopiracy cases from a legal and biological point of view
- to identify options for following legal actions
- to contribute to public awareness on this theme through videos, news and reports.

Development phases of the EWGPB

Two phases can be distinguished in the development of the EWGPB: establishment and commencement; consolidation and spread of information.

First phase: Establishment and commencement

One Ecuadorian NGO related to environmental law was the convenor of the first part of the working of the group. At the beginning, it was easy to find a group of colleagues working on access to GRs, TK protection and conservation who were interested in working pro bono on this topic.

From the onset, the EWGPB fortunately avoided the temptation of having its own definition of biopiracy for a number of reasons. Biopiracy has a lot of definitions. In the author's opinion, there exists an obvious gap of objective legal elements of analysis in these definitions and some of these definitions emanate from moral judgements. When both of these elements are confusing in a misappropriation case, the most likely thing that could happen is that the legal chances of prosecuting within the competent judicial or administrative authority are lost (Ribadeneira Sarmiento, 2007).

The EWGPB agreed that biopiracy is a violation of national access rules; in the case of Ecuador this rule is the Andean Decision 391 on Access to Genetic Resources. Of course, in the absence of national rules, or where they are not determined, it is hard to identify biopiracy as such. Ecuador had made two attempts to develop national rules at the time the EWGPB commenced its activities.

The first phase of EWGPB's work (between April 2005 and 2006) focused on the following issues: ·

Table 17.1 *Composition of the EWGPB*

Names and background of members	Institution	Expertise
Nestor Acosta Biologist, MSc	Catholic University	Natural science research
María Arguello Biologist, MSc	NGO Ecociencia	Biotrade
Ximena Buitrón Biologist	IUCN	Biodiversity
Alba Cabrera Agronomic Engineer	Ecuadorian Institute of Intellectual Property Rights (IEPI)	Patent system and breeders' protection
Martha Carvajal (*)	IEPI	Patent system
Rodrigo de la Cruz Lawyer	Responsible for Intellectual Property Rights and Traditional Knowledge (IEPI)	Indigenous Law
María Fernanda Espinoza Geographer, PhD	Latin American Social Sciences Faculty (FLACSO)	Indigenous public policies at national and international level
José Luis Freire Legal Bachelor	NGO Ecolex	Facilitator
Manuela González Bachelor of Arts	IUCN	International negotiation processes
Manolo Morales Lawyer, MSc	NGO Ecolex	Environmental Lawyer
Mónica Ribadeneira Sarmiento Lawyer, MSc	Private consultancy	Genetic resources and international law
Wilson Rojas Biologists	Ministry of Environment	National focal point of ABS CBD National focal point of Cartagena biosafety protocol
César Tapia (*) Agronomic Engineer	Ecuadorian National Institute of Agricultural Research (DENAREF – INIAP)	Focal point International Treaty on Plant Genetic Resources for Food and Agriculture (FAO Treaty)

(*) Ms Martha Carvajal and Mr César Tapia joined the EWGPB during the second phase

Source: Ribadeneira Sarmiento, M., September 2007

- discussion of terms of reference (ToR) of biological and legal criteria in case selection
- research findings through documentation, hemerographic research and interviews in each case.

The EWGPB had interest in the following cases: (1) ayahuasca (*Banisteriospsis caapi*); (2) *Horchata lojana* and other medicinal plants; (3) Andean bean (common name *fréjol nuña o reventón*, scientific name *Phaseolus vulgaris* subsp. *nunas* (formerly *Phaseolus vulgaris -Nuñas Group-*)); (4) marine micro-organisms, Galapagos Islands, also known as the Venter Case; (5) Amazonian frog (*Epipedobates tricolor*); and (6) medicinal plants in Awá Reserve.

In January 2006, the EWGPB participated in the Fourth Open Meeting Working Group on Access and Benefit Sharing – Convention on Biological Diversity (CBD) in Granada. During its intervention, the EWGPB presented a TV spot and its preliminary results of the research cases in a flyer for free distribution.

Also during this phase, in December 2006, the EWGPB gave to the NGO EcoCiencia, mainly Bio Trade Project and Horchata tea local producers, some recommendations on the procedure to register the brand and protect it, as well as develop their markets.

Second phase: Consolidation and spread of information

During 2007, under the coordination of the Ecuadorian Institute of Intellectual Property (IEPI), the EWGPB was very involved in cooperation within an academic initiative. For this purpose, all the members of the group were invited to give talks in conferences and lectures to the students of the Catholic University in Quito.

At the beginning of 2008, the EWGPB changed its coordination to IUCN.

Now the EWGPB is in the process of contracting an external coordinator to continue developing new activities.

What is and what is not biopiracy? Finding the cases

The selected cases for an in-depth study and analysis were: (1) marine micro-organisms, Galapagos Islands (Venter Case); (2) Amazonian frog (*Epipedobates tricolor*); and (3) medicinal plants in Awá Reserve. Information about these cases is available online at http://www.ecolex-ec.org/ecolex.htm, so the author will not present a summary of each one, but rather the lessons she learned as a legal researcher with the EWGPB until April 2008.

Is biopiracy over-aggrandized or under-aggrandized?

In answering this question, it is necessary to consider that there are common generalizations used to accuse all bioprospecting or biotechnology projects with commercial purposes of biopiracy, mainly when they come from the pharmaceutical sector, as well as food and cosmetic projects. The same happens to basic scientific projects when they involve collection and taxonomy involving DNA analysis.

But from a technical point of view, biological and legal, it is necessary to considerer objective arguments. Some bioprospecting or biotechnology projects (basic research or applied-oriented research), with or without commercial purposes, could involve irregularities; it is necessary to identify them.

Legal methodology: Some tips

Since we are looking at the legal situation, we should avoid coming up with definitions from moral judgements based on an ideal world. It is important to be emphatic about the need to analyse case by case and through this analysis establish the following:

- *The complete chronology of the case.* This is needed to collect and to identify some evidences relevant for the case.
- *The list of all administrative felonies, irregularities and misappropriation.* All cases could include biopiracy and other kinds of felonies, but it is necessary to distinguish all the legal situations. For that reason, it is necessary to identify each one of the incidents because sometimes what initially appears as an irrelevant detail could be the origin of an independent case about an administrative irregularity or some felony. Of course it is important to consider all testimonial evidences or witness statements available because they could be the origin of some new information.
- *Legal analysis.* Each incident should be put to legal test. It is necessary to determine what the legal resources in each case are and also which necessary proofs are available. Even starting administrative actions does not seem to be the most appropriate legal action. In fact, in some cases, due to the lack of options to track the main misappropriation, using other available ways to apply the law and generate public awareness might prove to be more efficient.

As mentioned above, the main problems behind the notion of 'biopiracy', legally speaking, are the inexistence of a legal category and the frequent use of biopiracy as a political denomination. Even when national laws regarding the protection of TK exist,[3] biopiracy is not a legal category. Hence, it is important to follow all the possible legal actions within the national and international legal framework regarding all the incidents in each case.

Some learned lessons from the Ecuadorian experience

International laws such as the CBD, the Convention on Indigenous and Tribal Peoples in Independent Countries (also known as Convention 169 ILO), or the International Treaty on Plant Genetic Resources for Food and Agriculture (known as FAO Treaty[4]) and others, recognize the national sovereignty rights over GRs and some international principles on access. The cases that do not follow these principles are misappropriation cases, which could also include biopiracy cases, administrative felonies, wildlife trafficking and so on.

There is some international level of consensus about the problem, but there are not enough legal instruments to prosecute and to avoid biopiracy cases.

Biopiracy damages sovereignty but also indigenous and local communities' rights, and these are vindications which could be used as arguments, but it is necessary to enforce the legal analysis and argumentation to chase all the irregularities. For chasing them, the author suggests a working group strategy because a working group is able to put together all necessary, qualified or competent expertise better than any other institution could do.

The main lesson to learn about biopiracy is that it could be a crime that can include other felonies. However, for it to be a crime, it is necessary to have an objective legal analysis of it. Regional and national legal instruments must be used as a legal argument during the analysis of the cases. Likewise, it is very important to distinguish the legal argument for a political position.

In working groups, the level of commitment is the keystone of the whole process. It is not only necessary to have lawyers and biologists, it is necessary that they themselves feel part of the team and work with the same level of responsibility. Regarding lawyers, it is relevant to incorporate in the team experts in intellectual property rights, as well as experts in environmental law and human rights.

Monitoring from a legal as well as a biological point of view is important; it is not a specific action, but should be a long-term process involving starting not only the legal actions, but also the public awareness about the topic.

Notes

1 Originally in Spanish: *Iniciativa Andino Amazónica para la Prevención de la Biopiratería.* Information available at http://www.biopirateria.org/spa/, accessed 18 May 2009.
2 Andean Amazon Biopiracy Prevention Initiative. Flyer, 2008.
3 The main legal instruments are the following:
   • *African Union:* African Model Legislation for the Protection of the Rights of Local Communities, Farmers and Breeders, and for the Regulation of Access to Biological Resources of 2000, also known as African Model Law
   • *Brazil:* Provisional Measure No 2186-16 of 2001, Regulating Access to the Genetic Heritage, Protection of and Access to Associated Traditional Knowledge, also known as the Brazilian Measure; original name *Medida provisória sobre o acesso ao patrimônio genético dispõe sobre o acesso ao patrimônio genético, a proteção e o acesso ao conhecimento tradicional associado, a repartição de benefícios e o acesso à tecnologia e transferência de tecnologia para sua conservação e utilização, e dá outras providências*
   • *China:* Patent Law of 2000 and Regulations on the Protection of Varieties of Chinese Traditional Medicine
   • *Costa Rica:* Law No 7788 of 1998 on Biodiversity, original name Ley de Biodiversidad de Costa Rica
   • *India:* Indian Biological Diversity Act of 2002
   • *Japan:* Unfair Competition Prevention Law No 47 of 1993
   • *Peru:* Law No 27,811 of 2002, Introducing a Protection Regime for the Collective Knowledge of Indigenous Peoples Derived from Biological Resources. Original name: *Régimen de Protección de los Conocimientos Colectivos de los Pueblos y Comunidades Indígenas Vinculados a los Recursos Biológicos*
   • *Philippines:* Indigenous Peoples Rights Act of 1997, also known as the Philippines Act
   • *Portugal:* Decree Law No 118 of 2002, Establishing a Legal Regime of Registration, Conservation, Legal Custody and Transfer of Plant Endogenous Material, also known as the Portuguese Law
   • *Republic of Korea:* Korean Unfair Competition Prevention and Trade Secret Protection Law No 911
   • *Thailand:* Act on Protection and Promotion of Traditional Thai Medicinal Intelligence, B.E 2542, also known as the Thai Act
   • *USA:* Indian Arts and Crafts Act of 1990 ('US Arts and Crafts Act'), Uniform Trade Secrets Act of 1979 with 1985 Amendments, also known as the US Trade Secrets Act.
4 Mainly in its chapter on Farmers' Rights.

References

Andean Community, Comunidad Andina CAN (2000) Decisión 391 sobre Acceso a los Recursos Genéticos, 2 July 1996. Available at http://www.comunidadandina.org/normativa/dec/D391.htm, accessed 4 February 2009

Andean Community, Comunidad Andina CAN (2000) Decisión 486 sobre el Régimen Común de Propiedad Industrial, 14 September 2000. Available at http://www.comunidadandina.org/normativa/dec/D486.htm, accessed 4 February 2009

Cháves, J., Vélez, J. y García, P. (2006) 'El acceso ilegal de recursos genéticos y conocimientos tradicionales – Estudio de Caso Colombia, Iniciativa Andino Amazónica de Prevención de la Biopiratería', Documentos de Investigación. Año II, Sociedad Peruana de Derecho Ambiental (SPDA), Lima

ECOLEX, Corporación de Gestión y Derecho Ambiental. Available at http://www.ecolex-ec.org, accessed 30 January 2009

Estrella, J., Manosalvas, R., Mariaca, J. y Ribadeneira Sarmiento, M. (2005) 'Biodiversidad y Recursos Genéticos: Una guía para su uso y acceso en el Ecuador', EcoCiencia, Ecuadorian National Institute of Agricultural Research (INIAP), Ecuadorian Ministry of Environment (MAE) y Abya Yala. Quito

Iniciativa Andino Amazónica para la Prevención de la Biopiratería. Available at http://www.biopirateria.org, accessed 30 January 2009

IUCN, International Union for Conservation of Nature (2005) Report 1er, Regional workshop about Biopiracy and Related Issues, Bogotá. Available at http://www.generoyambiente.org/admin/admin_noticias/documentos_noticias/biopirateria.pdf, accessed 3 February 2009

Ribadeneira Sarmiento, M. (2003) 'El Componente Intangible asociado a los Recursos Genéticos, Una Aproximación Económica', Specialist Thesis, Universidad Autónoma de Madrid, Universidad Complutense de Madrid y Universidad de Alcalá de Henares, Madrid – España. Available at http://www.uam.es/otros/fungobe/doc/Ribadeneira,%20Intangible.pdf (unpublished), accessed 4 February 2009

Ribadeneira Sarmiento, M. (2007) InWEnt Transfer Project: Ecuadorian Working Group on Biopiracy Prevention, Berlin

Ribadeneira Sarmiento, M. (2008) Conservación de la biodiversidad y política ambiental, 6ta. Convocatoria, Premio de Monografía Adriana Schiffrin 2007, Trabajos Premiados, FARN. Artículo 'Biopiratería: de una acusación política a una categoría legal', Buenos Aires – Argentina. Available at http://www.farn.org.ar/docs/p54.pdf, accessed 18 May 2009. Edición impresa Lottici María Victoria et al (2008). Buenos Aires, p87

WIPO (2003) Intergovernmental Committee on Intellectual Property and Genetic Resources, Traditional Knowledge and Folklore. Information on National Experiences with the Intellectual Property Protection of Traditional Knowledge, WIPO/GRTKF/IC/5/INF/2, Geneva

WIPO (2004) Intergovernmental Committee on Intellectual Property and Genetic Resources, Traditional Knowledge and Folklore. Protection of Traditional Knowledge: Overview of Policy Objectives and Core Principles, WIPO/GRTKF/IC/7/5, Geneva

Chapter 18

Sharing the Benefits of Using Traditionally Cultured GRs Fairly[1]

Christiane Gerstetter

Introduction

Maybe at some stage in the future a children's book will be written, telling tales of how medicinal and agricultural plants were discovered in all parts of the world and put to the use of humanity, as a result of the collective efforts of farmers, indigenous peoples and scientists all around the globe. It would contain the story of some tired Indian scientists marching through the Indian jungle and being offered a fruit unknown to them, called 'arogya-pacha', by their local guides – which to the scientists' surprise would relieve them of their fatigue.[2] It would also contain the narrative of a US citizen who bought a bag of yellow beans of a kind that he had never seen before, in a Mexican market, used them for breeding and later obtained various intellectual property rights (IPR) for the bean, which he called Enola, according to his wife's middle name (Tolan, 2001). As different as those stories may be, they raise essentially the same questions on who is allowed to use genetic resources (GRs) such as arogyapacha or the Enola bean and pertinent knowledge, for what purpose and under what conditions. This chapter deals with several issues concerning the protection against unwarranted use of GRs and sharing the benefits from their use.

The perspective of the chapter is that of traditional owners of GRs.[3] Of the different situations that the Convention on Biological Diversity (CBD) seeks to regulate, the chapter thus deals with one specific situation: A user seeks to gain access to and use a GR that is still in situ and is owned by indigenous or local communities in the sense outlined above. It thus deals with the situation *before* ABS takes place and seeks to clarify what legal rules provider countries should adopt for such situations, in order to make sure that benefit-sharing deals are fair and equitable, as mandated by the CBD. Moreover, the chapter focuses on issues surrounding the use of GRs, rather than TK.[4] It relates mainly to GRs that do not simply exist on the territory of indigenous or local communities, but have been used by them for certain purposes, in particular agricultural (such as in the case of the Enola bean) or medicinal (such as in the case of arogyapacha*).* The case of such GRs is, from a normative perspective, less complicated than the one of GRs found on indigenous territories. In the case of used or cultivated GRs, the point that indigenous or local communities should have the ultimate right to determine who is allowed to use the GRs under which conditions is much easier to make.

In the first part of the chapter, the disadvantages and advantages of public and private law approaches in protecting GRs from unwarranted use will be discussed. The second part of the chapter outlines some criteria of what makes a benefit-sharing deal fair and equitable and makes suggestions what rules provider countries should enact. The chapter adopts a theoretical rather than an empirical approach, but occasionally uses existing legal frameworks as examples and takes account of the current negotiations of an ABS regime.

Different approaches to fending off unwarranted use of GRs

It is far from settled what approach biodiversity-rich countries, mostly located in the global South, should adopt to best suit their interests when it comes to ABS from the use of GRs on their territories and used by the people on their territories. There are two main (rather complementary than alternative) models, which are currently debated. The first is restricting access through legal provisions of an administrative nature. Such rules oblige whoever seeks to use GRs in a certain country to obtain the prior consent of the competent state bodies and to sign a benefit-sharing contract for the use of GRs with state bodies. The CBD is based on this public law approach.

Many feel, however, that such national ABS provisions alone do not effectively protect provider states and local stakeholders against either unwarranted use of GRs or the lack of fair and equitable benefit-sharing arrangements (Martin, 2002; Posey and Dutfield, 1996, pp59ff). They suggest using new forms of IPR in order to make sure that GRs are not used against the will of their owners. This may be called a private law approach.[5] Many proponents of this approach opine that traditional IPR such as patents are, in most cases, not useful for protecting traditional owners of GRs. This is most evident in discussions on TK and its protection against misappropriation,[6] but the same rationales apply to GRs. There are several reasons why classic IPR do not constitute an adequate means for indigenous and local communities: Firstly, more often than not GRs and the associated knowledge have been developed collectively and over extended periods of time. Indigenous and local communities regard them as something belonging to them collectively. Individual ownership – which is a concept underlying classic IPR – does not go well with such concepts of collective ownership.[7] Moreover, an unaltered version of a GR may not qualify for patent protection – only minor changes, or, the extraction of certain substances, the description of certain components will earn the 'inventor' a patent – even though the initial knowledge of the usefulness of the GR may be the much bigger creative step (Brouns, 2004, p37). Finally, the process of applying for traditional IPR is frequently much too bureaucratic and expensive for communities from the countries of the global South.

In sum, forms of protection are needed that take into account the collective nature of GRs and are accessible to often poor communities throughout the world, which lack access to or do not wish to use classic IPR systems. The debate on such rights has been going on for at least 20 years (Posey and Dutfield, 1996, pp1–3) and many issues still need to be solved. As many indigenous communities reject the idea of individual, tradable property of GRs, the term 'traditional resource rights' is sometimes preferred in this context. This term also serves to capture the idea that many indigenous communities reject the idea of commodifying nature and its components.[8] In addition, it has also been pointed out that it may not be possible to develop *one* set of collective rights, as the customs, decision-making structures and needs of indigenous communities in different parts of the world vary considerably (Posey and Dutfield, 1996, pp60–61). Moreover, the development of new forms of intellectual property has focused much more on the protection of TK than on rights over GRs, and a wider range of instruments is discussed in this context.[9]

The debate on sui generis rights is difficult not only at the conceptual level, but also at the concrete legislative level. As of today, there are only a few (draft) legislative examples of community rights. For example, the draft Indian Karnataka Community Intellectual Rights Act[10] stipulates in Section 5 that a community 'shall have rights' to all its innovations, and that any user of a community innovation must pay to the local community, which is the custodian or steward of a certain innovation, a sum amounting to not less than 20 per cent of any product or process incorporating the community innovation. The system, according to the draft, would be implemented through a register of both communities and their innovations. The Philippines, in turn, have adopted an Indigenous Peoples Rights Act.[11] This Act uses a combination of techniques to protect indigenous rights. Traditional resource rights are mentioned in Sections 3 and 5 as part of the collective rights of indigenous people. GRs are, however, not cited in this context. In Section 34 indigenous communities are given a right to special measures to develop and protect their GRs. Section 35 makes their prior informed consent a prerogative for access to their biological and GRs. Comparing these two approaches, the Indian draft law provides for more concrete steps for the protection of indigenous communities' rights, while the Philippines' law essentially relies on an access clause as a means for securing the rights of indigenous communities. In both cases, it remains doubtful whether the rules would be legally effective at preventing others from claiming IPR over GRs. So far, there seems to be no fully fledged legal concept for property of GRs (Cabrera Medaglia and López Silva, 2007, p40) and much less for protecting traditional ownership of GRs through sui generis forms of IPR.

In the following, I will compare two situations: In the first one, a country adopts a public law approach. In the second one, a country also puts a system of sui generis IPR in place beside the public law ABS legislation. The former approach is currently the more widespread one. This is probably attributable to both the fact that legally viable concepts for sui generis IPR are not fully developed and the fact that only the public law approach is mandated by the CBD.[12] When comparing the pros and cons of a public law approach over combining a public law approach with a private law approach, it is useful to start from the perspectives of the different actors who are involved in ABS. The most important ones are the provider states, non-state actors in the provider states, that is, the indigenous and local communities mentioned in Article 8(j) of the CBD, and, finally, users of GRs – typically corporations and research institutions from the North.[13] The biodiversity-poor countries, which are home to the prospective users of GRs, normally do not play a major role in ABS agreements, but only in providing (classic) IPR on GRs to inventions based on GRs and in adding mechanisms, such as certificates of origin, to these systems that help to ensure that ABS agreements are con-cluded in the first place. The comparison is undertaken on the basis of the – as of now largely hypothetical – assumption that a type of legally enforceable, sui generis IPR exists, that is, that indigenous and local communities would, having obtained such IPR, have an effective mechanism for protecting themselves against illicit commercial uses and the appropriation of their GRs by others through patents and other classic IPR at their disposal.[14] Consequently, the following comparison is theoretical, rather than empirical.

Comparison from the perspective of traditional owners of genetic resources and advantages from their perspectives

Adopting first the viewpoint of traditional owners of GRs, an aspect which seems very often of primary importance to them, is the right to self-determination with regard to the use of genetic resources that they have traditionally used and cultivated. This is evident from a wide range of statements of farming and indigenous communities and their organizations from all over the world.[15] Those communities tend more often than not to be sceptical about claims of representation by the states in which they live.

In this context, the granting of sui generis IPR could play a positive role. IPR grant the right to decide about the use of GRs to the holder of IPR: in the case of sui generis IPR, the local or indigenous communities. They would thus lay the ultimate right to determine what does or does not happen with a GR in the hands of such communities and thus also give them certain leverage in negotiations. This would, however, not only entail rights for the respective communities, but also entrust them with the task of deciding among themselves whether they want to become holders of such IPR and to comply with necessary procedures to secure the respective IPR – whatever they may be (registering, writing down the existing knowledge on certain GRs and their uses, etc.). Privileges do not come without burdens. Nonetheless, given that the concept of IPR entails the freedom to choose whether or not one wants to become a holder of a certain IPR, it would seem that providing for the opportunity of soliciting such IPR – which entails the option not to claim them – would hold more benefits than disadvantages from the perspective of traditional owners of GRs.

Perspective of provider states

The perspective of provider states coincides to an extent, but in no way fully, with the perspective of the traditional owners of GR. Provider states will typically be interested in having effective mechanisms in place for making sure that they determine who will have access to the country's GRs and make sure that benefits are shared. It is probably not too bold an assumption that state bodies would also be interested in maintaining a degree of control over the entire ABS process, that is, would not want to leave it entirely to non-state entities. This is a more common position with regard to GRs – which states claim to be under their sovereignty – than with regard to TK, which is not amenable to any claims of state sovereignty.[16] On the other hand, a state may also be interested in securing for its population the optimal outcome concerning benefits from the use of GRs on its territory.

In this respect, an approach combining public law regulations on ABS with sui generis IPR would seem advantageous. It would enable a state to maintain a degree of control on who accesses GRs on its territory, while establishing a second mechanism to make sure that benefits are shared. Nonetheless, the state would lose its ultimate and final control over a benefit-sharing agreement. State officials may not like this in settings where relations between indigenous peoples and the state are particularly antago-nistic. Moreover, the respective state would have to invest some resources into establishing a system of sui generis IPR and of registering the holders, which is more than what it needs to do when only public law regulations on ABS are put in place.

Perspective of users

Prospective users of GRs will, typically, be interested in gaining easy access to GRs, in particular those that have been used traditionally and whose properties are therefore known to a certain extent. At the same time, users will usually seek to maximize their gain from the use of such GRs. Even though there may be some well-meaning, entirely altruistic individuals or institutions from user countries, it is reasonable to assume that mostly users would want the providers of GRs to obtain as small a share of the cake of benefits resulting from GRs as possible. Moreover, an important point from the perspective of users is to minimize transaction costs. This entails having to deal with as few parties as possible, knowing whom to address and, in general, having to overcome as few bureaucratic hurdles as possible (Liebig et al, 2002, p50).

From the perspective of users, sui generis IPR would thus rather seem to make things more complicated. First of all, users would not only have to deal with a variety of national ABS regulations, but also with IPR legisla-tion. Moreover, they would also not have to enter into negotiations only with state bodies, but also with the holders of sui generis IPR. Even though the CBD encourages states to involve non-state owners of GRs, it does not make this approach mandatory, so that a public law approach will not necessarily, even though optimally, provide for the participation of non-state actors in the negotiations on the provider side. A private law approach is thus likely to complicate things from the perspective of users of GRs.

Finally, on the basis of the hypothetical assumption outlined above, that sui generis IPR granted by a provider state to the owners of GRs on its ter-ritory would be legally effective in protecting the latter against the unwarranted use of GRs, users would probably be antagonistic towards such IPR. Such rights would, at least when formulated broadly, ex ante pre-vent users from claiming IPR over GRs and their components. Under a pure public law approach, in contrast, parties may agree on a benefit-sharing deal according to which users will not apply for (traditional) IPR over the GRs they seek to explore. Other than under a private law approach, where providers do have leverage to prevent misuse of their GRs, such agreement, depends, however, on the consent and later compliance of a prospective user. It is not mandatory. In summary, users can be expected to be rather hostile towards a private law approach.

How to make sure that an ABS deal is fair and equitable

The CBD contains several articles of relevance to the issue of benefit sharing. The most important one is, evidently, Article 15 CBD. Article 15.7 provides that each party shall take measures with the aim of sharing in a fair and equitable way the results of research and development, and the benefits arising from the commercial and other utilization of GRs. This section is dedicated to the question of what measures provider states can take to make sure that an ABS deal is fair and equitable. I will first discuss the meaning of the terms, followed by suggestions on which administrative rules states should adopt to ensure the fairness and equity of ABS arrangements.

The meaning of 'fair' and 'equitable'

Before discussing *how* to make sure that ABS deals are fair and equitable, it is necessary, of course, to know *what* is meant by these terms. Some scholars hold the view that the two words 'fair' and 'equitable' are synonyms (Henne, 1998, p174). Others attribute slightly different meanings to the two terms (Smagadi, 2006, p265). Some of them contend that 'fair' relates more to the process of benefit sharing, while 'equitable' relates more to its outcome (Herold, 2003; Tvedt and Young, 2007, pp89ff). Tvedt and Young (2007, pp89ff) identify some elements of equity which are known in international law. According to them, equity, as understood in international law, entails recognition of the historic contribution that provider countries and their population have made to the conservation of GRs. Moreover, it implies that no one should be able to earn a benefit that was essentially generated by another person. Even though these elaborations make the connotations of the terms clearer, the terms remain rather vague when one only considers the level of textual interpretation.

It is therefore helpful to take a look at the concept of justice – which is behind the concept of 'fair and equitable' – in the context of using GRs. What is just when it comes to sharing in an international context is, of course, a hotly debated topic, and solving it is much beyond the scope of this chapter. I will only offer some preliminary comments on the issue.

Many authors agree that justice has a procedural dimension to it. In simple words: the ones that are affected by a decision must have a say in what the decision looks like. This idea ultimately lies behind the prior informed consent requirement contained in the CBD. As the situation in which the sharing of benefits is discussed by definition involves negotiations and thus the participation of at least some of the people affected by a decision under which conditions GRs may be used, ABS negotiations will always fulfil certain standards of procedural justice. The concept would, however, require, that *everyone* affected by a decision would be able to participate in the negotiations. This also implies that people are allowed to speak for themselves; that is, indigenous communities are not represented by the state (Brouns, 2004, pp40–41). The concept of procedural justice also implies that this participation is meaningful; in other words, the negotiating parties are not only represented, but they also have a chance to actually influence the outcome of the negotiations. This will require mechanisms to level inequalities in knowledge and power. Some authors, in this context, have suggested that the requirement in Article 8(j) of the CBD to seek the 'approval' of indigenous and local communities when using their knowledge entails an obligation on state bodies to explain to them with a particular degree of diligence and care how, under which conditions and to what purpose their knowledge would be used (Henne, 1998, pp53, 155). Even though this norm only refers to the use of knowledge, not of GRs, a similar rationale applies when it comes to using GRs. Finally, procedural justice also entails the right of each of the parties involved to abort negotiations at any stage of the negotiations process. This also means that provider states and communities in those states should have the right to deny access in individual cases.

Justice does, however, also have another dimension to it: distributive justice.[17] This dimension relates to how much different parties are affected by the use of the GR and – in the context of benefit sharing – how much benefits the negotiating parties get from the use of GRs.[18] What is 'fair and equitable' from this angle is certainly much harder to capture than the procedural dimension. Certainly, there cannot be one standard agreement that is just in all situations, but every situation has its own standards of what is fair and equitable. The questions that need to be answered in each situation are what the benefits to be shared are and how they are distributed among the parties (Blais, 2002, p150).

Concerning what needs to be shared, the answer from a justice perspective would, obviously, be that *all* benefits resulting from the use of a certain GR should be shared. In this context, the idea that historic responsibilities and contributions have to be taken into account when marking out the equilibrium of rights and duties of actors in international environmental law has gained prominence. It is more explicitly written in some other multilateral environmental agreements than in the CBD,[19] but it is also reflected in the CBD, for example in the idea itself that benefits must be shared. Tvedt and Young identify the recognition of historic contributions as one element of equity in international law (pp89–90).

Concerning how the benefits are shared, an important idea that has been put forward is that the benefit-sharing process, in order to be fair and equitable, should be used to strengthen the weaker part in the process in a long-term perspective. This is an aspect of justice attributable to John Rawls as a general guideline for political decisions (Rawls, 2005, p14), but is also used in the context of ABS decision-making (Herold, 2003). Moreover, it has also been argued that environmental justice requires giving the rights of local communities priority over other interests in conflicts over natural resources (Sachs 2003, p39). In addition, even though a one-size-fits-all approach is certainly not possible when talking about substantive justice, it is possible to outline some of the aspects that need to be taken into account in concrete cases. Notably, the share of the benefits that providers should be given should be larger where their contribution to the ultimate usefulness of a GR (e.g. through cultivation) is larger, or where use by others reduces the usefulness for the traditional owners of the GR (e.g. because the price of the relevant resource increases on local markets).

Finally, one important element, to indigenous and local communities, is the recognition of their contributions towards preserving or developing a certain GR; awarding this recognition is considered by some as an important part of making a benefit-sharing agreement fair and equitable (Henne, 1998, p177). Recognition lies at the interface between procedural and distributive aspects of justice. Saying that procedural justice entails the right of someone affected by a decision to take part in the decision-making partly implies a recognition of that person. On the other hand, recognition may also be a compensation for someone's contribution and thus also contributes to substantive, distributive justice.

Altogether, it is obvious that rules seeking to make sure that a benefit-sharing deal is fair and equitable will necessarily have to contain a degree of vagueness when it comes to the substantive, distributive side of justice. What is more, over-regulation in this context may lead to injustice, as abstract rules may not be adequate to the concrete situation and the interests of the parties involved. This, however, makes procedural safeguards all the more important.

Which rules to adopt to reach fairness and equity in ABS arrangements

Which rules, then, should states that design a national framework on benefit sharing draw up in order to make sure that the criteria of justice laid out above are honoured in benefit-sharing deals?

Procedural justice
Concerning, first, the procedural side of justice, it is of primordial importance that the provider state does not decide about benefit-sharing

arrangements all by itself, but that it is made sure that indigenous and local communities are equal partners in benefit-sharing negotiations. It is not entirely clear if this is actually mandated by norms of the CBD other than the 'fair and equitable' requirement. Article 8(j), which mentions local and indigenous communities as stakeholders, only speaks of their approval concerning the dissemination of their knowledge, but is silent as to the use of GRs themselves. This is in line with the overall approach of the CBD which confers the sovereignty over GRs to provider states, not sub-state actors. Moreover, Article 8(j) is formulated rather as a guiding principle than as a binding imperative (Hahn, 2003, p298). That does not, however, speak against a regulatory approach which makes the consent of stakeholders within the provider states mandatory as a pre-requisite for a benefit-sharing deal. Some examples may be found in existing legislation. One example is the one from the Philippines, cited above, where the law sets forth that indigenous communities must give their consent on any access decision. A similar clause could be easily formulated with regard to ABS.

As pointed out above, the participation of traditional owners of GRs must be a meaningful and informed one. Thus, any national legislation should contain procedural rules that make sure that non-state actors participating in negotiations are informed about their rights and what is at stake, and thus have the full capacities to take part in negotiations. Depending on the country-specific situations, this may require rules providing for translation of relevant documents in indigenous languages, or for high-quality training or pro bono legal advisors to such communities. In this context, standard benefit-sharing agreements, such as the standard material transfer agreement provided for in the International Treaty on Plant Genetic Resources for Food and Agriculture (ITPGRFA), may help to level inequalities in power and knowledge when they serve as rather 'impartial' model agreements. They, however, would also have to be just themselves, that is, have to be concluded with the meaningful participation of relevant non-state actors, a condition which was not fulfilled by the ITPGRFA standard material transfer agreement (Herold, 2003).

Such rules are not problematic in the light of CBD provisions. This, however, is less clear for rules protecting the right of actors to say no and abort negotiations on a benefit-sharing deal, at least when a benefit-sharing arrangement is a pre-condition of access to GRs. As some authors have pointed out, Article 15.2 of the CBD imposes some limits in this respect. A general prohibition of accessing GRs of a certain territory or imposing prohibitive conditions would probably contradict the CBD's imperative to facilitate access (Henne, 1998, p148; Lochen, 2007, p120). However, the CBD does not prohibit provider states from denying access in individual cases (Henne, 1998, p201; Lochen, 2007, p123) – otherwise the requirement that consent of the provider state be given in advance and that benefit sharing must be on mutually agreed terms would not make sense. Countries that seek to make ABS deals more fair and equitable should thus seek to use the leeway granted by the CBD in favour of maintaining as much power of decision for themselves and communities under their jurisdiction, and explicitly including reference to the possibility of not granting access (Glowka, 1998, p9). This approach is entirely consistent with the CBD. As some authors have pointed out, Article 15.7 of the CBD can be read to imply that agreement on the sharing of benefits is a condition for providers to grant access (Lochen, 2007, p128). As Article 15.1 CBD allows members to delegate the right of deciding on access to non-state entities, such entities would also have the right to say no to the extent that representatives of the respective provider state could do so (Henne, 1998, p148).

Substantive justice

As pointed out above, developing general guidelines of securing benefit-sharing agreements that live up to the standards of distributive justice is much harder than identifying procedural safeguards, which should be incorporated into national law. This case-by-case approach is reflected by the fact that in national ABS legislation there are hardly ever any regulations on precisely what form a benefit-sharing deal is supposed to take. Equally, Annex II of the Bonn Guidelines is restricted to giving examples of how benefit sharing may take place. Moreover, there is also a major debate at the international level about differentiated obligations among user groups, mainly between commercial and scientific users. The rationale behind this debate, that scientific users researching for purposes other than profit making should have lesser responsibilities in benefit sharing than commercial users, should be taken seriously. Apart from the diversity of indigenous communities and socio-economic situations in different countries of the world, this rationale constitutes another reason why a one-size-fits-all approach is not appropriate in benefit sharing. Nonetheless, there are some overarching aspects which apply irrespective of the concrete situation.

A relatively easy aspect is the one of recognition. It is relatively easy because it usually does not cost users a lot. In order to make sure that the efforts of traditional owners of GRs are recognized, provider states should adopt rules that make the mentioning of the contribution of the respective indigenous peoples and local communities mandatory when the GRs are used for any public purpose, commercial or scientific. They should, of course, only be mentioned when it is their wish. In what form the mentioning takes place depends on the use to which the GRs are put. Examples would include references in footnotes of scientific publications or on the labels of products developed from a certain GR.

Concerning the question of how benefits need to be shared, states should and could make sure that there is a minimum sharing of benefits by, for example, incorporating clauses into national legislation which enumerate certain sub-categories of benefit sharing, and make it mandatory for parties to discuss each of these categories and provide evidence of the discussion in the agreement.[20] The Indian model cited above, where a certain, fixed percentage of gross sales would always have to be paid, seems less suitable for constituting a fair and equitable deal for all situations, in contrast. Finally, the idea that benefit-sharing agreements should serve to improve the position of the weaker part in the negotiations should lead to including clauses leading to long-term improvements in the provider country and, of course, for indigenous and local communities. A benefit-sharing deal that leads, for example, to better education for indigenous people can thus be considered to be closer to the ideal of fair and equitable than one which leads to the short-term transfer of sums of money. An ABS agreement which leads to long-term improvements for marginalized parts of the population in provider countries in turn would be preferable to one which benefits local elites. The idea is partially reflected in Articles 16 and 19.2 of the CBD. These give examples of types of benefit sharing that are likely to generate long-term effects. A suggestion for provider countries in this context would be to make mandatory the transfer of certain kinds of benefits for every benefit-sharing agreement; for example, to insist that some sharing of scientific knowledge and training of local scientists will take place.

Conclusion

This analysis has shown that there are some mechanisms which can be used to ensure that benefits from utilized GRs are shared fairly. It is possible to formulate some standards which national ABS rules should fulfil in order to ensure that ABS satisfy basic standards of justice: From a procedural perspective, it is indispensable that *all* relevant actors are involved in an informed, meaningful way and that they have the right to say no. From the perspective of substantive justice, it is important that a benefit-sharing deal contains mechanisms for empowering the weaker party. Moreover, it should be made sure that the contribution of providers of GRs is acknowledged publicly, in adequate forms. Nonetheless, the quest for justice is never an easy one and transforming principles of justice into concrete rules remains a difficult task. To what extent such mechanisms are established and used effectively will also decide what characters people such as the Indian scientists or the US farmer mentioned in the introduction will take in a children's book: heroes or villains.

Notes

1   I am indebted to Gerd Winter for having been the one to first stir my interest in the issues discussed in this book through a seminar held many years ago at Bremen University. Likewise I gratefully acknowledge his comments and those of Evanson Chege Kamau on this chapter. I also wish to express my gratitude to the members of the BUKO Campaign against Biopiracy (www.biopiraterie.de), with whom I have been discussing many issues to which this chapter is related over the past few years. Even though they may not support everything said in this chapter, they have definitely left their mark on my thinking on these issues. All shortcomings remain mine, obviously.

2   This story is told by Bijoy (2007). The author also reports the continuation of the story.

3   The term 'owner of genetic resources' in this chapter refers to any kind of right over GRs. It does not imply property in a strict legal sense. It thus denotes a set of very different rights to determine what happens to a certain GR, ranging from the sovereignty of a certain state over the GRs on its territories to the moral–political 'right' of traditional and local communities who have used, explored and developed certain GRs over long periods of time. The Indigenous Council on Bio-colonialism, a network of various indigenous organizations from all over the world, emphasizes that indigenous rights over GRs should include rights over GRs that are associated with indigenous knowledge, but more broadly over all GRs that originate in indigenous territories, lands and waters, whether or not associated directly with indigenous knowledge. See Collective Statement on an International Regime on Access and Benefit Sharing, www.ipcb.org/issues/agriculture/htmls/2007/unpfii6_ABS.html, accessed 5 January 2009. 'Traditional' refers to what the CBD calls indigenous and local communities, that is, individuals and groups of people who for extended periods of time have used certain plants or animals and accumulated a huge body of knowledge on how to use them.

4   Frequently, the GR itself will be the object of users' interest in the agricultural sector, as an input for breeding efforts. In the pharmaceutical sector, the second-largest sector where GRs are a valuable input, TK about the properties of medicinal plants plays a much larger role in comparison.

5   When using this label, it should be noted, however, that private law is in a deeper sense also public law, that is, state-created law which is used to create and break markets and provide the framework conditions for stable market transactions; in other words, capitalist economy.

6   See, for example, Ramsauer (2005, p61ff).

7   See Posey and Dutfield (1996, pp60–61) for a description of such systems of ownership.

8   See Halbert (2005, pp153–159).

9   See Posey and Dutfield (1996, Chapter 8), Martin (2002, pp68ff), and Glowka (1998, pp38–43), for overviews.

10 Available at www.grain.org/brl/?docid=916&lawid=1352, accessed 5 January 2008.

11 Available at www.grain.org/brl/?docid=801&lawid=1508, accessed 5 January 2008.

12 Article 15.1 in particular stresses the role of contracting parties in granting access to GRs.

13 Brouns (2004, p30) mentions imaginary 'citizens of the world' whose interest is the conservation of biodiversity as a fourth group of actors. This may be useful when the design of more general rules for the protection of biodiversity is discussed, but such citizens of the world do not sit at the table when a benefit-sharing deal is struck.

14 User state tort and IP law could be alerted to sanctioning breaches of the envisaged ownership of traditional GR. See Godt, in this book, for suggestions in this direction.

15 See, for example, the Declaration on Indigenous Peoples' Rights to Genetic Resources and Indigenous Knowledge, signed by about 50 indigenous organizations in 2007, at www.ipcb.org/resolutions/htmls/Decl_GR&IK.html, accessed 5 January 2008.

16 See Firestone (2003, p174), who cites Brazil as an example of a state which is particularly keen on exercising its sovereign rights over GRs.

17 See, for environmental law, Czarnecki (2008, in particular pp192–196), for the use of GRs.

18 Authors that invoke both a procedural and a substantive dimension of justice include Blais (2002, p154ff) and Tvedt/Young (2007). At a more general level, it is of course controversial if procedural safeguards are enough to make a decision just (as the Habermasian school would argue) or whether criteria of substantive justice must be satisfied in addition, as for example Czarnecki (2008, pp78–80) maintains. There are good arguments, in my eyes, for the latter position, at least in the absence of the empirical conditions for an 'ideal' discourse.

19 See, for example, Article 4 UNFCCC.

20 The idea of minimum requirements is also discussed in the current negotiations on an ABS regime; see UNEP/CBD/COP/9/6, Section III.2.

References

Bijoy, C. R. (2007) 'Access and benefit sharing from the indigenous peoples' perspective: The TBGRI-Kani "Model"', *Law, Environment and Development Journal*, vol 3, no 1, www.lead-journal.org/content/07001.pdf, accessed 31 May 2009

Blais, F. (2002) 'The fair and equitable sharing of benefits from the exploitation of genetic resources: A difficult transition from principles to reality', in Le Prestre, P. G. (ed) *Governing Global Biodiversity – The Evolution and Implementation of the Convention on Biological Diversity*, Hampshire/UK, Ashgate, pp145–157

Brouns, B. (2004) *Was ist gerecht? Nutzungsrechte an natürlichen Ressourcen in der Klima- und Biodiversitätspolitik*, Wuppertal/Germany, Wuppertal Institute

Cabrera Medaglia, J. and López Silva, C. (2007) *Addressing the Problems of Access: Protecting Sources, While Giving Users Certainty*, Gland/Cambridge, IUCN

Czarnecki, R. (2008) *Verteilungsgerechtigkeit im Umweltvölkerrecht*, Berlin/Germany, Duncker & Humblot

Firestone, L. A. (2003) 'You say yes, I say no. Defining community prior informed consent under the Convention on Biological Diversity', *Georgetown International Environmental Law Review*, vol 16, pp171–204

Glowka, L. (1998) *A Guide to Designing Legal Frameworks to Determine Access to Genetic Resources*, Gland/Cambridge/Bonn, IUCN

Hahn, A. von (2003) 'Implementation and further development of the Biodiversity Convention – Access to genetic resources, benefit-sharing and traditional knowledge of indigenous and local communities', *Zeitschrift für ausländisches öffentliches Recht und Völkerrecht*, vol 63, pp295–212

Halbert, D. J. (2005) *Resisting Intellectual Property*, London, Routledge

Henne, G. (1998) *Genetische Vielfalt als Ressource*, Baden-Baden/Germany, Nomos

Herold, B. (2003) 'Fair and equitable benefit-sharing within the International Treaty on Plant Genetic Resources for Food and Agriculture', Conference Presentation, http://www.syngentafoundation.org/db/1/414.pdf, accessed 31 May 2009

Liebig, K., Alker, D., Chih, K., Horn, D. and Wolf, J. (2002) *Governing Biodiversity – Access to Genetic Resources and Approaches to Obtaining Benefits from their Use: the Case of the Philippines*, Bonn/Germany, German Development Institute

Lochen, T. (2007) *Die völkerrechtlichen Regelungen über den Zugang zu genetischen Ressourcen*, Tübingen/Germany, Mohr Siebeck

Martin, P. C. (2002) *Providing Protection for Plant Genetic Resources – Patents, Sui Generis Systems, and Biopartnerships*, New York/The Hague/London, Kluwer Law

Posey, D. A. and Dutfield, G. (1996) *Beyond Intellectual Property: Toward Traditional Resource Rights for Indigenous Peoples and Local Communities*, Ottawa/Canada, International Development Research Centre

Ramsauer, T. (2005) *Geistiges Eigentum und kulturelle Identität*, Munich/Germany, Beck

Rawls, J. (2005) *A Theory of Justice*, Cambridge, MA, Re-print, Harvard University Press

Sachs, W. (2003) *Ökologie und Menschenrechte – Welche Globalisierung ist zukunftsfähig?*, Wuppertal-Papers no 131, Wuppertal/Germany, Wuppertal-Institute

Smagadi, A. (2006) 'Analysis of the objectives of the Convention on Biological Diversity: Their interrelation and implementation – guidance for access and benefit sharing', *Columbia Journal for Environmental Law*, vol 31, no 2, pp243–284

Tolan, S. (2001) 'A bean of a different color', www.biotech-info.net/different_color.html, accessed 24 December 2008

Tvedt, M. W. and Young, T. (2007) *Beyond Access: Exploring Implementation of the Fair and Equitable Sharing Commitment in the CBD*, Gland/Cambridge/Bonn, IUCN

Chapter 19

Streamlining Access Procedures and Standards

Evanson C. Kamau and Gerd Winter

Introduction

Whereas Article 15.1 of the Convention on Biological Resources (CBD) recognizes the sovereign rights of states over their natural resources, as well as their authority to determine access to GRs subject to their national legislations, Article 15.2 requires that contracting parties facilitate access to GRs and do not impose restrictions that run counter to the objectives of the CBD. Article 15 tries to engage both providers and users to collaborate in order to achieve mutual benefits for both parties, as well as benefits for the environment. Such collaboration, however, seems still far from being achieved. A mismatch of expectations has largely led to a deadlock; part of which is related to access procedures. In many cases, provider regimes have created too many constraints, making access to GRs extremely strenuous. In some instances, users have opted for synthetic raw materials in place of biological ones. Lack of legislative capacity has also contributed to the constraints. Users have also contributed to the stalemate. Until now, no user country has implemented the Article 15.7 obligation by putting in place 'legislative, administrative or policy measures with the aim of sharing in a fair and equitable way the results of research and development and the benefits arising from the commercial and other utilization of GR with the Contracting Party providing such resources'. To protect their interests, many provider countries have one-sidedly opted for stringent measures. As a result, they introduce often-insurmountable conditions for access.

This chapter investigates procedural and substantive requirements of access authorizations with a view to identify unnecessary transaction costs and suggest possibilities of streamlining procedures and criteria. Access regulation in Kenya is taken as an example, but represents many others. Upon analysis of the main shortcomings of the regime, the authors suggest how the situation could be improved by simplified procedures and criteria.

They recommend taking the integrated permit used in some countries in environmental law as a model.

In this chapter we suggest that access procedures could be simplified without reducing the provider states' control of the access and participation in benefits.

Sovereignty over GR: A right with obligations

Prior to the CBD, GRs were widely regarded as a common good (Sampath, 2005, p127). In other words, they were believed to be an inheritance of all mankind. The CBD made it clear that they fall under the territorial sovereignty of individual countries where they are found (Preamble; Article 15.1). Some authors argue that the CBD did not bring any significant change. Ruiz Muller (2003), for example, says that the CBD just reaffirmed and expressed 'in an unambiguous manner, a right that, theoretically, the States had always had and had never lost'. According to him, the quasi-erroneous, international customary notion that GRs were *res nullius* bred the impression that GRs were 'something over which everybody and, at the same time, nobody had rights' (Ruiz Muller, 2003). For one section of GR, however, there was a formal statement in the nonbinding International Undertaking on Plant Genetic Resources for Food and Agriculture (1983) that plant GRs are a common heritage of mankind, a provision which gave rise to a *res nullius* approach at least in crop GRs. However, also Ruiz Muller's statement seems to confirm that the CBD actually brought about a significant change. If the international customary notion gave everybody and nobody specifically the right to access and use GRs, then the state's right could only be understood within the context of everybody's right. That would imply that states in whose territories the GR are found have the right to use GRs, but not to impede any GRs connected activities[1] of other states in their territories. If the *res nullius* doctrine did not clearly delineate the rights of states over GRs, at least the CBD endorsed the sovereign right of countries possessing GRs (ten Kate and Laird, 1999, p15) to determine the rules of access and other conditions attached thereto subject to national legislation (Article 15.1), a right that had never been granted an international formal (legal) recognition before.

Was Article 15.1 meant to give provider states the right to arbitrarily deny others access to GRs found in their territories? Absolutely not; it merely allows states to make access subject to conditions in support of the CBD objectives, such as the fair and equitable sharing of benefits. Putting emphasis on the right to deny would be a wrong interpretation and one that is against the spirit of the CBD (Mugabe et al, 1997, p8; ten Kate and

Laird, p15f). The provision granting providers the right to regulate access cannot be interpreted in isolation from the rest of the provisions of the Convention. Article 15.2 in particular states: '*Each Contracting Party shall endeavour to create conditions to facilitate access to genetic resources for environmentally sound uses by other Contracting Parties and not to impose restrictions that run counter to the objectives of this Convention.*' This provision obviously addresses providers. It could, hence, also read: '*Each Provider State shall ....*' It implies that providers have a right to regulate access subject to national legislations, but they also have obligations to (1) *create conditions to facilitate access* and (2) *not to impose restrictions that are contra CBD objectives*. What are conditions to facilitate access? Which restrictions would be against the CBD objectives and which ones would not?

Legal and scholarly work has not yet ventured into trying to dissect these two obligations. It is especially difficult to distinguish them because the outcomes might often be the same. However, the concepts seem to differ. Whilst the first seem to revolve around procedural obstacles, which cause unjustifiable transaction costs, the second seem to anchor on the kind of substantive criteria the administrative agency is allowed to apply when supervising access. We will attempt to make this distinction in the next two sections.

Facilitation of access

Before looking at conditions for facilitation of access, it is good to first ask, what does the term 'facilitate' mean in the context of Article 15.2? To facilitate here could be interpreted to mean ease, enable or even assist (access). While designing access regimes, provider countries are thus expected to put in place legislative, administrative and policy measures that ameliorate and not impede access. The CBD does not offer a list of conditions that would be necessary to facilitate access. The Bonn Guidelines (BG) likewise did not attempt to elaborate this requirement. It is hard to imagine conditions that would be uniformly applicable to all or most providing states. However, since some experience exists showing why ABS regimes of many provider countries fail, we might get better results if we first ask the following question from the existing ABS regimes: Which conditions act counter to facilitation of access? Are there procedural requirements that do not serve the purpose of allowing the authorities to take a grounded decision, but unnecessarily cause or increase transaction costs?

An answer to the above question would be more vivid if an analysis of a complex ABS regime is made. Below we examine the access procedures of the Kenyan Regulations 2006, taking this as an example that exemplifies many others. Subsequently, we will discuss the possibilities of simplifying such procedures.

Access procedures: The case of Kenya

In full, the law that regulates ABS in Kenya is the Environmental Management and Co-ordination (Conservation of Biological Diversity and Resources, Access to Genetic Resources and Benefit Sharing) Regulations, 2006. The ABS provisions are found in parts III (Section 9–18) and IV (Section 19–20) of Regulation 2006. Part III states clearly that any person intending to access GRs in Kenya, whether an individual or a legal person, must be in possession of an access permit obtainable from the NEMA. Before an application is acceptable to NEMA, it must include other authorizations, as well as PICs.

The access procedure begins with a research authorization from the National Council for Science and Technology (NCST).[2] The procedure at the NCST takes approximately six weeks (Kamau, 2009, p83).[3]

In addition, a permit to enter into territories may be required. The Forests Act 2005 (FA) and the Wildlife (Conservation and Management) Act Cap 376 (WCMA) indicate clearly that any person intending to enter into territories placed under their jurisdictions, or collect or remove any type of biological resources, or carry out extraction for export, must be in possession of a licence or permit. The WCMA restricts access to and exploitation of wildlife resources (Mugabe and Otieno-Odek, 1997, p98). Any person seeking access to such resources or parts thereof must obtain a permit from the Minister for Tourism and Wildlife (Mugabe and Otieno-Odek, 1997).[4] The FA also stipulates that any removal of forest produce[5] without a licence or permit contravenes the Act (Section 52.1a) and is punishable by law (Section 52.2, 55.1). It is also illegal to extract, remove or cause to be removed, any tree, shrub or part thereof for export from any forest area (Section 54.8d). The Minister determines the circumstances in which licences/permits and other agreements are applied for, granted, varied, refused or cancelled, and the manner in which a person to whom a licence is granted may exercise a right or privilege conferred upon him by the licence (Section 59.2d). He also makes rules to control the entry of persons into forests (Section 59.2f) or nature reserves (Section 59.2h), how long they should remain there and under which conditions they may do so (Section 52.1b). Likewise, the Minister determines the amount of royalties or fees payable for any activities licensed under the Act (Section 59.2b). According to Section 4(j), such charges are collected by the Kenya Forest Service (KFS). Before an approval for a licence/permit is granted, a period of 90 days, after such an intention is published in the Gazette and in at least two newspapers of national circulation, is given to the public to make objections (Section 44.3). If there are any objections, 60 more days from the time of the receipt of the objection are needed to deliberate and deliver a decision to the objector (Section 44.4).

Concerning entry into forests and collection, harvesting, removal or extraction of forest produce, only activities undertaken within a management plan[6] are exempted from a licence/permit and an Environmental Impact Assessment Report (EIAR) in respect of the proposed activity (Section 44.1, 44.2). An application by a foreign institution (researcher) to conduct a basic research aimed at improving sustainable use and management capabilities, for example, might, hence, enjoy the ease created by this provision. Advanced research aimed at commercialization, on the other hand, would be caught by the provision.

PIC from local communities in Kenya might be complicated by the fact that there are few organized and issue-sensitized communities. It might be difficult, therefore, to trace the true representation of a local community. It could also imply that one might have easy access to PIC which is not representative and that might be challenged later by the legitimate local community.

It is only after all these hurdles have been overcome and the requirements above have been met that an application for an access permit is acceptable to the NEMA. The applicant has to seek all the clearances, licences and permits, even from government institutions, before applying for the access permit at the NEMA. Upon receipt of the application, the Authority shall, nonetheless, publish a notice in the Gazette and at least one newspaper with nationwide circulation, or in any other appropriate way (Regulation 2006, Section 10). This is meant to give the public an opportunity to bring representations or objections (Regulation 2006, Section 11). It takes 60 days from receipt of an application to the time the Authority decides to grant or refuse the permit (Section 13).

Drawing an example of a procedure that would be relatively short from Figure 19.1 if the applicant succeeds to get a research clearance from the NCST/MST within two months, PIC from KFS within 90 days and access permit from the Authority within 60 days, the duration of the process would amount to seven months. It is also very expensive as there are different fees to be paid, as well as other likely expenses to be incurred by the applicant. If an applicant succeeds in obtaining research authorization and access permits with the first attempt, he or she would have paid US$100–500 at NCST/MST and US$260–650 at NEMA as administrative fees. But this still does not include the fee(s) of the lead agency(ies) (LA) under whose jurisdiction the resources are to be found and without whose PIC NEMA cannot issue an access permit. Assuming the applicant needs a permit from only one LA with a fees estimate to that of NCST/MST or NEMA, the applicant will have paid a total of US$460–1650 or US$620–1,810.

This is the shortest access procedure one can imagine under Regulations 2006. If the applicant requires PIC from more lead agencies or ex situ

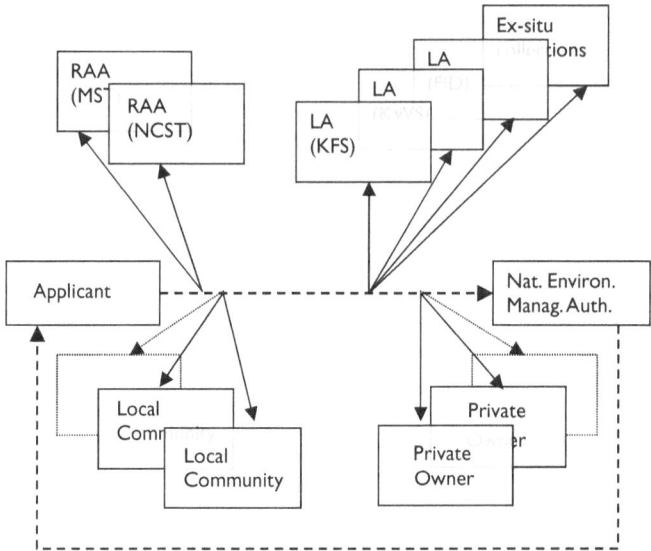

FiD Fisheries Department; KFS Kenya Forest Service; KWS Kenya Wildlife Service; LA Lead Agency; RAA (MST) Research Authorizing Authority (Ministry of Science and Technology); NCST National Council for Science and Technology; NEMA National Environment Management Authority

Figure 19.1 *Current access procedure in Kenya*

collections and perhaps one or two local communities, the procedure becomes extremely complicated and expensive.

It should also be kept in mind that an applicant has no assurance that the application will succeed at all (Section 11) and if it does, after how many attempts. In addition, the validity of the permit after such a great effort lasts only one year (Section 14.1). The renewal provision (Section 14.2) does not mitigate the situation, but creates more uncertainty. First, by stating that 'an access permit may be renewed', it gives the impression it might not. Second, it allows for new terms and conditions to be imposed, which might force the researcher/bioprospector to give up a project that had already been started. Third, the second renewal also lasts for only one year. Fourth, a new fee for renewal has to be paid.

From the three regulations analysed above, the following conditions are easily identifiable: (1) lengthy procedures; (2) cumbersomeness; (3) high costs; (4) multiple costs; (5) overlapping procedures; (6) long delays; (7) vagueness; (8) uncertainty; and (9) ambiguity. Such procedures would most likely discourage basic researchers. Likewise, they might not be capable of attracting potential commercial bioprospectors.

Table 19.1 illustrates in a condensed form the negative characteristics (for access) identified in Regulations 2006 – which are also prevalent in

ABS regimes of many other countries, including those of forerunner countries such as the Philippines (Executive Order 247) and Brazil (Provisional Measure No 2.186-16) – and the negative impacts they are likely to produce.

In light of the outcome of the analysis above, it is justifiable to conclude that the above ABS regime does not facilitate, but rather impair, access to GRs. Hence, it does not comply with Article 15.2 of the CBD and needs to be revised.

Simplifying access procedures

There are two different possibilities to improve the situation and simplify the procedure: coordination by an LA, and the concentration of licensing. Coordination by an LA is the simplest way of easing the citizen's struggle with the multitude of administrative bodies. It is also called procedural integration by European Union (EU) environmental legislation. For instance, Article 7 of the Integrated Pollution Prevention and Control (IPPC) Directive[7] states:

> *Member States shall take the measures necessary to ensure that the conditions of, and procedure for the grant of, the permit are fully coordinated where more than one competent authority is involved, in order to guarantee an effective integrated approach by all authorities competent for this procedure.*

Table 19.1 *Characteristics identified in and impacts created by Regulations 2006*

Identified (negative) characteristics	Possible negative impacts						
	Delay	Expensive	Complicated	Cumbersome	Uncertainty	Ambiguity	High transactional costs
1 Long procedure	✓					✓	✓
2 Multiple permits	✓	✓	✓	✓	✓	✓	✓
3 Multiple PICs	✓	✓	✓	✓	✓	✓	✓
4 Multiple fees		✓			✓		✓
5 Other likely fees		✓			✓	✓	✓
6 Overlapping procedures	✓	✓	✓	✓	✓	✓	✓

This means that the different administrative agencies that are competent to provide a permit are asked to coordinate their procedures and conditions of granting the permit. This shall avoid consecutive decision-making, a time-consuming practice where the single agencies presuppose that the permit of another agency must first be obtained before it deals with the matter. The citizen shall be entitled to file all applications at one time, and the agencies addressed shall handle them simultaneously. Moreover, the agencies must coordinate their decision and the conditions they attach to the permit. In this way, contradictions shall be avoided which may arise in cases of overlap of the material objectives and criteria for which different agencies are competent. For instance, the agency providing a research permit may wish to reject the application or impose very strict conditions, while the agency in charge of access to GRs may follow a more generous line. They should be coordinated in order to take a harmonized decision. Such coordination must be organized. The most appropriate way to do this is to designate an LA and provide it with competences to coordinate and even combine the publication of the application, the receipt of comments, the holding of hearings and the drafting of decisions.

The concept of an LA was, for instance, introduced by the German provision, which transposed Article 7 IPPC Directive. Section 10(5)(2) of the German Federal Immission Control Act (Bundesimmissionsschutzgesetz, BImSchG) says:

> *Insofar the project ... requires authorization according to other laws, the authority competent for the authorization [according to the Immission Control Act] must ensure a full coordination of the licensing procedures and conditions.*

If applied to the Kenyan ABS regime, procedural integration would mean that one agency – possibly NEMA – is entrusted with the function of an LA. NEMA would then be obliged and entitled to coordinate procedures, decisions and conditions of KWS, KFS, NCST and so on.

Even more simplification of procedures is possible by what is called material integration. This appears in two variants: the concentration of permits; and the full integration of permits. Only the first shall be discussed here.[8]

Concentration means that the various permits are consumed in one. Only one permit is required for an activity, and this permit comprises all other permits, which would otherwise have to be obtained. An example of such concentration can be found once more in the BImSchG. In doing so, the German Act goes further than the IPPC Directive, which settles for procedural integration:

The licence shall include other official decisions with a bearing on the installation, in particular public-law licences, approvals, grantings, permits and authorizations – with the exception of plan approvals, ... and authorizations under water law pursuant to Articles 7 and 8 of the Federal Water Act.[9]

This concept, if applied to ABS, would mean that – taking Kenya as an example – only one permit would be necessary, most probably the one provided by NEMA. The person seeking access to GRs would have to file only one application. However, he or she would have to submit all data relevant for the different permits according to legislations other than Regulations 2006. NEMA would have to consult the agencies normally responsible for the other permits. The comments of other agencies could either be framed as recommendations or even as consent requirements.[10] It would have to respect all material criteria of the other permits. But NEMA would have the exclusive competence to take the final decision. However, following the proviso of the cited German provision, if the access activity involves larger works such as the construction of a building or road, the permit requirements related to this would remain separated from the concentrated permit.

Non-imposition of restrictions that run counter to the CBD objectives

The CBD requires that restrictions imposed in regulating access do not run counter to the objectives of the Convention. The objectives of the CBD are listed in Article 1. They are the conservation of biological diversity; the sustainable use of its components; and the fair and equitable sharing of the benefits arising out of the utilization of GR. It is restrictions that hinder the realization of these objectives that Article 15.2 forbids and not all restrictions in general. Which restrictions are these?

Neither the CBD nor the BG give a clue as to what such restrictions might entail. The BG simply reproduced CBD's Article 15.2 wording by stating that '*Providers should strive to avoid imposition of arbitrary restrictions on access to genetic resources*' in Article 16(c)(ii).

Again, restrictions that are likely to run counter to the three objectives of the CBD would apparently emanate from laws and/or regulations. They are especially easier to understand in regard to basic research, which tends to be conveniently disadvantaged (Erdos, 1999; Ruiz Muller, 2003, p195ff; Swiderska et al, 2001), in spite of its utmost importance for conservation, as well as sustainable use of biodiversity.[11]

Let us try to conceive some likely restrictions in regard to foreign researchers/bioprospectors. They are not uniformly applied by provider countries, but appear with some frequency. They are presented here with comments on whether they are compatible with the three objectives of the access regime. They may include:

- prohibition of collection of samples of a degraded species – protection of biodiversity
- issuance of access permit on condition that a defined benefit-sharing agreement is reached forthwith – benefit sharing; means to coordinate permit and contract
- issuance of access permit on condition that the researcher/bioprospector employs some or a certain number of local collectors for the duration of collecting – benefit sharing
- issuance of access permit on condition that the researcher/bioprospector collects not more than a certain amount of sample specimens – protection of biodiversity
- issuance of permit on condition that the researcher/bioprospector will continue paying a standing fee during the course of the research – benefit sharing
- issuance of access permit on condition that ensuing research will only take place in the resource state – benefit sharing
- issuance of access permit on condition that the researcher will make available/reveal the results of the research before publication – benefit sharing.

The wording of Article 15.2 CBD (*not to impose restrictions that run counter to the objectives of this Convention*) must be understood to mean that there can be regulatory objectives, which do not run counter to the CBD objectives because they do not belong to the realm of the Convention. For instance, the CBD does not address questions of military use of lands. Therefore, preventing access to these areas is a legitimate objective.

What would be examples of reasons that 'run counter to the objectives'? For instance, legitimate objectives may be pursued in a too strict and possibly counter-productive manner. This would be the case if the access permit was issued on condition that the researcher/bioprospector collaborates only with local partners or scientists recommended by a national authority, thus prohibiting collaboration with a partner of his or her own choice. This could jeopardize independent, high-quality research.

Let us once more take a look at the Kenyan example in order to identify actual and better practices for alerting licensing to the CBD objectives. Two aspects must be distinguished: the scope of application of access requirements; and the criteria for access permits.

Scope of activities subject to licensing
The Kenyan Regulation 2006 states in Section 9(1): 'Any person who intends to access GR in Kenya shall apply to the Authority for an access permit'. 'Access' is defined by Section 2 to mean 'obtaining, possessing and using GR conserved, whether derived products and, where applicable, intangible components, for purposes of research, bioprospecting, conservation, industrial application or commercial use'.

It appears that the scope of the permit requirement is too broadly and vaguely delimited by this provision. 'Obtaining' GRs would include, for instance, the purchase of a plant on a market. Shall this really be subject to authorization? 'Possessing' GRs would include the growing of plants on any property in Kenya. Must the owner really ask for authorization for this? 'Using' would include eating, burning and crafting. Is all this to be authorized? 'Purposes of research' would cover any research related to the genetic characteristics of the resource, but are not all properties of an organism related to its genetics? 'Bioprospecting' is not defined in the Regulation, although it is a technical term not used in ordinary language. The definition of 'conservation' is again very broad. For instance, if someone plants a rare tree on her farmland in order to conserve the species, shall this be subject to authorization? The examples show that the realm of activities that fall under the permit regime must more narrowly be designed in order to become manageable and be directed to the objectives of the Convention. Otherwise, legal certainty lacks and authorities may use their powers of requiring permits arbitrarily.

It is core that the law properly defines what is meant by 'genetic resource'. The definition given by the CBD, that GR is 'genetic material of actual or potential value' (Article 2 CBD), must be specified in order to draw a line between the value of the genetic material and the value of the organism as such (of its bulk use, as it is sometimes called). Those activities, which aim at researching and using the immediate value of the organism, should not be covered by the ABS regime. Examples include the carving, carpeting and burning of wood, the growing and collecting of plants and the catching of animals for food and feed, and the consumption of organisms. Unfortunately, the CBD does not make clear what the value of genetic material means. After all, even the nutritive value of corn is a value of the functional units of heredity of corn. Nevertheless, a line must be drawn, and it is up to the national state to fill the gap of the CBD and decide where to draw the line. It is here suggested that a state should distinguish between an immediate and an elaborate use of a resource. The mere consumption of corn is not making use of the genetic material. The latter starts if an organism's hereditary traits are identified and exploited for specific purposes, thus if the genetic material is developed further in order to improve its usability beyond the immediate use of its organism of origin.

Criteria for the access permit
Section 11(1) of the Kenyan Regulation 2006 states:

> *The Authority shall, on receipt of representations or objections to the proposed access permit from the public, review the application and if satisfied that the activity to be carried out shall facilitate the sustainable management and utilization of genetic resources for the benefit of the people of Kenya, issue an access permit to the applicant.*

The yardstick for the granting or refusal of the permit is thus that the activity 'shall facilitate the sustainable management and utilization of genetic resources'. It appears that this is very broad language that does not give much guidance, thus allowing the authorities a very wide margin of discretion. Authorities may be tempted to grant or refuse permits arbitrarily or even ask for bribes. Although the yardstick (sustainable management and utilization) is related to the three CBD objectives (conservation, utilization and benefit sharing), it does not fulfil the task of national implementation of international agreements, that is, to specify the general principles of such agreements. For instance, 'conservation' could be defined as protection of wild fauna and flora, rare as well as common, 'utilization' as research in and development of the genetic characteristics plants, animals and microorganisms, and 'benefit sharing' as the provision of employment for local people, joint ventures of research and development activities, communication of scientific results, crediting of authorship and shares in monetary income from using GRs.

In addition, guidance may be laid down about conditions of permits. This may be done on the kind and calculation of fees and down payments, on time limits, on allowed uses, on come-back clauses for new uses, on the transfer of material to third persons, on employment of locals, on joint ventures, and so on.

Criteria and conditions must also be clarified concerning other permits. For instance, as stated above, in Kenya as in many other legal systems a permit for conducting research in the country is required. The relevant law should clearly state the purpose of this requirement and specify it by yardsticks to be applied by the competent authority. Is it ensuring participation in research and technology activities and the sharing of results? Is it the generation of state income from research results? Or might it simply be due to traditions of authoritarian states to closely supervise societal activities? The authors suggest that the latter reason is not defendable in a liberal state. Still, if the sharing of research and research results as well as of monetary benefits is the objective and yardstick for research permits, would this not amount to an overlap with the access regime, if the research concerns GRs?

In cases of such overlap the concentration principle as introduced above would allow that one permit includes others, and is only provided if the material yardsticks of the other permits are also respected. In the case of the research permit, the law on access to GRs could provide that the access permit also includes the research permit, and that the agency responsible for the access permit must fully consider the yardsticks concerning the research permit. The administrative agency responsible for the access permit would be required to invite the research agency to comment on the project. But a separate permit would not be required in such cases.

Likewise, there must be clarity about the criteria and conditions if the GR to be accessed is located in a protected area. In Kenya, besides the general permit provided by NEMA, a permit must be obtained from KWS, the agency responsible for the protected areas. The objective for the second permit could be that due regard must be given to the specific value and vulnerability of the protected area. Once again, the concentration principle could be introduced in this case.

Finally, a special permit must be obtained from the Forest Service if the GR is located in a forest. Once again, clarity is necessary as to the objectives and criteria of this permit. Is it to ensure that environmental damage is prevented? Is it the generation of state income from a public resource? Is it the protection of local communities? It is submitted that in the case of access to GRs this permit too could be concentrated in the access permit, and that the Forest Service must be heard in the decision-making process?

Conclusion

As the Kenyan example shows, and a broader study of many more legal systems would reveal, time has come for an evaluation of existing ABS legislations of provider states. The first round of laws was heavily – and legitimately – influenced by policy considerations. How to establish full control of the access process was the primary concern. This led sometimes to over-ambitious and loosely framed concepts that caused legal uncertainty and bureaucratic overkill. There is priority need and indeed potential for bringing legal doctrinal scrutiny into play. After all, access procedures are administrative law and must correspond to its general concepts and ambitions. Clear and parsimonious criteria and procedures not only reduce transaction costs, but also further the purpose they shall serve, that is, the sustainable use of GRs. Our concrete suggestion is that the procedures and conditions for access permits should better be coordinated, most appropriately by one agency entrusted with such a lead function. Even better would be to comprise the different permits in one concentrated permit.

Notes

1 IU-PGRFA: www.fao.org/ag/cgrfa/iu.htm, Article 1: 'This Undertaking is based on the universally accepted principle that plant genetic resources are a heritage of mankind and consequently should be available without restriction'.
2 The NCST is under the Ministry of Science & Technology.
3 Generally for access procedure before and after 1999, see Kamau (2009).
4 The WCMA is not clear concerning regulation of access to flora (in parks and reserves). Provisions to this effect can only be hypothetically derived from sections 13 and 16, which forbid a variety of other activities against both fauna and flora without authorization, and empowers the minister to make entry regulations, as well as establish the fees to be paid for such entry. Now a draft bill, the Wildlife (Conservation and Management) Bill 2007, which incorporates research and bioprospecting concerns, has been developed and is pending in Parliament for approval. If it is adopted, the law will establish a clear requirement for basic and commercial researchers to seek for an access permit and pay the required fee before any activities are conducted. Bioprospectors would still have to possess PIC, material transfer and benefit-sharing agreements from stakeholders whose interests are involved before a permit can be issued by the wildlife department. A copy of the bill is available at http://www.fankenya.org/downloads/wildlife-conservation&managementbil2007.pdf, accessed 29 October 2008.
5 According to section 2 of the Act, 'forest produce' includes bark, creepers, fibres, fruit, grass, gum, honey, leaves, limestone, plants, rubber, sap, seeds, spices and wax.
6 Section 2 defines a 'management plan' as a systematic programme showing all activities to be undertaken in a forest or part thereof during a period of at least five years, and includes conservation, utilization silvicultural operations and infrastructural development.
7 Council Directive 96/61/EC concerning integrated pollution prevention and control (IPPC).
8 Full integration means that only one permit is required for a certain activity, not any other, even not in the form of inclusion in the concentrated permit.
9 Section 10(5)(2) BImSchG.
10 The PIC of the local community in cases of access to TK would certainly be a binding requirement.
11 Note that Article 12(b) of the CBD places an obligation upon contracting parties to '[P]romote and encourage research which contributes to the conservation and sustainable use of biological diversity'.

References

Erdos, J. E. (1999) 'Current legislative efforts in Brazil to regulate access to genetic resources', *Executive Report* (December), available at http://www.planeta.com/planeta/99/1199brazil.html, accessed 29 October 2008

Kamau, E. C. (2009) 'Sovereignty over genetic resources: Right to regulate access in a balance. The case of Kenya', *Revista Internacional Dereito e Cidadania*, no 3, February, pp73–88

Mugabe, J. and Otieno-Odek, J. (1997) 'National access regimes: Capacity building and policy reforms', in Mugabe, J. et al (eds) *Access to Genetic Resources. Strategy for Sharing Benefits*, Washington DC, IUCN Environmental Law Centre & ACTS, Kenya & World Resources Institute, pp95–114

Mugabe, J., Barber, C. V., Henne, G., Glowka, L. and La Viña, A. (1997) *Access to Genetic Resources. Strategy for Sharing Benefits*, Washington DC, IUCN Environmental Law Centre & ACTS, Kenya & World Resources Institute

Ruiz Muller, M. (2003) 'The international treaty on plant genetic resources and decision 391 of the Andean community of nations: Peru, the Andean region and the international agricultural research centers' (June), available at http://www.cipotato.org/library/pdfdocs/AN65154.pdf, accessed 29 October 2008

Sampath, P. G. (2005) *Regulating Bioprospecting: Institutions for Drug, Research, Access and Benefit Sharing*, Tokyo, New York, Paris, United Nations University Publisher

Swiderska, K., Dano, E. and Dubois, O. (2001) 'Developing the Philippines' executive order No. 247 on access to genetic resources', available at http://www.cbd.int/doc/case-studies/abs/cs-abs-order-ph-en.pdf, accessed 29 October 2008

Ten Kate, K. and Laird, S. (1999) *The Commercial Use of Biodiversity: Access to Genetic Resources and Benefit Sharing*, London, Earthscan

Chapter 20

Capacity Development in a Changing World – Three Years of the ABS Capacity Development Initiative for Africa: Achievements and Perspectives

Peter Munyi, Fabian Haas, Andreas Drews and Suhel al-Janabi

Introduction

The third objective of the CBD relates to the fair and equitable sharing of the benefits arising from the utilization of GRs, in short, ABS. It was included in the text of the CBD at the insistence of the developing countries, which possess approximately 80 per cent of global GRs. Functioning ABS regulations at national, regional and international level are essential for biodiversity conservation. Since the use of biodiversity is almost as diverse as biodiversity itself, such regulations can only be negotiated by competent partners, including a wide range of stakeholders, including governments, business, and local and indigenous communities. Building on this insight, the Dutch–German ABS Capacity Development Initiative for Africa started in 2005 to support African stakeholders. Guided by the elevator principle – connecting all levels bottom-up and top-down, and 'stopping on request' – the Initiative links the local level with UN negotiations using regional and sub-regional activities as kick-off platforms. To date, the Initiative has established three sub-regional forums for exchanges on ABS issues that take into account particularities of languages in Africa and the diversity of legal systems: (1) eastern and southern Africa; (2) central Africa; and (3) West Africa, Maghreb and the Indian Ocean Islands.

National sovereignty and the 'commons'

The CBD[1] was agreed in Rio de Janeiro, Brazil, at the United Nations Conference on Environment and Development (UNCED) in June 1992.[2] The CBD, which has been ratified by an overwhelming majority of countries,[3] reaffirms the sovereign rights of states over their biological resources. This principle of international law evolved after World War II, culminating with the UN Declaration on Permanent Sovereignty Over Natural Resources of 1962.[4] During colonial times biological resources were perceived and treated as property of the respective countries and colonial powers; for example, as illustrated by the rubber monopoly of Brazil, which fell after the British secretly collected and smuggled seeds of the Pará rubber tree (*Hevea brasiliensis*) in the late 19th century from Brazil to Kew Gardens for propagation, and from there to their Southeast Asian colonies.

State sovereignty over natural resources is acknowledged in Principle 21 of the Declaration of the United Nations Conference on the Human Environment,[5] which took place in 1972 in Stockholm. This principle was followed by the report of the Brundtland Commission[6] stating that national sovereignty is increasingly 'challenged by the realities of ecological and economic interdependence ... in shared ecosystems and in "the global commons" – those parts of the planet that fall outside national jurisdictions'. It further reveals that 'collective responsibility for the common heritage would not mean collective international rights to particular resources within nations. This approach need not interfere with concepts of national sovereignty'.

In a separate line of thought and discussion, by the late 1970s political concerns were raised by developing countries over control, ownership and access to plant GRs for food and agriculture (PGRFA). At that time PGRFA were generally considered as common heritage of humankind, that is, free access for further improvement of the genetic material. Against this political background in 1983, governments created the International Undertaking on Plant Genetic Resources (IU) under the Food and Agriculture Organization of the United Nations (FAO). The entering into force of the CBD required the IU to take into consideration the unresolved issues of the Farmers' Rights and the status of the ex situ collections in existence prior to the CBD. The initiated renegotiation of the IU resulted in the International Treaty on Plant Genetic Resources for Food and Agriculture (ITPGRFA) which entered into force in 2004. The ITPGRFA established a CBD-compliant multilateral system (MLS) for access to PGRFA (listed in Annex 1 of the Treaty) and benefit sharing (BS) if exclusive intellectual property rights are evoked on a specific PGRFA accessed under the MLS.

Therefore, the CBD did not bring a real paradigm shift away from common heritage of humankind to state sovereignty over biological resources as often stated by CBD-critical voices, but reaffirms existing international legal principles. In this context the IU was renegotiated to be compliant with the BS requirement of the third objective of the CBD without giving up the common heritage principle for PGRFA, where in many cases ownership cannot be traced back.

The third objective of the CBD

Taking on board the political recommendations of the Brundtland Commission about the need to link conservation and sustainable development,[7] the CBD became the first multilateral environmental agreement to explicitly link biodiversity conservation with sustainable development. This is clearly reflected in the three objectives of the Convention: the conservation of biological diversity; the sustainable use of its components; and the fair and equitable sharing of the benefits arising out of the utilization of GRs.

The third objective was included at the insistence of developing countries, where approximately 80 per cent of the global GRs are found. It underscores that owners as well as users, who are often different actors, must enjoy benefits from the sustainable use. Indeed, it gives biodiversity a practical or monetary value which provides a driver for biodiversity conservation, by incentivizing governments to conserve biodiversity in order to benefit from bioprospecting, and thus protect indigenous and local communities who live on the land. Klaus Töpfer, the former Executive Director of UNEP, summed up this argument as 'use it or lose it'. ABS further addresses this issue by providing local communities with livelihood generating potential from the commercial use of their biodiversity.

Because of their relative complexity, ABS issues have been discussed primarily by experts, at occasions including Conferences of the Parties (COP), Ad-hoc Open-ended Working Groups and numerous workshops in industrialized and developing countries, focusing mainly on bioprospecting, marketing and biopiracy. Although the CBD clearly asks Parties in several articles to cooperate and facilitate access to GRs for environmentally sound uses, most national ABS regulations developed so far restrict rather than facilitate access.[8] This has to be seen from two sides. First, many users perceive any regulation as a restriction particularly as a hindrance to academic research. Secondly, provider countries are concerned about how to monitor and track the further use of their GRs after granting international access. Both these views are understandable and highlight the regulatory uncertainties that characterize the relationship between providers and users of GRs.

So far, however, the potential for ABS mechanisms to contribute to poverty alleviation has rarely been utilized, partly because of a lack of international guidance (Henne et al, 2003). The Bonn Guidelines (Decision VI/24A)[9] are merely a voluntary code of conduct providing a set of best practice guidelines for the various stakeholders in any bioprospecting activity. Beyond this, recommendations have been put forward in various international conferences and workshops dealing with issues such as the cross-border transfer of GRs, disclosure of information, mechanisms for participation, stakeholder cooperation and the documentation of TK. Despite widespread recognition of the importance of bringing the respective sectoral approaches in harmony, of enacting national level regulations and forging local approaches, efforts by the ABS community have not yet resulted in a mutually supportive relationship with the relevant key issues of the global trade and intellectual property rules. Nevertheless, global debate has reaffirmed the ethical principle of BS, with the outstanding issue being how equity is established on sharing the benefits arising from utilization of GRs. Other outstanding matters include how to operationalize enabling mechanisms, such as on access and monitoring and reduction of transactional costs.

At the World Summit on Sustainable Development (WSSD) in 2002 at Johannesburg, South Africa, the ABS debate gained more momentum, with political leaders requesting an international regime to ensure the implementation of the third objective of the CBD. Since the WSSD, there has been an ongoing negotiation of an international regime on ABS, and at the last (ninth) COP in 2008 a clear roadmap was established for the negotiation of the international ABS regime by 2010 – to coincide with the WSSD call for a significant reduction in the rate of loss of biodiversity: the so-called '2010 biodiversity target'.[10]

If ABS mechanisms are to have a positive impact on sustainable development, poverty reduction and biodiversity conservation, functioning ABS regulations at national, regional and international level are essential. These regulations should include provisions on land and property rights, access to resources, national and international market mechanisms, profit sharing and technology transfer, capacity building and the recognition of TK and intellectual property, including disclosure obligations in the user countries of GRs.

As the use and complexity of biodiversity is almost as diverse as biodiversity itself, such regulations require input from competent partners, including a wide range of stakeholders. So far the ABS negotiations have involved only a small community of experts, many of them distant to real-life problems of sustainable use of biological and GRs and to life sciences. Governments need to establish a broad participatory process at the national level, whether in preparation for international ABS negotiations or drafting national ABS legislation, to ensure that all stakeholders are involved and that awareness is raised among those who are affected positively or otherwise by these processes. This is very critical, particularly in Africa, where potential exists for ABS to contribute to poverty alleviation.

To enable African stakeholders to actively participate and contribute to the ABS negotiations, debate and legislative process, capacity development is necessary. Concerted capacity development measures would empower stakeholders not only to identify and articulate their own interests, but also to understand positions of their partners in negotiations at all levels – international to contractual – in order to find fair and mutually agreed solutions. Key stakeholders in this context are policy makers, legislators, implementing agencies, national and international research institutions, the private sector and, last but not least, representatives of local and indigenous communities. Only consensus of the groups mentioned on principles, procedures, rules and regulations will lead to the generation of sustainable benefits in the framework of ABS.

The ABS capacity development initiative for Africa

Building on these insights, the Netherlands Directorate General for International Co-operation (DGIS) and Deutsche Gesellschaft für Technische Zusammenarbeit (GTZ) GmbH (the latter acting on behalf of the German Federal Ministry for Economic Co-operation and Development, BMZ) joined forces to build human and institutional capacity in developing countries to deal with these complex ABS issues. Priority was set on Africa because by international comparison this is where the need for capacity development is greatest: from supporting national legislative processes to strengthening African positions on ABS at the relevant international negotiations of the CBD, but also within the WTO and WIPO.

DGIS and GTZ, in cooperation with Ethiopia, held an initial sub-regional multi-stakeholder orientation and needs assessment workshop in Addis Ababa in 2005. The workshop's recommendations were presented at the meeting of the CBD Working Group on ABS in Spain in January 2006. Together with African stakeholders, DGIS and GTZ staged a joint information session at the eigth COP in Brazil in March 2006 – to present and discuss the results of an ABS Capacity-Building Needs Assessment carried out in the interim, along with the possible instruments for an effective ABS capacity development programme.

On the strength of the outcomes of the workshop in Ethiopia in 2005 and its follow-up activities and, no less importantly, the positive evaluation by the African partners, the joint *Dutch–German ABS Capacity Development Initiative for Africa* was set up in 2006. Considering the results so far and the identified gaps of the Needs Assessment, the Initiative was designed to implement the *CBD Action Plan on Capacity-Building for Access to Genetic Resources and Benefit Sharing* (adopted by the seventh COP in February 2004) in a manner tailored to the African needs. This was not only seen as a procedure serving CBD obligations: instead in the long run considered as an iterative process that supports the vision of ABS as a powerful instrument to reduce poverty in Africa. With the co-funding support of DGIS for the next three years, the GTZ programme 'People and Biodiversity – Implementing the Biodiversity Convention' became responsible for the concept and implementation of the Initiative on the basis of regular steering meetings.

From the very beginning, the Secretariat of the Convention on Biological Diversity (SCBD) became a cooperation partner to the Initiative, providing substantive inputs on the ongoing negotiation process and logistical support for the delegates' briefings. In order to further the integration of francophone African countries into the Initiative's activities, specific support was provided by the Government of Quebec (Canada) and the *Institut de l'énergie et de l'environnement de la francophonie* (IEPF). In 2008, IEPF joined the Initiative as a permanent partner and consequently the Initiative was renamed the ABS Capacity Development Initiative for Africa (the 'Initiative').[11]

The cooperation of GTZ, DGIS, SCBD, Government of Quebec and IEPF formed the first steps towards developing the initial Dutch–German partnership into a multi-donor initiative for a concerted ABS capacity development within Africa as an entire global region. Thus, following the spirit and the requirements of the Paris Declaration on Aid Effectiveness, the Initiative with its joint Steering Committee is contributing to an improved donor harmonization, aid effectiveness and better valorizing the potential for poverty alleviation at the interface of natural resources management, trade and governance.

Through its support the Initiative aims to create a win-win situation for poverty alleviation, conservation and sustainable use of biological diversity by the implementation of the third objective of the CBD. The latter requires the private sector to invest in bioprospecting activities (in cooperation with national and international research institutions) and share generated benefits, including profits, with national governments as well as

Box 20.1 *Approaches and Instruments*

Guided by the elevator principle – connecting all levels bottom-up and top-down, and 'stopping on request' – the Initiative implements its activities using the following approaches to achieve its objectives:

- Capacity development of relevant stakeholders
- Preparation and follow-up of CBD meetings on ABS
- Support to national implementation.

The activities under each of these approaches may include one or more of the following eight instruments:

1. multi-stakeholder workshops to discuss emerging ABS issues and to define priorities for capacity development in a participatory manner
2. Thematic and stakeholder – focused training courses based on needs
3. Peer-to-peer knowledge transfer at national and local level between African countries as well as globally; for example, between private-sector and governmental decision-makers
4. ABS best practices with the private sector, including lessons learned in order to identify additional participants and potential investors
5. Information exchange and knowledge management within the different stakeholder groups at national level and on a pan-African and global basis
6. Technical papers/studies in order to set priorities, stimulate substantive discussion giving support for decision making
7. Communication, Education and Public Awareness (CEPA) for ABS
8. Active participation of and/or substantial inputs by African representatives to important ABS meetings at UN level.

local communities. The Initiative supports directly the main steps towards achieving this win-win situation:

1. the International ABS Regime, which provides a norm setting for benefit-sharing that contributes to poverty alleviation, that is, inclusion of local and indigenous communities as custodians or owners of genetic or biological resources and associated TK

2 ABS policies and poverty alleviation approaches, which are linked to other relevant policies and incentives to prevent or counteract marginalization within countries, such as land tenure, ownership and rights of use for biological resources, food security, support for small and medium-sized local enterprises and access to (micro) credits

3 national ABS legislation, which may establish the legal basis for ownership of GRs, associated TK and identifies the roles and responsibilities of indigenous and local communities in the ABS process, thus contributing to poverty alleviation

4 market opportunities, which enhance the support for the private-sector development (including public–private partnerships) and the prolific potential of economically poor but knowledge and biodiversity-rich communities by improving entrepreneurial capacities with respect to biological and GRs, as well as associated TK.

Working in a developing political process, unable to predict what kind of international regime will emerge by the year 2010, the Initiative had to provide a high level of flexibility to a variety of regional and national political settings and different socio-cultural contexts in providing training and capacity development for all stakeholders involved in the ABS process: from village chiefs to CBD negotiators, business representatives and even ministers. Therefore, the implementation of the Initiative is taking place in a learning loop, that is, as an ongoing process that requires iterative monitoring and adaptation for continuous improvement.

Objectives

The objectives of the Initiative are to: (1) increase awareness of African policy makers and legislators on ABS matters, especially their cross-sectoral nature and their potential for poverty alleviation; (2) foster meaningful participation of all relevant stakeholders at all stages of the negotiation, development and implementation of ABS regulations – at the international, national and local level; (3) improve regional cooperation on ABS issues among African countries; and (4) support the development of partnerships for business opportunities.

The Initiative's instruments offer a range of options for the transfer of best practices and lessons learned to and among the various stakeholder groups and, by achieving these objectives, the short-term effects of the Initiative include increasing the knowledge of all relevant stakeholders on ABS matters by offering different platforms for experience exchange (workshops, peer-to-peer learning exchange, Clearing House Mechanism); and by providing a facility for the documentation of, for example, case studies, information on biopiracy, conclusion of benefit-sharing agreements, legislative and regulatory approaches, cross-border harmonization and national governance.

In the long term, the Initiative will make a positive impact on food security and a contribution to the achievement of the Millennium Development Goals (MDGs) through fair and equitable sharing of benefits generated from the use of GRs. Towards these objectives, the Initiative considers the private sector to be a key actor.

Sub-regional targeting

Guided by the 'elevator principle' (see Figure 20.1), the Initiative targets its objectives by linking the local level with the CBD negotiations process using regional and sub-regional activities as kick-off platforms. Until 2008, the Initiative established three sub-regional platforms, taking into account different levels of ABS implementation, diversity of language and legal systems: (1) eastern and southern Africa; (2) central Africa; and (3) West Africa, Maghreb and the Indian Ocean Islands. The political vision provided the Initiative's starting point from which to provide comprehensive information, and to prepare all stakeholders at political and technical level as a basis for national or sub-regional ABS approaches leading to a sub-regional harmonization of environmental legislation with ABS as a pilot theme.

In all three sub-regions, analysis of case studies has played a critical part in the Initiative's capacity development agenda. Participants present examples from their countries of origin. These real-life cases, including panel discussions with all stakeholders involved in the particular case, evolved to be a standing element of the major gatherings conducted by the initiative. Wherever possible, the Initiative also organized field visits to locations of the case studies. Examples of subjects of field visits include gluten-free Teff cereal in Ethiopia and enzymes found in herbivore's dung in Kenya. Participants have reported that this approach has assisted to technically and socially 'ground' the often politically driven discussions on ABS in Africa, including the discussion of the particular role of TK. Additionally, the Initiative has examined the often slim distinction between ABS and biotrade, focusing on the market chain of argan oil (Morocco) or *Prunus africana* (Cameroon).

The Initiative's kick-off workshop was held in Cape Town in November 2006 and resulted in the development of a common ABS vision, and a roadmap of priorities for Africa. More than 50 participants representing mainly southern and eastern African countries were drawn from the spheres of environmental policy, administration, science, development

Level of intervention	Preparation national ABS authority	Other stakeholders	Main instruments used at different levels
UN level	Negotiation International Regime	ABS authorities of other countries	• African inputs on ABS to relevant meetings • Technical papers and studies • CEPA for ABS
(sub-)regional level in	Harmonization of legislation Regional cooperation	ABS authorities of other countries Academia Private sector Communities	• Multi-stakeholder WS • Information exchange/CHM • Issue based Trainings • Technical papers/studies • Best practices with private sector • CEPA for ABS
National level	Laws and regulations	Academia Private sector Communities	• Best practices with Private Sector • Peer-to-peer knowledge exchange • CEPA for ABS
Local level	Implementation and monitoring	Academia Private sector Communities	• Best practices with Private Sector • Peer-to-peer knowledge exchange • CEPA for ABS

(Diagram annotations: "Developing recommendations" / "Developing and implementing recommendations")

❶ Multi-stakeholder workshops; ❷ Issue or stakeholder-focused trainings; ❸ Peer-to-peer knowledge exchange; ❹ ABS best practices with private sector; ❺ Information exchange and knowledge management; ❻ Technical papers and studies; ❼ Communication Education and Public Awareness (CEPA); ❽ African inputs on ABS to relevant meetings.

Source: ABS Capacity Development Initiative for Africa, 2008

Figure 20.1   *The elevator principle for ABS capacity development through the Initiative*

cooperation, non-governmental organizations, local communities and (pharmaceutical) industry representatives. Representatives from French-speaking African countries also attended this Anglophone workshop, as well as the second Africa-wide workshop in December 2007 in Nairobi. In the latter workshop, participants from 19 African countries gathered and shared their experiences in developing products for the national and international markets, and the relationships of local producers and national and international private sector businesses.

The methodology of the Initiative and the results of the Cape Town, South Africa, workshop were presented at the constitutional meeting of the Working Group on Biological Diversity of the Central African Forest Commission (COMIFAC), held in December 2006 in Sao Tome. The ten COMIFAC countries[12] requested a complementary ABS capacity development process for the central African region. Considering the objectives of the COMIFAC convergence plan and with financial contributions of the Agence Française du Développement (French Development Agency, AFD) and BMZ, this process started as a sub-regional part of the Initiative, focusing on a common ABS approach in policy and implementation for the COMIFAC members.

Co-financed by the Institut de l'Energie et de l'Environnement de la Francophonie (Francophone Institute for Energy and Environment, IEPF), a formal ABS process of the Initiative for West Africa and the Indian Ocean Islands started in Marrakech, Morocco, in November 2007. It aims at

---

Box 20.2 *Sub-regional milestones*

- Eastern and southern Africa:
  ABS vision for Africa; African position on the 'Certificate of Origin'; workshop on ABS, business and commercial research; ABS negotiation skills trainings; agenda setting on the interface of ABS and protected areas.
- Central Africa:
  Study on relevant legislation in the COMIFAC countries; definition of cornerstones for sub-regional ABS framework and national ABS regulations.
- West Africa, Maghreb and francophone Indian Ocean Islands:
  Roadmap for common ABS process; establishment of an ABS communicator's network, priority setting on national awareness-raising on ABS.

Source: ABS Capacity Development Initiative for Africa, 2008

network building and exchanging experiences on bioprospecting cases and regulatory approaches in the francophone African Countries. Whereas in the other African sub-regions national ABS regulatory processes are comparatively advanced (eastern and southern Africa) or have strong political backing (COMIFAC), this regional process requires more awareness-raising and information elements.

Feedback loops and pan-African team building

On the one hand there is an increasing demand to support regional capacity-building processes addressing specific needs, framework conditions and language particularities. On the other hand, there is a political necessity for a common African position in the current negotiations of the International ABS Regime. In order to facilitate the development of this common position, viewpoints regarding the potential elements of the International Regime on ABS have been elaborated in all sub-regional processes. Prior to the ABS Working Groups (ABS-WG 5, Montreal, October 2007 and ABS 6, Geneva, January 2008), the Initiative provided a meeting platform with simultaneous interpretation of the African ABS Focal Points, enabling them to discuss the results of the sub-regional workshops and to elaborate a common position. Technical briefs on the agenda items of the ABS negotiations were prepared in French and English for the African delegates by the independent Centre for International Sustainable Development and Law (CISDL).

Box 20.3 *Milestones of pan-African team building*
- ABS delegates' briefing:
  Common position of francophone and anglophone countries; improved preparedness of the African Group at the ABS negotiations; results of multi-stakeholder processes integrated into African ABS negotiation positions; increased collaboration and harmonization of standpoints within the African negotiator's group – also regarding other Programmes of Work of the CBD.
- Mahé, Minister briefing:
  First high-level dialogue on ABS in Africa ever; significantly raised awareness of ABS at African decision maker's level; elaboration of joint African ministerial position on ABS for the ninth COP High-Level Segment.

Source: ABS Capacity Development Initiative for Africa, 2008

Addressing the high demand from the African Group also prior to the ninth COP in Bonn, Germany, a briefing was conducted, which was attended by both the CBD Secretariat and the African Group ABS Focal Points. Furthermore, the Initiative invited 15 African Ministers responsible for national CBD implementation to the ninth COP preparatory meeting in Mahé, Seychelles. This high-level gathering[13] also provided simultaneous interpretation, focused strongly on ABS, while also tackling other important COP issues on the sidelines. The meeting included hands-on presentations of bioprospecting cases during a field trip, a standard element of the events under the Initiative. The demand for regional preparation and coordination has continued to be met by supporting the preparation of a coordinated submission of the African Group to ABS-WG 7, Paris, April 2009, by providing the final coordination platform during the third Pan-African ABS Workshop in Antsiranana (Diego Suarez), Madagascar, in November 2008.

Reinforced efforts for Africa are crucial for success

At the sixth meeting of the Ad Hoc Open-ended ABS Working Group in Geneva, January 2008, the ABS negotiations reached a turning point. For the first time all parties agreed upon initial basic elements for the International ABS Regime. The negotiating dynamic shifted from impasse to a concerted effort to make progress. This has led to a need for parties to engage more technical expertise to evaluate all aspects of the elements under negotiation, including potential overlaps and linkages with other international regimes, treaties and conventions.

Having these challenges in mind, it is imperative that ABS capacity-development measures in Africa be intensified at all levels, including communication, education and awareness-raising of issues. If African countries do not contribute effectively and infuse their perspectives towards the shape of the International ABS Regime, in the long run this would probably compound disadvantages for the region, negatively impacting conservation of biodiversity and economic development. For these reasons, the Initiative is supporting efforts to sharpen the depth and breadth of the knowledge of key African actors in the process and substance of ABS issues.

Perspectives for the future

Whilst the Initiative continues to support the African Group in the negotiations in the ways detailed above, it is imperative that all actors engaged in the process engage with the challenge of implementing the regime beyond the 2010 deadline. There is a critical need to consider how best the incumbent regime can be implemented at the regional, national and local levels. This task requires ABS legislation to be crafted in a way that dovetails with existing regional and national frameworks on conservation and the rights of indigenous and local communities. This constitutes recognition of the broader CBD context from which ABS has emerged.

The Initiative's focus on developing the capacity of stakeholders during the negotiations feeds into the subsequent implementation of an ABS regime. The specific knowledge, skills and partnerships forged through the Initiative will be crucial for effectively implementing the regime at the international as well as subsequently at the national level. Moreover, by focusing on developing capacity among Africa's diverse ABS stakeholders, it is hoped that the results will also be diverse, including the conservation of the region's biodiversity and increased African biotrade, and lead to empowered local communities better able to generate culturally and environmentally appropriate livelihoods.

Notes

1 See The Convention on Biological Diversity, http://www.cbd.int, accessed 13 January 2009. The Handbook of the CBD, www.cbd.int/doc/handbook/cbd-hb-all-en.pdf, accessed 13 January 2009, and the decisions of the CBD, which can be found at http://www.cbd.int/decisions, accessed 13 January 2009, are also useful resources.
2 The other two 'Rio Conventions' are the United Nations Framework Convention for Climate Change (UNFCCC) and the United Nations Convention to Combat Desertification (UNCCD).
3 As of May 2008, 191 parties had ratified the Convention, including the European Union.
4 General Assembly resolution 1803 (XVII) of 14 December 1962, 'Permanent sovereignty over natural resources', http://www.unhchr.ch/html/menu3/b/c_natres.htm, accessed 4 February 2009. See also Schrijver, N. J. (1995) *Sovereignty Over Natural Resources: Balancing Rights and Duties in an Interdependent World*, Dissertation, Rijksuniversiteit Groningen, 482pp.
5 Declaration of the United Nations Conference on the Human Environment, http://www.unep.org/Documents.Multilingual/Default.asp?DocumentID=97&ArticleID=1503, accessed 4 February 2009.
6 Our Common Future: Report of the World Commission on Environment and Development, http://worldinbalance.net/agreements/1987-brundtland.php, accessed 5 February 2009.
7 For example, as reflected in the proposed legal principle 7 for environmental protection and sustainable development: 'States shall ensure that conservation is treated as an integral part of the planning and implementation of development activities and provide assistance to other States, especially to countries of the global South, in

support of environmental protection and sustainable development'. See Our Common Future: Report of the World Commission on Environment and Development. Annex 1: Summary of proposed legal principles for environmental protection and sustainable development adopted by the WCED Experts Group on Environmental Law, http://www.worldinbalance.net/agreements/1987-brundtland. php, accessed 5 February 2009.
8   For a detailed discussion, see Kamau and Winter, 'Streamlining access procedures and standards', in this book.
9   The Bonn Guidelines on ABS, http://www.cbd.int/abs/bonn.shtml, accessed 13 January 2009.
10 Homepage of the '2010 target' to significantly reduce the rate of loss of biodiversity by 2010, http://www.cbd.int/2010-target, accessed 13 January 2009.
11 ABS Capacity Development Initiative for Africa, http://www.abs-africa.info, accessed 13 January 2009.
12 Burundi, Cameroon, Congo, Democratic Republic of Congo, Gabon, Guinea Equatorial, Central Africa, Rwanda, Sao Tome and Principe and Chad.
13 See the statement, for example, that was delivered at the meeting of the Executive Secretary of the CBD by Dr Ahmed Djoghlaf, http://www.cbd.int/doc/speech/2008/ sp-2008-04-10-seychelles-en.pdf, accessed 13 January 2009.

References

More information on the ABS Capacity Development Initiative for Africa is available at www.abs-africa.info
Henne, G., Liebig, K., Drews, A. and Plän, T. (2003) *Access and Benefit-Sharing (ABS): An Instrument for Poverty Alleviation. Proposals for an International ABS Regime*, German Development Institute (GDI), Bonn
Le Prestre, P. G. (ed) (2002) *Governing Global Biodiversity: The Evolution and Implementation of the Convention on Biological Diversity*, Aldershot, Ashgate
Secretariat of the Convention on Biological Diversity (2002) *Bonn Guidelines on Access to Genetic Resources and Fair and Equitable Sharing of the Benefits Arising out of their Utilization*, Montreal, Secretariat of the Convention on Biological Diversity
Secretariat of the Convention on Biological Diversity (2005) *Handbook of the Convention on Biological Diversity Including its Cartagena Protocol on Biosafety*, 3rd edn, Montreal, Secretariat of the Convention on Biological Diversity

PART FIVE

USER COUNTRIES' MEASURES

Chapter 21

Disclosure Requirement – A Critical Appraisal[1]

Evanson C. Kamau[2]

Introduction

Very little generally has been done in the European realm to bring national laws into compliance with the ABS regulations of the 1992 Rio de Janeiro CBD,[3] as required by Article 15.7.[4] The first move taken within Europe was to create an enabling basis by developing the so-called *Pan-European Biological and Landscape Diversity Strategy* in 1994.[5] This was intended to introduce a coordinating and unifying framework that could strengthen and expand existing initiatives that were supportive to the implementation of the CBD.[6] In 1998, the European Community (EC) adopted a biodiversity strategy,[7] which recognized the need for the Community to promote multi-lateral frameworks for ABS, encourage the development of voluntary guidelines for bilateral cooperation in ABS and support countries of origin of GRs to develop national strategies on bioprospecting and access, while considering relevant multilateral frameworks and instruments.[8]

The 1998 EC Biodiversity Strategy defined a precise framework for action by setting out four major themes and specifying sectoral and horizontal objectives to be achieved.[9] Among the four themes around which it was developed are conservation and sustainable use of biological diversity and sharing of benefits arising out of the utilization of GRs.

In the same year, the European Parliament approved the proposal for a European *Directive on the Legal Protection of Biotechnological Inventions* (Directive 98/44/EC)[10] (Leskien, 1998, pp16–19), also known as the 'Biotechnology Directive'. The Directive took ABS considerations into account.

This chapter examines how the CBD benefit-sharing (BS) provisions were implemented in the EC through Directive 98/44/EC, as well as their transposition in individual Member States. It then gives an evaluation of the disclosure requirement and makes suggestions on other means of achieving the BS objective and obligation of Article 15.7 of the CBD.

Implementation of the **ABS** provisions through Directive 98/44/EC

Directive 98/44/EC encourages patent applications to include information on the geographical origin of biological material.[11] It also requires Member States to give particular weight to, among others, Articles 8(j),[12] 16.2[13] and 16.5[14] of the CBD when bringing into force their laws, regulations and administrative provisions necessary to comply with the Directive.[15] In addition, it embodied the observation of the third Conference of Parties (COP)[16] concerning existing need to clarify the relationship between intellectual property rights (IPRs) and the CBD, in particular on issues relating to technology transfer, conservation and sustainable use of biological diversity and the sharing of benefits arising from the utilization of GRs, and the protection of relevant knowledge of local communities.[17] These considerations did not make any amicable contribution to the question of BS, for two main reasons. One, all three were taken into the recitals, thus making them legally non-binding on the Member States.[18] Two, disclosure (i.e. including information) of geographical origin of biological material was left to the application and registration procedure of individual Member States.[19] The recital did not place any consequences on applicants who claimed the geographical origin was unknown. It also did not demand further requirements to prove their claims before rights were granted.[20] With this approach, the Directive produced undesirable results in relation to the thorough implementation of the ABS provisions of the CBD, the desires of provider countries to have their measures enabled in user countries and the possibility of the formation of a uniform and effective ABS regime in Europe.

As a follow-up to the EU's Biodiversity Strategy of 1998, the EC adopted four biodiversity action plans in 2001 (March 28).[21] These (sectoral action) plans aimed at supporting integration of biodiversity protection into EU policies in natural resources, agriculture, fisheries and economic development.[22] They defined concrete actions and measures for fulfilling the objectives spelled out in the strategy and specified measurable targets.[23] The action plans aimed at stopping the loss of biodiversity in wildlife, ecosystems, crop varieties, animals and fish.[24] They are also a fundamental basis for the sixth Environmental Action Plan, which targets at halting further biodiversity loss by 2010.[25] Like the recitals, these action plans were non-binding and consequently offered no solution to the core problem of ABS.

The pace and approach of adoption of the CBD in the EC in relation to third countries was also hampered by the fact that adoption was focused on preservation of EC Member States' environment. From the perspective of the EC, this was the kind of approach expected of resource states. But this trend took a different turn after the enactment of the so-called Bonn Guidelines[26] by COP 6[27] in the Hague in April 2002.

The call to promote the implementation of ABS contracts by users received another boost from the vote of the WSSD (World Summit for Sustainable Development) environment summit in Johannesburg in August 2003. The WSSD environment summit made recommendations that the Bonn Guidelines be made into a legally binding instrument.[28] Towards the close of the summit, an agreement was reached to call for an international regime[29] to promote and safeguard the fair and equitable sharing of benefits arising from the utilization of GRs.[30] Another positive output of the summit was the restatement that World Trade Organization (WTO) rules shall not override global environmental treaties.

These developments prompted the European Commission to reconsider its responsibility in the process of creation of a legal order in international transfer of genetic materials.[31] Consequently, it undertook two measures.[32] First, it initiated a discussion on: (1) possible introduction of a disclosure requirement for patent applicants in the EC legal order, as well as making it a formal condition of patentability; and (2) the possible development of a certificate of origin for genetic materials as evidence of prior informed consent (PIC). Second, it placed the obligation of observing the access regulations of provider (resource) states on the European enterprises and institutions.[33] The Environment Commissioner concretized the latter in a press release in January 2004.[34]

The next section looks at the course and options the implementation of the ABS provisions of the CBD has taken in the EU and the Member States.

Implementation of CBD ABS provisions in EU Member States

Application of CBD in individual EU states

Article 15.7 CBD reads: '*[E]ach Contracting Party shall take legislative, administrative or policy measures, ... with the aim of sharing in a fair and equitable way the results of research and development and the benefits arising from the commercial and other utilization of genetic resources with the Contracting Party providing such resources ... upon mutually agreed terms.*' States are required to perform their international obligations in good faith, having the liberty to decide on the modalities necessary within their legal systems to achieve such performance. There is also a general duty on states to bring domestic law into conformity with obligations under international law, but the method of achieving this result is left to the domestic jurisdiction of states.

In practice, there is lack of uniformity in the different national legal systems as to how international obligations should be translated into internal law and which legal status they should have domestically. This variation is mainly influenced by the distinct legal traditions based on dualism and monism.

Dualists uphold the existence of two distinct sets of legal order, which regulate different subject matter. In case of a conflict between international law and municipal law, the dualist would assume that a municipal court would apply municipal law.

Monists uphold the unity of the two legal systems; both international and municipal law form part of the same legal order. There are, however, different schools of thought here. One school upholds the primacy of municipal law over international law with the argument that it is states that possess the power to enforce law – international law included. The second school upholds the primacy of international law over municipal law. The most radical among the latter argue that any law that conflicts with international law is void. The less radical monists say that municipal law applies within the state, regardless of whether or not it conflicts with international law. Accordingly, it is up to the state to decide to what extent municipal law must conform to international law. Hence, if a state passes legislation that conflicts with international law, it will be valid within the jurisdiction of the state. However, at the international level, international law would have primacy over municipal law. Hence, if a conflict arises before international law courts, a national regulation will inevitably be inapplicable. This is the same as saying that international law and domestic law are superior each in its relative sphere and hence seems to conform to the dualist position.

Due to these practices, the CBD would not apply directly in countries with a dualist tradition. Provisions that are recognized as 'self-executing', that is, clear and unconditional, are usually transposed into national law depending on varying constitutional approaches either through 'incorporation', 'adoption', 'transformation' or 'reception'. In Germany and Austria, publication of the act of consent or the act of publishing the treaty in the relative organs will automatically make the treaty effective internally. This of course is under the assumption that the treaty has already come into force internationally. This procedure is often referred to as general transformation or adoption of international law rules. Special transformation refers to the practice according to which special internal legislation must be passed by the law-making bodies of a state for international treaties, which are self-executing and have entered into force, to be legally effective or binding internally, that is, within the state. The UK adheres to such a practice.

The question as to whether individuals can derive rights and obligations directly from international law does not arise in cases where treaties are not self-executing or in cases where special transformation is envisaged. Where self-executing international norms become automatically part of national law, the question then arises as to whether such norms become part of municipal law (general transformation), or whether they become binding internally but in their capacity as international law norms (adoption).

General transformation or incorporation makes the international norm part of municipal law. The act of consent decides which status they are to have. Some authors maintain that such norms will prevail over earlier municipal or internal laws. Likewise, later internal laws and decisions can repeal them. However, the state will still be bound by the international rules and will not be able to rely on provisions in its internal laws as a defence to a claim based on international law. As a rule, the internal laws transforming the international law norm will only be valid after entry into force of the international treaty in question and will lose validity once the international norm expires. Since they have become part of the internal laws, national courts will interpret such rules in the light of national law.

According to the theory of adoption, the international rule becomes part of municipal law, but does not lose its character as an international law norm. Such a law will be subject to international law rules, especially in questions pertaining to validity, expiry and interpretation.

It is not clear whether Article 15.7 of the CBD is a self-executing norm, but it is true that no user country has conformed to this provision's obligation (Tvedt and Young, 2007, p99ff). The standard set by Directive 98/44/EC is not sufficient to achieve such compliance. EU states, especially where self-executing norms of international law do not automatically become part of municipal law, must take individual initiatives to meet this obligation. Deficient corporate (EU) compliance would still not justify failure by individual states to fulfil their obligations under international law.

Implementation of ABS provisions in selected EU states

The implementation approach of ABS provisions in the European Member States vary. A cross section of implementation procedures in some of the EU countries though reveals that the options have been highly influenced by recital 27 of Directive 98/44/EC. This has resulted mostly in the adoption of measures of a non-binding character.

The adoption of the 'Biotechnology Directive' through the amendment of Article 34a of the patent law of 16 December 1980 facilitated implementation in Germany.[35] The amendment text embraced the non-binding formulation of recital 27 of the Directive.[36] This requires that, in patent applications, information of the geographical origin of biological material used in an invention be disclosed, if known. However, whether the applicant discloses the origin, or claims that it is unknown, the processing of patent applications or the validity of rights arising from granted patents is not prejudiced.

The initial implementation approach in Belgium differed from that of Germany. It sought to appropriate the patent law *ordre public* and morality maxims to disqualify any invention developed from biological material that was collected or exported in breach of Article 3,[37] 8(j),[38] 15[39] and 16[40] of the CBD. To this effect, Article 4(4) of the 1984 Belgian Patent Act was modified by the draft proposal of 2000,[41] which required that an application contain the geographical origin of biological material on the basis of which the invention was made. This approach, however, was strange and lacked a legal basis as no precedent for such a broad interpretation of *ordre public* or morality existed (Dutfield, 2003, p34 and note 74; van Overwalle, 2002, pp233–236). The draft (amendment) was not adopted, but rather a new approach, similar to the German one, was used. Article 5 of the patent law was modified to adopt recital 27 of the Biotechnology Directive in 2005. With this, the formal disclosure requirement in patent application was anchored in Belgian law. Consequently, 'unknown' geographical origin of biological material used in an invention has no consequences for the patent application procedure and the rights arising therefrom. However, as would be seen later, other consequences might arise if violation was established.

In response to the EC Biotechnology Directive, Sweden also implemented a new rule under the Patents Act, which came into effect in May 2004. This provision, Rule 5(a) (SFS 2004: 162), adopts the formulation of recital 27 of the biotechnology Directive. Accordingly, patent application involving an invention from biological material of plant or animal origin shall include the geographical origin of such material, if known. Missing information on geographical origin does not seem to affect the processing of the application or the validity of rights arising from a granted patent. The law demands though that the applicant indicates or declares that the origin is unknown.

Denmark implemented the Directive 98/44/EC disclosure requirement[42] by enacting a disclosure of origin clause in the IPRs legislation in 2000.[43] With it the Danish Patent Act was amended,[44] and subsequently the ministerial regulations on patents (Regulation 374 19/6/1998) also, by adding the new provision on disclosure to §3.[45] The provision[46] essentially adopted the biotechnology Directive's formulation. It requires that geographical origin of biological material of vegetable or animal origin on the basis of which an invention is made be disclosed in patent application, if known. If the origin is not known, the applicant has to indicate this in the application, although such a declaration and ignorance concerning such information have no effect on the application procedure or the validity of granted rights.

The implementation regulation of the Romanian Patent Law 64/1991, rule 14, point (1)(c), requires that if TK is used in an invention ('state of the art'), this should be clearly indicated in the description including its source, when known.[47] This regulation was republished on 15 October 2002 and approved on 18 April 2003. Romania recognizes the information requirements of patent applications for GRs used in inventions regardless of technology used or nationality of applicant.[48] The law, however, does not foresee any consequences for non-compliance under patent or civil law.

Apart from the five countries examined above, there are more countries in the EU with ABS legislations in place, but they are resource-state oriented. Three of them are examined below.

In 2002, Portugal passed a law, Decree law no 118/2002,[49] to protect plant varieties and TK. Indigenous plant material not protected by IPRs may be registered by application of the entity representing the interests of the geographical area where they exist and included in the Directory of Registration of Plant Genetic Resources.[50] The entity is responsible for in situ maintenance of the plant material.[51] Authorization of access to germ plasm of plant material and use of plants or parts thereof for industrial or biotechnological purposes is granted subject to prior consultation with the entity owning the registration.[52] It also shares in the benefits arising from utilization of the registered variety.[53] TK is also protected against reproduction, commercial or industrial use through registration.[54]

Bulgaria's Biological Diversity Act (State Gazette No 77/9.08.2002)[55] is very clear on its intention to regulate access to GRs and BS. According to Article 66, the state may provide access to the GRs of its natural flora and/or fauna prior to a written agreement. The terms and manner of sharing benefits are to be included in the agreement and must be profitable to all parties involved. The agreement shall show the natural origin of the material, provide for provision by the 'user of results of the research and technologies obtained from, related to, or derived from accessed resources', 'recovery of part of the resources obtained in use of the material, as well as of derivatives or studies for commercial purposes' and 'participation in joint scientific studies'.

The state may permit gratuitous access to GRs for non-commercial purposes. Where they are 'subject to patents and other intellectual property rights, such access shall be provided in compliance with the provisions of legislation specific to this sphere'. Whichever the case, the user of GRs is under an obligation to seek the written consent of the provider (owner) prior to allowing third-party utilization. A regulation adopted by the Council of Ministers shall establish the terms and a procedure for provision of access to GRs.

The Medicinal Plants Act[56] regulates the management, conservation and sustainable exploitation activities with regard to medicinal plants and the collection and buyout of herbs obtained therefrom,[57] irrespective of their ownership (Articles 1 & 2). The Act establishes the procedure for acquisition of access permits and payment of utilization fees depending on the location of the plants (Articles 21–24). All medicinal plants to which the Act applies are annexed to the Act. Obtainment of genetic material from artificially propagated medicinal plants shall be provided for in the Seed Stock and Planting Stock Act, and also in the Protection of New Plant Varieties and Animal Breeds Act (Article 20(2)).

Lithuania adopted the Law on National Plant Genetic Resources No IX-533 of 9 October 2001 to regulate the collection, conservation and use of plant GRs. It allows free access of GRs for non-commercial purposes listed in Article 14 that include plant breeding and research, scientific research, reproduction, exchanges and satisfaction of rational human needs. The law possesses no provisions on PIC, BS or monitoring mechanisms. Article 17, however, contains enabling provisions on liabilities for violations.

The UK has no specific legislation on access to GRs and BS. To deal with issues pertaining thereto, relevant rules in other areas of law viz. access and trespass, property, IPRs, national parks and overseas territories are used.[58] However, the government's focal point, the Department for Environment, Food and Rural Affairs (DEFRA), has undertaken a number of measures in two stages in order to clarify issues pertaining thereto.

In 2002, the so-called Wilding Report[59] carried out a review of policy on GRs for food and agriculture. Its results recommended the development of a new and comprehensive policy on conservation and sustainable use of GRs for food and agriculture, linking biodiversity and agriculture. The report, however, concentrated on national policy on GRs; it did not attend to the CBD perspectives concerning PIC, mutually agreed terms (MAT) and BS.

In 2004, two years later, a second stage was undertaken to review the experience of implementation of UK stakeholders of ABS arrangements under the CBD. It was also important to investigate the issue of biopiracy, especially following multiple implications of UK users, in order to address the question of reputation of British organizations.

The report reveals, inter alia, that the CBD and ABS provisions are better understood and more welcomed by public and semi-public institutions.[60] Awareness of the Bonn Guidelines, on the other hand, is low except in large ex situ collections.[61] At the implementation level, putting CBD into practice is generally proving difficult and complex: the understanding of ABS provisions with regard to scope and implications are often not properly understood.[62]

Finally the report gave recommendations, which include the need to: identify any best practice necessary to ABS; include ABS issues in UK law as common law and other statutory provisions relevant thereto subject to constant change; encourage main organizations to develop ABS arrangements; and create a wider stakeholder network.[63]

Assessment

User measures have been implemented only very sparingly in the EU. From the nine countries examined, only five – Germany, Belgium, Sweden, Denmark and Romania – have attempted to do so. Three – Portugal, Bulgaria and Lithuania – have only provider measures, and one – the UK – none. The five did not go beyond the requirement of the Biotechnology Directive, which just requires disclosure of the geographical origin of biological material used in an invention during patent application, if the origin is known. Declaration by the applicant that the origin is unknown or the applicant's ignorance concerning such information has no consequences for the patent application procedure or obtained rights. This is also true for Sweden and Denmark, which in addition require a clear indication or declaration by the applicant in the application that the origin is not known.

Value of disclosure requirement

The requirement to disclose the (geographical) origin of genetic material used in an invention during patent application was proposed as one of the possible tools for enabling provider measures in user countries. It is built on a number of assumptions. One of it is that, once the source is known, the provider's burden in pursuing his rights is eased.[64] That, however, might be a hasty conclusion. There are a number of issues to consider before any conclusion can be drawn as to how useful a disclosure clause in patent law would be, especially in easing BS. First, can and how can the disclosure requirement be enforced? Second, does it create any obligations for the user country? Third, does it yield any rights for the source country? Fourth, does it oblige the user to share benefits? Another important question, which will not be discussed in this chapter, is whether it is in conflict with the Trade-Related Aspects of Intellectual Property Rights (TRIPS) Agreement (Barber et al, 2003; de Carvalho, 2003; Dutfield, 2003).[65]

Can it and how can it be enforced?

As seen above, Directive 98/44/EC did not place any obligation on the EC Member States to demand mandatory disclosure of the origin of genetic material upon which an invention is based in patent application (also Garforth et al, 2005). So far, there is no appropriate and/or effective mechanism of enforcement. This requirement relies wholly on voluntary disclosure by users.

The Directive does not hinder the imposition of stricter measures such as stand-alone or enhanced disclosure[66] requirement. A number of countries have instituted mandatory disclosure requirement within their patent laws (Hoare and Tarasofsky, 2006), mostly through a stand-alone disclosure requirement (Hoare and Tarasofsky, 2006). Norway and Denmark have instituted such a requirement: non-compliance is punishable under their penal codes (Hoare and Tarasofsky, 2006; also Tvedt and Young, 2007, p35), but is not treated as sufficient legal ground to invoke refusal or invalidation of a patent (Hoare and Tarasofsky, 2006; also Tvedt and Young, 2007). Belgium has instituted a formal requirement whereby non-compliance could risk the processing of the patent application (Hoare and Tarasofsky, 2006). Already-granted patents, on the other hand, would remain valid, but attract a fine if disclosure was wrongfully made (Hoare and Tarasofsky, 2006). This is considered as an occurrence that is unlikely to happen, as the Belgian patent office does not check compliance, not being a searching authority (Hoare and Tarasofsky, 2006).

Enhanced disclosure imposes strict compliance and attaches grave consequences to non-compliance. It results in rejection or revocation of a patent and loss of patent rights in case of violation (Hoare and Tarasofsky, 2006; Tvedt and Young, 2007, p35). None of the EC countries have instituted such measures. The Andean Pact,[67] Costa Rica, Brazil and India apply such measures for failure to disclose the geographical origin and/or PIC.[68] Decision 486, Article 75 of the Andean Community, for example, states the following:

> *The competent national authority may, either ex officio or at the request of a party, and at any time, declare a patent null and void, where:*
>
> *...*
>
> *(g) when pertinent, the products or processes in respect of which the patent is being filed have been obtained and developed on the basis of genetic resources or their by-products originating in one of the Member Countries, if the applicant failed to submit a copy of the contract for access to that genetic material;*

(h) when pertinent, the products or processes whose protection is being requested have been obtained or developed on the basis of traditional knowledge belonging to indigenous, African American, or local communities in the Member Countries, if the applicant has failed to submit a copy of the document certifying the existence of a licence or authorization for use of that knowledge originating in any one of the member Countries.

Until now, enhanced disclosure is known to apply only within the territories of those countries that have enacted such measures and in connection to patents of inventions based on their own GRs. In addition, they are so limited to offer a feasible solution to the challenges faced in BS. Also, these measures have not been applied widely and, therefore, do not provide a lead on how BS can be ensured through enforcement of disclosure requirement.

Does it oblige a user('s) country?

There are presently no mechanisms, which oblige a government to oversee the BS interests of another government on its territory. Disclosure requirements as such do not oblige the state in whose territory a patent application, based on foreign genetic material, is made to verify whether information given by the applicant is correct. Likewise, they do not commit governments to inform source countries when patent applications based on their genetic material are made (also Tvedt and Young, 2007). Apart from that, even where voluntary disclosure is made, there's no sure way of ascertaining the exact source of the genetic material (Tvedt and Young, 2007).

Does it create rights for a source country?

Assuming that the user obtains genetic material legally, discloses voluntarily and informs the provider, which post-access rights would the latter have gained and which mechanisms would he use to claim them? Arguably, there are no additional rights gained by the provider. Disclosure does not give entitlement to new rights, but attempts to pave a way for the provider to quasi-receive or -claim, from the user, the rights accruing from the user's utilization activities. The rights of the provider remain as spelled out, for example in Article 8(j), 15.7 and 16.3 of the CBD, and as hopefully adopted in all existing national ABS legislations. After allowing access, the provider is entitled to a fair and equitable share of the benefits from utilized genetic material, as well as any technology that utilizes such material – on mutually agreed terms. It implies that, first, there might not be any known benefits at the level of disclosure that the user country could inform the provider about. Second, it is not always obvious whether and when benefits will accrue. Third, in spite of disclosure, the provider will still have to track and make demand on the user who is located in a foreign jurisdiction and not simply rely on some rights. Like disclosure, the contractual or moral rights of the provider to receive benefits from the utilization of genetic material or TK are highly dependent on good faith from the user side.

Does it oblige the user to share benefits?

Each contract party is expected to fulfil its duty(ies). Once the provider has allowed access, as well as export of genetic material, the remaining duty is for the party that has acquired such materials to share any accruing benefits with the provider.[69] Based on this, can the disclosure requirement oblige a 'contract-bound' user to honour his contractual obligations? For GRs acquired irregularly, it is a question of morality whether a user would after all want to play it fair with the provider. For legally obtained GRs, the disclosure requirement has no ability to oblige the user to share benefits as well. Whether the resources are obtained through a regular or irregular manner, the deciding question is, how does the provider enforce his BS rights in a foreign jurisdiction? The most the user's (EC) government can do under the disclosure requirement, as already seen, is to penalize a patent applicant who does not comply with the formal (patent) application procedure by giving wrong information. It cannot ensure that users in its territory share benefits with providers. What happens from there and throughout the working of a patent (or any other usage of the material) is the parties' business. Therefore, in the absence of user measures, disclosure requirement becomes a very abstract notion and does not avail much for a provider to pursue his BS rights once the resources leave his territory. If the user is unwilling to collaborate with the provider, the short way paved by disclosure turns into a cul-de-sac (dead-end street) for the provider, thus making BS based on disclosure, be it for illegally or legally obtained genetic material, an issue of morality (Posey and Dutfield, 1996).

Interim remarks

Disclosure requirement as an enabling tool for BS suffers from numerous limitations, for a number of reasons. First, Directive 98/44/EC, which adopts it into EU law, neither makes disclosure a strict obligation nor imposes consequences that have earnest effects on its violators. Second, it depends on a legal system outside the jurisdiction of the provider to track the usage of the resources and compel the user to share benefits. Third, it relies on the same system, which is meant to protect the rights of the patent holder, to enforce the rights of the provider of GRs and TK. Fourth, it relies on the party who is protected by the (patent) law, the user, to initiate legal action on behalf of the provider.

Some authors consider the latter assumption that the intellectual property system functions through legal action initiated by the protected party as one of the major fallacies of the patent disclosure idea (Tvedt and Young, 2007, note 165). Indeed, the third and fourth points are at a crossroads with the essence of patent rights, which is to deter possible infringements against the invention (including challenges) for a stipulated period of time (Kamau, 2004, pp37, 69ff). According to some economic theories, patent protection is justifiable because it encourages inventors to invent more, thus enhancing technology growth, to disclose the invention, to disseminate technology, to innovate and to ensure that they recover resources invested in research (Kamau, 2004, p47ff). Therefore, it is unlikely that patent law would initiate, or would be used to initiate, an action that contradicts its essence or renders it unreliable (in fulfilling its purpose). This makes patent law of little help in enforcing BS. The next section proposes ways of encouraging compliance.

Which way out? Searching for alternatives

Obviously, as already seen, the disclosure requirement in its current state is not sufficient to achieve the BS objective and obligation of Article 15.7 of the CBD. At the moment, there are virtually no user-side measures available to this end. Ideas for improving the situation vary from analysis of mandatory or binding measures to softer measures in the form of voluntary approaches and incentives.[70] There are feelings that the latter are likely to be more useful in the present situation – especially if mixed with the former – due to the unwillingness of most parties to adopt mandatory measures, as well as hardships in achieving enforcement.[71]

Voluntary measures

Voluntary measures are of two distinct kinds: motivational and permissive. Permissive measures are often not useful because they normally consist of recommendations to act in a certain manner without benefits; for example, disclosure in patent application. Motivational measures, on the other hand, offer benefits for defined actions willingly taken. Such benefits include protection from liability, increased legal certainty, access to special government services, priority treatment (Tvedt and Young, 2007, Chapter 6) and tax benefits (Ruiz Muller and Lapena, 2007, Chapter 4). Nonetheless, even these have limitations because the user will most likely opt for them if their value is greater than the costs of compliance.

Incentives and other motivators

There are a number of incentives and motivators that user countries may use to enforce compliance. These include elimination of perverse incentives, creation of disincentive for non-compliance and removal of disincentives for compliance.

Elimination of perverse incentives aims at levelling the ground for compliant and non-compliant users. Currently, the disclosure of origin requirement in practice favours the non-compliant user and places the compliant user in comparison at a disadvantage. A user who accesses GRs and/or TK illegally and declares their origin unknown saves on access expenses and time, and escapes the obligation to share benefits for commercialization and/or any other usage of GRs and/or TK. In contrast, the compliant user who incurs expenses to meet the administrative and regulatory requirements of the source country, including PIC and MAT, as well as complies with the user disclosure measures, is still the one expected to share benefits fairly and equitably with the provider. This provision, which creates reverse effects, needs to be redesigned by obliging all users to comply with the requirements of source countries and their BS obligations, with or without ABS contracts. In addition, a compliant user should enjoy some benefits as motivation for compliance.[72]

User measures should also discourage non-compliance by facilitating *post facto* negotiations with source countries and BS according to statutory terms and standards defined by responsible agencies or courts. Such measures should yield less favourable results than if the user had negotiated an ABS contract *ab initio* and hence serve as disincentives for non-compliance.

Finally, disincentives that discourage users to comply should be eliminated. An example is the legal uncertainty of obtained rights, created when users who participate openly in ABS negotiations are targeted for claims of 'biopiracy'.

Conclusion

The EU states have taken minimal measures, probably as a formal gesture, to assist source countries' ABS measures. They are not effective and supportive to ABS regimes of provider countries. A clear lack of commitment to undertake more serious measures exists. There is also an indication that the reluctance is influenced by a conflict of interests of providers of GRs and users.

Without proper user ABS enforcement measures, the requirement to disclose the origin of the GRs used in an invention during patent application remains a toothless tool for securing benefits from utilized genetic materials. Most user countries have not taken seriously the obligation of Article 15.7 of the CBD. More proactive user measures are needed. Probably mandatory measures, criminal and civil procedures might accelerate this process, but incentives and motivators for users are expected to improve the results.

Notes

1 I would like to thank the DFG for funding the Forschungsstelle für Europäisches Umweltrecht (FEU)-ABS project within which this chapter has been written. Many thanks also to Prof. Gerd Winter, the project supervisor, for his invaluable comments.
2 The views expressed herein are solely those of the author.
3 The CBD was adopted at the Earth Summit in Rio de Janeiro, Brazil, in June 1992, and entered into force in December 1993. As the first treaty to provide a legal framework for biodiversity conservation, the Convention established three main goals: the conservation of biological diversity; the sustainable use of its components; and the fair and equitable sharing of the benefits arising from the use of GRs. Contracting Parties are required to create and enforce national strategies and action plans to conserve, protect and enhance biological diversity. They are also required to undertake action to implement the thematic work programmes on ecosystems and a range of cross-cutting issues that have been established to take forward the provisions of the Convention. As of December 2002 there were 187 Contracting Parties to the Convention. For a free copy of the CBD, see http://www.biodiv.org/doc/legal/cbd-en.pdf, accessed 3 January 2009.
4 The Article states that '[E]ach contracting party shall take legislative, administrative or policy measures, as appropriate, ... with the aim of sharing in a fair and equitable way the results of research and development and the benefits arising from the commercial and other utilization of GRs with the Contracting Party providing such resources. Such sharing shall be upon mutually agreed terms'.
5 http://www.chm.org.uk/cats.asp?t=402, accessed 3 January 2009.
6 Ibid.
7 COM (98) 42.
8 Ibid.
9 Ibid.
10 [1998] OJ L 213/13, http://eur-lex.europa.eu/LexUriServ/site/en/oj/1998/l_213/l_21319980730en00130021.pdf, accessed 3 January 2009.
11 [1998] OJ L 213/13, recital 27, http://www.europabio.org/documents/offdocs/199844.pdf, accessed 21 April 2008.
12 Article 8(j) requires each contracting party, as far as possible and as appropriate, subject to its national legislation, to respect, maintain and preserve knowledge, innovations and practices of indigenous and local communities relevant for the

conservation and sustainable use of biological diversity, and promote their wider application with the approval and involvement of the holders of such knowledge, innovations and practices, and encourage the equitable sharing of benefits arising from the utilization of such knowledge, innovations and practices.

13 Article 16(2) requires contracting parties to provide and/or facilitate access and transfer of technologies that are relevant to the conservation and sustainable use of biological diversity or that use GRs (without causing significant damage to the environment) to developing countries under fair and most favourable terms, including on concessional and preferential terms where mutually agreed, and, where necessary, to provide financial support and incentives in accordance with the established financial mechanism (Article 21) so as to enable developing countries to achieve the objectives of the CBD.

14 Article 16(5) requires that contracting parties cooperate to ensure that intellectual property rights do not run counter to, but rather are supportive of, the objectives of the Convention.

15 Above, note 10, recital 55.

16 See COP 3, Decision III/17.

17 Above, note 10, recital 56.

18 Recitals do not create legally binding obligations on Member States. See also Garforth et al (2005).

19 This was taken in recital 27, which reads: '[W]hereas if an invention is based on biological material of plant or animal origin or if it uses such material, the patent application *should, where appropriate*, include information on the geographical origin of such material, *if known*; whereas this is without prejudice to the processing of patent applications or the validity of rights arising from granted patents.'

20 Save the non-binding nature of recitals, the term 'should' does not suggest an absolute legal requirement. For more on the voluntary nature of this requirement, see Dutfield (2003, p33 & note 71).

21 http://biodiversity-chm.eea.europa.eu/stories/STORY1016812291/, accessed 21 April 2008.

22 Ibid.

23 Above, note 5.

24 http://biodiversity-chm.eea.europa.eu/stories/STORY1016812291/, accessed 3 January 2009.

25 Ibid.

26 In full: Bonn Guidelines on Access to Genetic Resources and Fair and Equitable Sharing of the Benefits Arising out of their Utilization. This is an international law instrument of a non-binding, voluntary character. It lays down a detailed process for access and benefit-sharing (ABS) with the participation of local and indigenous communities, that is, based on prior informed consent (PIC) and mutually agreed terms (MAT), and proper arrangements on benefit sharing. For further reading, see Tully, S. (2003) 'The Bonn Guidelines on access to genetic resources and benefit sharing', *RECIEL* 12(1), pp84–98. For the shortcomings of the Bonn Guidelines, see SEARICE notes of May 2002, 'The Bonn Guidelines on access to genetic resources. Another false hope against biopiracy?', http://www.searice.org.ph/pdf/Other%20Pub/Bonn%20Guidelines.pdf, accessed 21 April 2008.

27 CBD COP, Decision VI/24 A (2002).

28 UN WSSD, Johannesburg, 26 August–4 September 2002, Plan of Implementation, §42(o). The argument by megadiverse countries in the WSSD Summit that lack of clear rules on access to GRs might prompt them to restrict access for researchers, business and private investment is seen as a decisive contribution towards the agreement.

29 To be negotiated within the framework of the CBD and its Bonn Guidelines.

30 UN WSSD, Johannesburg, 26 August–4 September 2002, Plan of Implementation, §42(o).

31 See Communication from the Commission to the European Parliament and the Council, 'The implementation by the EC of the "Bonn Guidelines" on access to genetic resources and benefit-sharing under the Convention on Biological Diversity', COM (2003) 821, http://eur-lex.europa.eu/LexUriServ/site/en/com/2003/com2003_0821en01.pdf, accessed 20 March 2009.

32 Ibid.

33 The Commission's Communication to the European Parliament and the Council (COM (2003) 821) read in part: 'Article 15.5 of the CBD provides that access to genetic resources shall be subject to the PIC of the Contracting party providing such resources, unless otherwise determined by that Party. Therefore, companies/institutions conducting bio-prospecting activities are expected to require the PIC of the provider countries. … The Commission strongly encourages stakeholders from the EC to respect the prior informed consent requirements of provider countries'.

34 IP/04/21. In the press release, the environment Commissioner, Margot Wallström, stressed the need for companies and research institutes utilizing GRs from developing countries to guarantee providers a fair share of the profits arising from the use of these resources. These resources should be used with the consent of the provider. The Commissioner pointed to the Commission's willingness to ensure that companies and research institutes acted responsibly by sharing gains with developing countries. This was seen as a win-win situation for trade and the environment where the benefits were used to foster nature conservation. Companies and research institutes were encouraged to use standard agreements in ABS arrangements between them and providers. Agreements should set terms and conditions on ABS. Users should also develop their own Codes of Conduct in response to CBD. The Commission's responsibility was also spelled out in measures envisaged in the Commission's action:

  • awareness raising – undertaking measures to make users' obligation under CBD clear, for example, by creating a European network to provide information on existing ABS laws at international, European and national level. To this effect, the EC has developed its own 'clearing-house mechanism' (that is, an agency that brings together seekers and providers of goods, services or information, thus matching demand with supply);

  • open debate on disclosure requirement – vis-à-vis source of GR and whether TK of indigenous or local communities was used;

  • action in international forums – organize a fully coordinated EU action within all international fora relevant to ABS (e.g. biodiversity, agriculture, trade and intellectual protection) in order to ensure a consistent international regime on ABS.

35 BR-Drs. 546/03 vom 15.8.2003; 'Gesetz zur Umsetzung der Richtlinie über den rechtlichen Schutz biotechnologischer Erfindungen'. Available online at:

http://217.160.60.235/BGBL/bgbl1f/bgbl105s0146.pdf, p148, accessed 3 February 2009; 'Blatt für PMZ', 2005, Issue 3, http://www.heymanns.com/servlet/PB/menu/1119800/index.html, accessed 3 February 2009.

36 This law came into force on 28 February 2005.

37 The sovereign right of states to exploit their own resources pursuant to their own environmental policies, and their responsibility in ensuring that activities within their jurisdictions or control do not adversely affect the environment beyond their national jurisdictions or that of other states.

38 Above, note 12.

39 Refers to access to GRs: national governments' authority to determine access based on national legislation; need to ensure access for environmentally sound uses and without restrictions that are counterproductive to the objectives of the CBD; only GRs provided by contracting parties that are countries of origin of the GRs or by parties who have acquired GRs in accordance with the CBD; acquisition on MAT; subject to PIC of contracting party, unless otherwise determined by the party; development and research based on GRs with full participation of, and where possible in, contracting parties; and fair and equitable sharing of benefits on MAT.

40 Above, note 13.

41 Queen Mary Intellectual Property Research Institute, Report on disclosure of origin in patent applications, prepared for the European Commission, DG-Trade, http://trade.ec.europa.eu/doclib/docs/2005/june/tradoc_123533.pdf, accessed 21 April 2008.

42 This concerns only resources of other provider countries and not its own. According to Dutfield (2003, p36), Denmark has resolved not to have any ABS regulations for its own GRs.

43 See UNEP/CBD/WG-ABS/2/INF/1 (30 September 2003), p82, http://www.biodiv.org/doc/meetings/abs/abswg-02/information/abswg-02-inf-01-en.pdf, accessed 3 February 2009.

44 For the full text of the Act (Patent Act 926 22/9 2000), see UNEP/CBD/WG-ABS/2/3 (20 October 2003), Annex, para 3.

45 This resulted to the new ministerial regulations on patents, Regulation 1086 11/12/2000. For an unofficial English translation of the provision, see UNEP/CBD/WG-ABS/2/INF/1 (30 September 2003), p82, http://www.biodiv.org/doc/meetings/abs/abswg-02/information/abswg-02-inf-01-en.pdf, accessed 3 February 2009.

46 See unofficial English translation; see ibid.

47 UNEP/CBD/WG-ABS/2/INF/4 (23 October 2003), paras 66 & 72, http://www.osim.ro/index3_files/laws/patents/lg20302en.htm, accessed 3 February 2009.

48 Ibid.

49 WIPO/GRTKF/IC/5/8 (28 April 2003), para 99.

50 Ibid.

51 Ibid.

52 Ibid.

53 Ibid.

54 Ibid.

55 Text available online at http://www.gti-kontaktstelle.de/toolkit/task_2.html or http://www.internationalwildlifelaw.org/BiodiversityActBulgaria.pdf, accessed 3 February 2009.

56 Promulgated, State Gazette No 29/7.04.2000; Amended: SG Nos 23/1.03.2002, 91/25.09.2002, http://www.gti-kontaktstelle.de/toolkit/task_2.html, accessed 3 February 2009.

57 According to Article 20, exploitation of medicinal plants shall be the exploitation of their resources and it shall include: (1) collection of herbs of wild and artificially propagated medicinal plants; (2) purchasing of herbs intended for primary or further processing; and (3) collection of genetic material from wild medicinal plants for the purpose of artificial propagation, for conservation out of the natural environment of medicinal plants or for re-introduction into other areas.

58 DEFRA, Review of the experience of implementation by UK stakeholders of ABS arrangements under the CBD, http://www.defra.gov.uk/farm/policy/geneticre-sources/abs-cbdreview.pdf, accessed 3 February 2009.

59 Ibid.

60 Ibid.

61 Ibid.

62 Ibid.

63 Ibid.

64 Other assumptions that are considered in this article as not presenting any problem are: (1) that users will concur that their inventions are based on GRs of foreign countries; (2) that users will disclose the origin of those resources; and (3) that the usage of diverse terminologies (for biological resources) in national legislations will not present a problem. For hardships that these factors are likely to present, see Tvedt and Young (2007).

65 For a discussion on this, see J. Sarnoff, 'Compatibility with existing international intellectual property agreements of requirements for patent applicants to disclose origins of GRs and traditional knowledge and evidence of legal ABS. Memorandum to public interest intellectual property advisors', Inc., 23 June 2004, http://www.piipa.org/DOO_Memo.doc, accessed 3 February 2009.

66 Stand-alone disclosure requirement does not hinder patentability or enforceability of a patent in case of non-compliance. Any legal consequences of non-compliance lie outside the ambit of patent law, but may be punishable under other codes. Enhanced disclosure, on the other hand, may jeopardize the processing of a patent application or result in withdrawal of a patent, as well as patent rights in case of non-compliance.

67 The Andean Community of countries comprises of Bolivia, Colombia, Ecuador, Peru and Venezuela. It has its origins in the Andean Subregional Integration Agreement (Cartagena Agreement).

68 For the Andean Pact and Brazil, see also Kathryn Garforth et al.

69 Generally on how the CBD user commitment could be achieved, see Tvedt and Young (2007).

70 See Tvedt and Young (2007, Chapter 6).

71 Ibid., also Chapter 3.

72 The Japanese Guidelines, for example, include an incentive measure for users who comply with the Guidelines, whereby the government of Japan offers to assist the user if some difficulty arises in negotiations or discussions with source countries. See JAPAN: METI/JBA, 2006, Guidelines for Access to Genetic Resources for Users in Japan, Ministry of Economy, Trade and Industry (adopted March 2005, published in English in 2006), http://www.jba.or.jp/english/pdf/081126_Japan%27s_ABS_Guidelines.pdf, accessed 3 February 2009.

References

Barber, C. F., Johnston, S. and Tobin, B. (2003) *User Measures: Options for Developing Measures in User Countries to Implement the Access and Benefit-Sharing Provisions of the Convention on the Biological Diversity*, 2nd edn, UNU/IAS, Tokyo

Carvalho, N. P. de (2003) 'Requiring disclosure of the origin of genetic resources and PIC in patent applications without infringing the TRIPS agreement: The problem and the solution', *Washington University Journal of Law and Policy*, vol 2, pp371–401

Dutfield, G. (2003) 'Protecting traditional knowledge and folklore. A review of progress in diplomacy and policy formulation', UNEP/CBD/WG-ABS/2/INF/2 (29 September)

Garforth, K., Noriega, I. L., Medaglia, J. C., Nnadozie, K. and Nemogá, G. R. (2005) 'Overview of the national and regional implementation of access to genetic resources and benefit-sharing measures', 3rd edn, CISDL, http://www.cisdl.org/pdf/ABS_ImpStudy_sm.pdf, accessed 3 February 2009

Hoare, A. and Tarasofsky, R. (2006) 'Disclosure of origin in IPR applications: Options and perspectives of users and providers of genetic resources', IPDEV-Project, Work Programme 8: Final Report (May), http://www.chathamhouse.org.uk/files/6409_iprreport0506.pdf, accessed 14 March 2008

Kamau, E. C. (2004) *A Hard Patent System: An Impediment to Technological (Economic) Development in Less Developed Countries*, Baden-Baden, Nomos Verlag

Leskien, D. (1998) 'The European Patent Directive on Biotechnology', *Biotechnology and Development Monitor*, no 36, pp16–19

Posey, D. A. and Dutfield, G. (1996) *Beyond Intellectual Property. Toward Traditional Resource Rights for Indigenous Peoples and Local Communities*, Ottawa, Canada, International Development Research Centre

Ruiz Muller, M. and Lapeña, I. (eds) (2007) *A Moving Target: Genetic Resources and Options for Tracking and Monitoring their International Flows*, Gland, Switzerland, IUCN

Tvedt, M. W. and Young, T. (2007) *Beyond Access: Exploring Implementation of the Fair and Equitable Sharing Commitment in the CBD*, Gland, Switzerland, IUCN

Van Overwalle, G. (2002) 'Belgium goes its own way on biodiversity and patents', *European Intellectual Property Review*, vol 24, pp233–236

Chapter 22

Enforcement of Benefit-Sharing Duties in User Countries

Christine Godt

Introduction

ABS is the third goal of the Convention on Biological Diversity (CBD, Article 1). The relationship of *access* on the one hand and *benefit sharing* on the other hand can be described as a quid pro quo (Article 15 CBD): The precondition for access should be the agreement about the sharing of benefits; and vice versa, benefits have to be shared only when access has been granted.

To date, the legal discussion has focused on the legislation of ABS in provider countries. The common perception prevailed that the implementation of ABS is primarily a task of the provider states. The reason is partly technical. Applicable legislation is to be adopted in the provider states. In this regard, the international community supported with capacity building and consultations on the national and international level. On the other hand, pushing for benefit-sharing rules has been perceived as being (only) in the interest of provider states. It would (only) be a matter of justice to share profits being generated in industrialized countries (so-called user states) on the basis of GRs or TK accessed in other countries (commonly called 'provider states') (Gerstetter, in this book). However, there is more to it. Connecting both goals provides for a central functional mechanism of the CBD as a whole. The convention was inspired by the regulatory economic thinking of the 1980s. It gave way to the rationale that sharing sbenefits from the use of GRs and TK would set an incentive for their conservation. From this perspective, the third goal of the convention provides for the instrument to achieve the first goal, namely conservation of biological diversity. This is the very reason why benefit sharing is not only in the interest of the provider states, but also in the interest of the user states.

Therefore, it is incompatible with the CBD to relinquish the ABS implementation to countries of origin. It has become evident that effective enforcement of both statutory and contractual benefit sharing will depend on supplementing enforcement by and inside user countries. Those can be purely self-regulatory and non-binding, such as internal corporate guidelines supporting employed bioprospectors in complying with local ABS regulations or public research-funding organizations, which support researchers in obeying the law of host countries. This chapter, however, will deal with coercive judicial enforcement. Its focus is on tort liability for non-consented access in respect of the use of GRs and TK.

Bioprospecting – A transnational activity

Typically, bioprospecting is a transnational endeavour. Field research and collection occur inside the provider states, whereas screening and commercialization is conducted in the user countries. This inherent transnational constellation demands the application of transnational legal instruments.[1] Up to now, the focus has been on intergovernmental cooperation (consultations for implementing provider states' legislation and capacity building[2]). However, effective enforcement of statutory and contractual duties depends on a transnational coordination of legal regimes. Four instruments have been discussed recently with regard to such coordination: mandatory disclosure rule in patent procedures,[3] certificates of origin,[4] border measures (Stoll, 2004, p83) and regulations of research foundations.[5] In contrast to such prior research, this contribution looks beyond regulating access and raises the question of how benefit-sharing claims can be enforced in user countries. Can a lawsuit of a provider community or a provider state be brought in front of a court in an industrialized state? Can a case be successfully argued that a company that is incorporated in an industrialized country, for example Germany, is liable in torts for violating ABS legislation in the course of its bioprospecting activities in a provider country? Can a share in benefits be claimed in damages?

In order to answer these questions, we need to know which country's law applies. The applicable law is not self-evident in transnational constellations. In contrast to public law, a civil law court does not necessarily apply domestic law. Private law is not in the same way bound to the territoriality principle, which stems from the principle of sovereignty. Private law relies on the principle that the law which suits the case best and which is in the interest of both parties is to be applied.[6] In fact, each single question can be governed by a different country's law.[7] Which law should apply is decided by the (domestic) body of conflicts of law.

Therefore, three different questions are to be distinguished: (1) Which court is internationally competent ('has jurisdiction')? (2) Which country's law applies for which question of a given case ('conflict of laws')? (3) What does the applicable law say (*lex causae*)? These questions will be answered consecutively in this chapter. Its conceptual focus is on how to conceive the right to GRs. Is the right in GRs 'material' or 'immaterial', comparable to 'material' or 'immaterial' property? The answer determines the applicable law and the determination of damages.[8] In the following, the relevant legal questions will be analysed departing from 'the standard' situation: Bioprospecting was conducted in violation of domestic ABS rules (no contractual benefit-sharing agreement was concluded). The commercial activity is executed inside an industrialized country; the subsequent product is primarily marketed on Western markets.[9]

For the sake of the analysis of the legal question involved, remaining factual problems such as the application for a necessary visa (purpose: conducting a law suit) and costs (travel, hotel accommodation and lawyer's fees) will not be discussed.[10]

International and local jurisdiction

The judicial competence (jurisdiction) of courts in international conflicts was harmonized in the European Union in 2002 by the so-called Brussels I Regulation (Regulation 44/2001).[11] Article 2 Regulation 44/2001 promulgates as a general rule that the court of the defendant's place of permanent residence enjoys jurisdiction. This rule corresponds to most continental rules;[12] for example, §13 German Civil Procedure Act (Zivilprozessordnung, ZPO), including interim injunctions, Article 2 Regulation 44/2001, §934 ZPO. It is applicable to all contractual claims and damage claims resulting from torts. Additional competences are available for tort claims under Article 5 Section 3 Regulation 44/2001, §§13, 32 ZPO (the country's courts where the effects of the tortious act occur). Exclusive jurisdiction is stipulated for claims of invalidity of a registered immaterial right, Article 22 Abs. 4 Regulation 44/2001 ('patents, trade marks, designs and other similar rights'). For those claims, only the country's courts in which the right is valid enjoy jurisdiction. This rule does not only govern the primary law suit which challenges the validity of the right, but also the (in)validity argument as a defence.[13] However, it does not govern damage claims stemming from the violation of patent.[14] As this chapter will not focus on patents as such, but on benefit sharing (thus claims for financial or other means of remuneration), the rule of Article 22 Abs. 4 Regulation 44/2001 will be neglected in the further course of the chapter.

Following the example of a manufacturer registered and residing for example in the city of Karlsruhe in Germany as a defendant, the civil court in Karlsruhe enjoys international and local competence to decide a claim against the company for the violation of contractual duties and for damages raised by any person or entity based in a provider state. With regard to this question, there is no legal uncertainty.

Standing

A first legal incertitude may arise when the plaintiff is a collective entity, such as an indigenous community. Some years ago, the literature argued that these entities would lack standing as Western legal systems do not acknowledge rights vested in groups, unless the group is incorporated.[15] However, under the influence of the modern human rights discourse, the opinion has recently been voiced that communities enjoy standing if their legal status is acknowledged by the home country (Fikentscher, 2005, p14; Godt, 2007, p279ff; van Hahn, 2004, p320). Therefore, it is of central importance that the provider state adopts rules that identify its traditional communities and formulate a procedure which allows identifying any rights to GRs and TK in a given case.

The next question that arises is whether a governmental entity, for instance a ministry of a provider state, enjoys standing according to civil procedure rules. As far as property, contractual and tortuous civil law claims are concerned, governmental institutions are not exempt from standing. In principle, the public hand is allowed to pursue its civil law claims as any other person does. However, any financial claim of a foreign government, which is vested in its sovereign rights, would be exempt, such as taxes and duties. Therefore, it is a relevant question whether a certain monetary amount is rooted in a 'civil' or a 'public' norm. This ambiguity arises in two situations. First, the situation is ambiguous in countries that acknowledge the special category of 'public property' (as many African countries do, e.g. Cameroon) (Godt and Nde Fru, 2008, p61). Second, the legal quality of the remuneration claim is unclear when lump sums are regulated by law.[16] Then the conceptual environment will decide about the legal quality (private property or public duty).[17]

Finding the applicable law

At the focus of this chapter is the question: Which country's law will govern the suit? The answer to the question is provided by the conflict of law rules of the country where the lawsuit is filed. These rules differentiate according to the nature of the claim raised (contract, torts, quasi-tort, intellectual property, material property). Therefore, the same amount of money can be raised under different laws if rooted in different legal subsystems. In addition, each question, which forms the basis of each claim, can eventually be judged differently. Consequently, a single claim can be decided on the basis of a mosaic of laws.

Conflict of laws on contracts

Although this chapter will focus on tort claims, it will briefly touch on the conflict rule of contractual claims. Applicable rules in this regard are in the process of harmonization. The so-called Rome I Regulation 593/2008[18] (implementing the Convention of the Law Applicable to Contractual Obligations of 1980[19]) was adopted on 17 June 2008 and will be applied to contracts concluded after 17 December 2009. Until then, German law will apply. The basic rule is the same in both regulations. Parties can choose the law, which shall apply to the contract (Article 3 Regulation 593/2008; Article 27 German Conflict of Law Code, EGBGB[20]). In the absence of choice, the methodologies of the regulations differ. Under Article 4 Regulation 593/2008, rules are stipulated for a set of standard contracts. For example, a sales contract shall be governed by the law of the country where the seller has his/her habitual residence (Article 4 Section 1 lit a Regulation 593/2008). Licence contracts are not among this list. For contract types which are not included, Article 4 Sections 2–4 provide for general rules. Primarily, the contract shall be governed by the law of the country where the party required to effect the characteristic performance of the contract has his/her habitual residence. Only where the law applicable cannot be determined otherwise shall the contract be governed by the law of the country with which it is most closely connected. Under German law, Article 28 phase 1 EGBGB stipulates as a general rule that the applicable law is to be the law of the country with which the contract is most closely connected. The rule of doubt is that this is the country in which the party who is to effect the performance, which is characteristic of the contract, has his/her habitual residence (Article 28 Section 2 EGBGB). For licences, the application of these rules has been contested (Groß, 2007, p229). Generally, the characteristic duty is performed by the licensor (in our example usually resulting in the application of the law of the provider country). However, if the licence is exclusive and the licensee is obliged to perform, then the characteristic duty is performed by the licensee (usually the law of the user country) (Pfaff and Osterrieth, 2004, p25, note 114).

The question is crucial in cases where benefit-sharing contracts stipulate that only non-commercial research is authorized. This agreement gives rise to the duty of the bioprospector to re-negotiate the contract as soon as the development matures and the research phase enters the development phase. The violation may give rise to contract penalties or damages.

Conflict of laws in torts

The 'statute' of the delict

With regard to the recovery of profits, claims can be based on various norms, which belong to different subsets of the law. Conflict of laws rules distinguish between torts, quasi-torts and property. In general, tort claims are governed by the law of the place where the delict occurred, the so-called *lex loci delicti* rule. In theory, this can be the place where the 'wrong' was performed *or* where the right was violated. According to traditional German conflict law, it was the victim that selected the law that suited the victim the best (Article 40 EGBGB). The principle was called 'the most favourable rule'. However, the victim lost this privileged position due to the Rome II Regulation (Regulation 864/2007),[21] coming into force in 2009 and applicable to cases occurring after 11 January 2009. The Regulation is applicable to all claims raised in the EU states; it applies universally (Article 3 Regulation 864/2007). The primary *lex loci delicti* in the EU will be the country's law where the violation of the right occurred (Article 4 Section 1 Rom II Regulation, *lex loci damni*). The place where the actual damage occurs does not matter (Kegel and Schurig, 2004, p730). With regard to a tort claim, it can also have several places where behaviour violates a right. If the alleged right was violated in Germany *and* in Kenya, a European Court has to apply both laws and attribute separately the damages, which resulted from a violation of the right in each country.

A claim for recovery of profits can also be based on *negotiorum gestio* (necessity of agency). In this regard, Article 11 Regulation 864/2007 stipulates a set of rules to determine the applicable law. In case of a parallel tort claim, the claim arising from *negotiorum gestio* shares the *lex loci delicti*. In case no tort can be identified, the law of the country in which the act was performed applies (Section 3). In cases of illegal bioprospecting, the unauthorized agency would be the use and the commercialization of the protected right. A precondition, at least under German law, means that there is a potential consent with the activity of the 'user', which was not negotiated in the given case. If commercial activities (development and commercialization) occurred inside the industrial country, it can be argued that the user state's law is to be applied.

In addition, different rules apply with regard to infringements of an intellectual property right (IPR). In this regard, Article 8 Regulation 864/2007 codifies the traditional *lex protectionis rule* according to which the law of the country is to be applied for which protection is claimed.

Therefore, it is important to identify what kind of damages occur in (and after) bioprospecting activities. What kinds of rights are violated? Do damages occur in the country of residence of the right holder or in the country where the damage occurred in financial terms? There is a broad legal uncertainty which is due to the unclear legal characteristics of GR and TK. These will be explored in the following.

Material or immaterial property?

Legal uncertainty stems from the fact that it is unclear whether a 'right to genetic resources' is to be qualified as *material* or *immaterial*.[22] However, this qualification is central. All questions relating to property are governed by a special 'property statute' – independent of the *lex causae* applicable according to the tort rule. The conflict of law rule is different for material and immaterial property. In the case of a material property right, all questions related to property (identifying the owner and the scope of property) are governed by the so-called *lex rei sitae* (in Germany: Article 43 EGBGB). The rule stipulates that these questions are governed by the law of the country where the object is located and as long as it is there. Is the property right in question immaterial, one has to distinguish IPRs and autonomy rights: IPRs are, in principle, governed by the territoriality principle (*lex protectionis*, recently codified in Article 8 ROM II Regulation).[23] In contrast, claims resulting from the violation of personal autonomy right (in most jurisdictions so-called personality rights) share the destiny of the *lex loci delicti* (Staudinger-von Hoffmann, 2001, Article 40, note 54).

What is the character of GRs? Are they *material* or *immaterial*? More important, is the violated right a *material* or an *immaterial* property right? For GRs, the immaterial or material quality has ever been discussed. The first discussions came up when the CBD was negotiated (Stoll, 2004). During the 1990s, the discourse shifted to patent law.[24] More recent discussions refer to data protection in tissue banks (Schulte in den Bäumen, 2008). The question is most probably not easily answerable, but only with due account of the circumstances. Therefore, we need to look for the prerequisites and into the consequences of each qualification.

Material property

What qualifies a damage claim for a violation of a material property right? When are GRs corporal (and not incorporal) information? I argue that the answer depends on the facts of each individual case. The material qualities prevail when, for example, fruits, wood and roots *as such* are concerned. 'As such' they are, for example, objects of a sales contract. In contrast, immaterial values prevail when their 'genetic make-up' is concerned. The specific plant, then, appears only to be the carrier. However, neither are the categories of distinction clear yet, nor are the consequences.

If the title to GR was to be aligned with material property, three issues have to be considered.

First, the conflict of law rule is the *lex rei sitae* rule. Following the example of Kenya and Germany, as long as the plant (as material) is situated in Kenya, property questions are to be determined according to Kenyan property law. After transport to Germany, for example, the property statute shifts to German law. Unless no statutory loss of title occurs due to good faith or adverse possession, a traditional community claiming 'biopiracy' could argue that their material property right was stolen if they bring evidence that the (identical) 'material' is owned by them and the bioprospector took it without permission.

Usually, however, traditional communities refer to something other than 'property' in the (Western) sense; for example, §903 German Civil Code. They claim a right to decide about what is done with 'their' biological resources. Their claim is not directed to the economic value of the single plant (damages), nor do they claim vindication of the material. In essence, this alleged right is rooted in the right to self-determination (a human right), as recognized by Article 8(j) CBD. As far as Western property is concerned, the right to self-determination is in principle embodied in property as the most comprehensive exclusion right. However, material property is limited to a specific object. Only if the community claims the mere vindication of or remuneration for a specific object would the property be protected.

The second consideration relates to the transfer of a plant to a third country under violation of an export prohibition. The question might arise if good faith or prescription could be prevented by arguing that the plant was illegally exported. Similar rationales have been applied to cultural heritage. Some countries stipulate that illegal (not permitted) export of cultural heritage goods will result in a loss of the acquired title.[25] National regulation stipulates to whom the property title will be attributed. In that case, the traditional community being the plaintiff can argue that it remained the owner (irrespective of potential acquisition in good faith). However, the regulatory recovery of title will not necessarily be acknowledged by European courts. The recognition depends on the domestic conflict of law rules with regard to the recognition of foreign regulation (so-called mandatory rules).[26]

Third, the scope of property protection is limited. In (material) property law, protection of prior informed consent (PIC) is granted by the prevention of statutory loss of title. This measure secures the owners' claim for vindication and injunction. Tort claims with regard to (material) property violations are, in principle, restricted to the economic damages (referring to the monetary value of the very good in question and lost profits for German law: §§249, 252 German Civil Code). With regard to some plant leaves, the economic value is often small, not to say marginal. The immaterial value of material property is acknowledged in only a few cases (for the German law: pets in §251 Section 2 German Civil Code; proprietary environmental goods in §16 Section 1 Civil Environmental Liability Act). A claim for recovery of profits generated by the violator of the right is, in principle, not acknowledged as an 'economic loss' and can only be claimed under *negotiorum gestio* (unauthorized agency of necessity).[27]

In consequence, the determination of damages is the biggest obstacle for benefit-sharing claims resulting from a violation of a material property right. The economic loss is minimal. Recovery of profits is in principle not possible; the actual use value has no market price.

Immaterial property

Two kinds of immaterial property. GRs and TK are often referred to as 'immaterial property'. However, immaterial property encompasses intellectual property and the autonomy right (in civil-law countries referred to as 'personality' right). The former protects various forms of human creativity. The latter protects the prior consent of a person and relates to corporal material and to information alike. IPRs are, in principle, governed by the *lex protectionis* rule; the right to personal autonomy shares the *lex loci delicti*.

With regard to GRs and TK, both sorts of rights have been discussed. Nothing is settled or cleared. No agreement about the qualities of a sui generic right has been achieved yet, neither in the framework of the CBD nor in the Intergovernmental Committee on Intellectual Property and Genetic Resources, Traditional Knowledge and Folklore at the World Intellectual Property Organization (IGC-WIPO). The difficulties in defining a sui generic right are not only due to a lack of political will, but also to the complexities of the topic. Precise qualifications of any right attributed to GRs and TK depend on the factual circumstances.

Those who stress the ingenuity, for example, of traditional healers or creators of designs and melodies draw the parallel to intellectual property (Coombe, 1998; van Hahn, 2004; van Overwalle, 2007, p355).[28] It gives credit to individual and collective creativity – concealed or shared with the community. A parallel to intellectual property is also drawn to useful natural substances for three reasons. First, traditional communities know about their effects and how to employ them. This knowledge would have an equivalent value of the costly search for a new 'leading target substance' (Balick, 2007, p280; Godt, 2007, p369).[29] Second, a comparison is drawn to the isolation theorem in patent law which justifies the grant of a patent with regard to the novelty requirement. Third, the informational value of the genetic make-up creates an economic value.

In contrast, others stress the right of communities to decide about who may access their resources and their knowledge, for which purposes, under which conditions (Fikentscher, 2005; Ramsauer, 2005; Rosenthal, 2008, p373). At the very centre is the recognition of autonomy and self-determination (Xanthaki, 2007, p131).[30] The model is an autonomy right which is acknowledged in the form of a 'personality right' in Germany by the High Court,[31] or as rights entrusted to the person in common law countries.[32] In acknowledging a sui generis right to GR and TK by analogy to the autonomy right, the court of the user country could transfer the (international law) recognition of traditional communities in Article 8(j) CBD into an actionable legal property title of (binding) national law.

Intellectual property

Assuming the community claims an IPR, the conflict rule to apply is the territoriality rule (*lex protectionis*). Whereas for registered rights the choice of law is explicitly aligned with jurisdiction (Article 22 Regulation 44/2001), Article 8 Regulation 864/2007 stipulates that the law applicable to non-contractual obligations arising from infringement of an IPR shall be the law of the country for which protection is claimed. The norm does not distinguish between registered and non-registered rights. Reading both rules together, two rationales can be identified. First, the territoriality principle as such does not require the alignment of jurisdiction. Therefore, the validity of non-registered rights and their violation can be argued in any court which is internationally competent/has jurisdiction. Both are to be determined in applying each respective territorial law.[33] This means that, for damages with regard to intellectual property, the *lex protectionis* replaces the general *lex loci delicti* rule. The consequence is that the territorial law decides not only about the violation of the right, but also about the infringing behaviour which results in a violation. Therefore, both the behaviour and the violation must occur in one and the same state.

Transposing these principles to a transnational GR and TK dispute, a European court would adjudicate both the violation of an IPR in the provider state *and* the IPR of the user state, each according to the applicable domestic laws. With regard to the applicability of a provider state's law, one would require that the IPR is recognized by the state and that the disputed body of knowledge is attributed to the community. I argue that the application of provider country's protection of communal right does not amount to a violation of the European or the German *ordre public*, which could block the application of the foreign law (Article 34 EGBGB).

With regard to the violation of an IPR applying, for example, German law as *lex loci protectionis*, one would find that *de lege lata*, statutory German law, does not recognize a sui generis (intellectual) property right in GRs or in TK of traditional communities as such.

However, it can be argued that the recognition of such a 'property' right (sui generis) is required by the CBD.[34] Two constructions of the claim are possible. Either one argues for an autonomous interpretation of what 'property' means (in Germany) or one argues for the recognition of such a right as an equivalent of 'absolute rights' protected by tort law (in German terms: 'ein sonstiges Recht' in the sense of §823 German Civil Code). The reason why such a right is to be acknowledged by parties to the CBD can be argued in two different ways. First, one can argue in a straightforward way that the CBD requires recognition directly (arguing for a direct obligation under the CBD). The autonomy right is a universal right, which exists independently from the political will of the sovereign where a traditional community is hosted. Second, one can argue that recognition stems from comity. The duty to recognize is derivative from the recognition of another party to the convention. Therefore, the duty arises under the condition that the provider state acknowledges and identifies such a right. This way of argumentation would give credit to the sovereignty and internal decision making of the provider state. The latter seems to better fit with contemporary concepts of a generalized conflict of law approach influenced by systems theory (Teubner and Fischer-Lescano, 2008, p17). The former is sensible to existing conflicts inside provider states.

Autonomy

As noted above, often it is the claim for self-determination which is formulated in terms of a property language. At the very heart is the PIC requirement. It does not commodify any good, but it attributes an exclusivity right – a precondition for the PIC mechanism to function. A parallel *de lege lata* is the autonomy right (personality right), which also refers to parts of the person which are *res extra commercium* (blood, organs). In common law, interests with regard to the person that are not primarily related to economic property are equally framed as a property right. The applicable conflict of law is the general rule *lex loci delicti*.

Let us assume the GR right in question is an autonomy right: the plant was bioprospected in a provider state and brought to an industrialized country, where research and development was performed and profits generated: Where does the violation of the right occur? Immaterial autonomy rights in general are not bound to a state (even less than copyright). Being rooted in human rights, their quest for recognition is universal. Therefore, various additional 'connecting factors' are discussed (Staudinger-von

Hoffmann, 2001, Article 40, note 60): the territoriality rule (creating a mosaic of applicable laws), the *lex fori* and the place of residence of the plaintiff.[35] The facts of each case are decisive.

Applied to the example, we can argue as follows: If the nexus is close to the provider state, the *lex causae* of the provider state has to be applied. A German court will acknowledge a collective right, especially when domestic regulation in the provider state is in place recognizing the traditional community as a legal entity, and when their right to the specific GR and TK is unquestionable. I argue that the recognition of such a right does not give rise to doubts with regard to the German Ordre Public (Article 6 EGBGB). If the case is closer connected with the user state (e.g. when TK was used for advertising the product on the Western market, catering for the argument that the violation of the right occurred in the user state), then the user state law applies. With regard to the management of such a suit, the application of the user state law will have pros and cons. The advantage of applying user state law is the autonomous characterization of a right which opens up the possibility to interpret the law with due regard to the CBD[36] and other instruments with regard to the recognition of Traditional Communities.[37] The decision depends on the individual constellation of the case.

Damages

In contrast to damages for the violation of a material property, the determination of damages caused by a violation of an immaterial right follows a different rationale. Whereas the damage to a material property right is the lessened monetary value of the item itself, damages for infringements of immaterial property rights are primarily determined by their use value. It is not the protected piece or the information as such which is taxed. Nor is it limited to the actual use value. In intellectual property, the use value is traditionally valorized by three different doctrines: the lost profit on the part of the right holder; reasonable royalties; and unjust enrichment on the part of the infringer. This methodology was first developed for intellectual property law.[38] Later, it was transferred to competition law violations (Micklitz and Stadler, 2003, p258f; Stoll, 2000, p101f), and then, when the right to autonomy evolved, it was employed in cases of a violation of an autonomy right by analogy (Stuhlmann, 2001, p309f; Wachs, 2007, p342f). The recovery of profits, however, was especially regulated by *lege speciales*,[39] as an exception to the general rules limiting damage claims.[40] However, Article 13 EC-Enforcement Directive 2004/48/EC now requires the universal account for profits generated by the infringer. Only in cases where the infringer did not knowingly engage in the infringing activity are member states allowed to alternatively attribute either recovery of profits *or* damages.[41]

With regard to the violation of an immaterial right, which relates to GR and TK, it is sensible to employ the same doctrines. A benefit-sharing claim that is based on biopiracy is to be calculated with regard to lost profits, reasonable royalties and/or profits generated by the infringer. In the case of intention, the new directive demands that national norms are interpreted in such a way that damages encompass the recovery of profits.

Interim conclusion

With regard to tort claims, the applicable law can be the provider *or* the user country's law. The applicable *lex causae* depends very much on the construction of the case. With regard to illegal bioprospecting, both the applicability of the provider country's law and the user country's law can be argued. In essence, the law cannot be determined in the abstract in advance. The analysis reveals, however, that a meaningful law suit for damages has to be based on a violation of an immaterial property right, be it an IPR or an autonomy right.

General tort requirements

Fault

The national law of the *lex loci delicti* determines the general tort requirements. A claim for damages is usually fault based, thus requiring the violation of a duty. In the case, for example, where Kenyan tort law is to apply, one may ask how the duty of care is to be determined. Do the domestic ABS rules of the provider state constitute the relevant duty of care with regard to the right holder (in contrast to the general public)? Again, things will depend on how the law is formulated and on the facts of the case. I argue that the domestic laws are not to be disregarded just for the reason of being public law which only requires respect inside the relevant country. However, a precondition is that the law's goal is the protection of the respective right in question. If the domestic law transposes the CBD and the user state (e.g. Germany) shares the values expressed by them, it is to be argued that these national norms concretize the duty vis-à-vis the right holder (as installing rights to GR is the functional core of the CBD).[42] In contrast, if German law is to apply, one might refer to the Bonn Guidelines in order to identify the duty vis-à-vis the right holder. They formulate duties of bioprospectors, scientific and commercial users – especially for cases of transnational cooperation (Godt, 2004, p205f). Ignoring one of these duties amounts to a violation of the duty of care, which one owes to the right holder.

Causality

In order to successfully argue a case it is important to establish a clear link between the tortuous bioprospecting activity and the defendant, namely the company registered in the court's state. The careful establishment of this link is significant in cases where the bioprospecting activity was outsourced or executed by a contractor. It has to be shown that the company acted under the defendant's knowledge, its control and, ideally, that the company had issued internal bioprospecting guidelines but deliberately did not enforce them. It might also be possible that the company knew about violations of the provider country's rules, but did not do anything about them. It is important to establish a causal breach of a respective duty *by the defendant*.

Conclusion

A claim for benefit sharing can be raised on the basis of contracts and torts alike. The analysis shows that it is possible to litigate a (meaningful) benefit-sharing claim for biopiracy in a user country's court. Prospects for success are better with regard to immaterial property than to material property. Against common wisdom, it is not the applicable law that forms an obstacle. Applying foreign law might be inconvenient, but it can ease the burden of argumentation. There is a high degree of flexibility with regard to the argumentation about which country's law is applicable with regard to tort and property-related questions. This is due to the hybrid character of GR – being both material and immaterial. The most important result of the analysis is the following: it reveals that different rationales apply to how damages are determined for violations of material and immaterial property. With regard to material property, it is the economic value of the good as such which is taxed. In contrast, it is the rule to recover a share in profits or the equivalent to the licence fee for the violation of an immaterial right. This rationale is better suited for cases of illegal bioprospecting and is applicable to IPRs and autonomy rights alike. After previous lawsuits against biopiracy have focused on patents, a civil lawsuit for damages based on immaterial rights sui generis is another promising route worth exploring.

Notes

1   Coined as 'user measures', see Barber et al (2003).
2   With regard to the CBD Draft International Regime for Access and Benefit Sharing, see only Kongolo (2008, p73f).

3 The mandatory disclosure requirement intends to create a nexus between the provider country's ABS regulations and the very first point in time when commercialization is about to start, enacted, for example, in India in §10(a)(4)(d)(ii)(D) Indian Patent Act (see Godt, 2007, p330f; de Carvalho, 2007, p241).

4 Certificates aim at raising and conserving information about the resource (country, source) (Glowka, 2001).

5 For example, regulations of the German Research Foundation, 2008.

6 Conceptualized as the 'theory of the "*seat*" of a legal relation' (see Kegel and Schurig, 2004, p183). In common law, justice and comity are put in the forefront (North and Fawcett, 1999, p4). For an account of how conflicts of laws and modern tort law have changed under the conditions of globalization, see Halfmeier (2009).

7 Referred to as 'connecting factors' in common law (Hayward, 2006, p3); 'Statuten' or 'Anknüpfungsnormen' in continental theory (Kegel and Schurig, 2004, p300ff).

8 Keeping in mind that benefits can emerge at any 'incremental step' of research (Tvedt and Young, 2007, p70).

9 A typical scenario is 'Umckaloabo', broadly discussed prior and in the course of the last CBD Conference of Parties (COP 9) in Bonn, May 2008. The active component of this medicament (strengthening the immune system) is taken from the root of *Pelargonium sidoides* (from South Africa). Its physiological function was first alleged by an English traveller (Charles Stevens, born 1880), who was sent in 1897 to South Africa for a cure of tuberculosis and where he met a local healer. As 'Stevens' Cure', the treatment with this root became a widely administered therapy for tuberculosis at the beginning of the 20th century (http://de.wikipedia.org/wiki/Umckaloabo). Today, it is used as a stimulus for the immune system in general. The pharmaceutical product in territoriality, licensed for Germany to Schwabe and sub-licensed to ISO (registered in Karlsruhe, Germany). About 4.1 million packets are sold each year, with a value of €55 million (http://www.arznei-telegramm.de/zeit/0303_a.php3).

10 Should be adopted as an intergovernmental issue. National focal points should cooperate in this regard. Financial means for access to justice should be provided upon request by the Global Environmental Facility (GEF).

11 EC Regulation No 44/2001 of 22 December 2000, OJ L 12 of 16.1.2001, 1–23, in force since 1 March 2002. The Regulation applies to all international law suits, not only to those which involve EU member states. It replaces the Brussels Convention 1968 for all EU Member States except Denmark (for which the provisions of the old Brussels Convention will continue to apply).

12 In contrast, the UK used to determine jurisdiction in close relation to the applicable law. For the common law, the regulation brought great change (see Briggs, 2008, p53f; North and Fawcett, 1999, p179f; for a comprehensive comparative account from the copyright perspective, Peinze, 2002, p375f).

13 Forcing the courts to set the law suit aside, see ECJ in C 4-03, *GAT/LuK* (2006) ECR I-6509 (as in the USA, see *Voda v. Cordis*, 476 F.3d 887 (Fed. Cir. 2007)). This rule has been criticized by Hess et al (2007, p405).

14 Explicitly LG Düsseldorf in its decision *Schußfadengreifer*, GRUR Int. 1999, 455 (at 456). The underlying rationale is that the legal effects of the court ruling are limited to the territory of Germany.

15 This argument was merged with the one that property can only be held by individuals. For an elaboration of this argument, see Gerstetter in this book (note 3). The

discussion about sui generis rights for communities with regard to GRs and TK was especially intricate due to the initial rejection of any further commodification of natural resources by critics of the CBD (see C. Godt, 2004, p206).

16 The mere fact that lump sums are regulated does not necessarily qualify them as 'public'; compare, for example, tables of lump sums with regard to immaterial damages in cases of lost body parts in German civil law procedures.

17 From the perspective of conflicts of law, it is an open question whether the conceptual framework of the *lex fori* or of the *lex causae* would govern the determination of the legal quality.

18 OJ L 177/6 of 4 July 2008. The original rule proposed by the European Commission (Article 4 lit. f of COM (2005) 650 (final)) was fiercely debated (Procedure Code COD/2005/0261), resulting in a compromise formula, published by the Council and the EP, now Article 4 Section 2–4 Regulation 593/2008.

19 http://www.jus.uio.no/lm/ec.applicable.law.contracts.1980/doc.html

20 Einführungsgesetz zum Bürgerlichen Gesetzbuch (EGBGB).

21 Regulation 864/2007, OJ L 199/40 of 31.7.2007.

22 For TK, the answer seems to be clear: The property right is related to immaterial knowledge. However, what kind of immaterial rights are suitable is a much discussed question.

23 There is a remarkably high degree of legal uncertainty; for an overview, see Kegel and Schurig (2004), p729; this is especially true for non-registered rights such as copyright (see Peinze, 2002).

24 Attempting to downplay the distinctiveness of the informational quality, see Straus (1998, p314).

25 For example, Indian cultural heritage protection laws; see the facts of the Swiss High Court, Decision of 8 April 2005 (No BGE 131 III 418).

26 Germany has been very restrictive in acknowledging them: Article 34 EGBGB (there is only one case, BGHZ 59, 82 – *Nigerian Masks*); more open is the Swiss law in Switzerland, both in the jurisdiction (see Swiss High Court Rationale in the decision of 8 April 2005, ibid.) and the written law: Article 19 Section 1 Swiss Conflicts of Law Code provides that instead of applying the law which should govern the case according to a specific provision of this law, a rule of a foreign state can be applied under three conditions: the foreign law wants to be applied; prevailing interests of the parties demand so; and the facts of the case are closely connected to the law of the given state.

27 In broader terms, there are three sets of norms which allow for the claim for profits under the German Civil Code (BGB) (in German: 'Verletzergewinn'): (1) delict requiring fault and economic loss, §§823, 249, 252 BGB; (2) unjust enrichment, being either limited to what was 'objectively' received, therefore *not* profits: §§812, 818 II BGB, or to positive knowledge: §§819, 687 II, 681, 667 BGB. Therefore, the courts extended the rules of (3) agency of necessity, §§687 II, 681, 678, 667 BGB. However, jurisdiction remained strict.

28 Inter alia, contributions in Graber and Burri-Nenova (2008) and von Lewinski (2008).

29 Ethnobotanical knowledge substantially accelerates the finding of a lead substance.

30 Notwithstanding that self-determination 'is a thorny topic'.

31 A concise introduction to this jurisdiction is provided by Brüggemeier, G. (2009).

32 For a comprehensive and comparative account, see Resta (2005).

33 In reality, for copyright the principle of reciprocal recognition of Article 5 Bern Convention had left no room for applying this rule (Peinze, 2002, p115). Notwithstanding that inside Europe, other principles may prevail: it was proposed that the EU principle of origin prevails if the state where the violation occurred is not identical with the country where a service/a product was first placed (Baetzgen, 2007; Wild, 2007).

34 This is close to what Fikentscher (2005) and Godt (2007, p279) propose.

35 Fikentscher, who departs from the territoriality rule (as he aligns the autonomy right with intellectual property), comes to similar results. He autonomously qualifies the *lex causae* and applies provider state law.

36 Especially in those cases, where the political situation of the provider country is characterized by tensions between the state and the communities (which renders, in the end, legal recognition and the assignment of rights by domestic laws improbable).

37 International law material calling for recognition is collected inter alia by Gibson (2005), Rosskopf (2004) and Xanthaki (2007).

38 Already in the 18th century (Münchener Kommentar-Seiler (2009), §687, note 27; Staudinger-Schiemann, 2005, §249, note 199). The first German decision was the Aniston case of the Reichsgericht in 1895 (Wachs, 2007, p342). The rationale, however, is different with regard to each right. In patent law doctrine, the rationale is rooted in the process of innovation stretched over time. Initially, the new invention has no market value. In order to capture the economic value being generated over time (and to render the patent system functioning), it is the *future* value which needs to be attributed to the inventions along the time line. This is also the very base of the concept of dependency in patent law, which allows the pioneer (holding the so-called dominant patent) to participate in the profits being generated by those later ones who refine the technology (holding so-called dependent patents) and bring a product to a wider market. In copyright, the doctrine mainly responds to the difficulties in determining the damage value.

39 For example, §§15, 128 German Trademark Law and §97 I 2 Copyright Act.

40 For the general rules which also apply to immaterial property, see supra note 36. In addition, the amount is usually strongly debated – and court rulings differ. Which costs can be subtracted? GRUR 2001, 329: not general costs as wages and rents, and according to BGH GRUR 2004, 532: not damages to be paid to business partners in the distribution chain; concurring however: OLG Düss GRUR 2004, 53 (see Klüber, 2007, p267; Tilmann, 2003).

41 Transposed into German intellectual property laws (however, only *lege speciales*) in July 2008, Bundesgesetzblatt I No 28, 1191–1211 (after the deadline of transposition elapsed on 29 April 2006). With regard to the recovery of profits, courts will need to turn to the doctrine of directive conform interpretation when applying the general tort rules.

42 Applied to the CBD; see an earlier approach by Godt (2003).

References

Baetzgen, O. (2007) *Internationales Wettbewerbs – und Immaterialgüterrecht im EG-Binnenmarkt: Kollisionsrecht zwischen Marktspaltung ('Rom II') und Marktintegration (Herkunftslandprinzip)*, Köln, Heymanns

Balick, M. J. (2007) 'Traditional Knowledge: Lessons from the past, lessons for the future', in McManis, C. R. (ed) *Biodiversity and the Law – Intellectual Property, Biotechnology and Traditional Knowledge*, pp280–296

Barber, C. V., Johnston, S. and Tobin, B. (2003) 'User measures – options for developing measures in user countries to implement the access and benefit sharing provisions of the convention on biological diversity', 2nd edn, http://www.ias.unu.edu/binaries/UNUIAS_UserMeasures_2ndEd.pdf, accessed 28 January 2009

Briggs, A. (2008) *Conflict of Laws*, 2nd edn, New York, Oxford University Press

Brüggemeier, G. (2009) '"Du sollst Dir kein Bildnis machen…"– Der I. Zivilsenat des BGH und die Paradoxien des Persönlichkeitsrechts', in Calliess, G. C., Fischer-Lescano, A., Wielsch, D. and Zumbansen, P. (eds) *Soziologische Jurisprudenz*, Berlin, de Gruyter, pp231–248

Carvalho, N. P. de (2007) 'From the shaman's hut to the patent office: A road under construction', in McManis, C. R. (ed) *Biodiversity and the Law – Intellectual Property, Biotechnology and Traditional Knowledge*, pp241–279

Coombe, R. J. (1998) 'Intellectual property, human rights and sovereignty: New dilemmas in international law posed by the recogniton of indigenous knowledge and the conservation of biodiversity', *Indiana Global Legal Studies Journal*, vol 6, pp59–115

Fikentscher, W. (2005) 'Geistiges Gemeineigentum – am Beispiel der Afrikanischen Philosophie', in Ohly, A. et al (eds) *Perspektiven des geistigen Eigentums und Wettbewerbsrechts*, München, Beck, pp3–18

German Research Foundation (May 2008) 'Leitfaden für die Antragstellung von Forschungsvorhaben, die unter das Abkommen über die Biologische Vielfalt (Convention on Biological Diversity, CBD) fallen', http://www.dfg.de/forschungsfoerderung/formulare/download/1_021.pdf, accessed 28 January 2009

Gerstetter, C. (2009) 'How to effectively protect traditional owners from unwarranted use of genetic-resources and make sure benefits are shared fairly and equitably – some suggestions for provider countries', in this book

Gibson, J. (2005) *Community Resources: Intellectual Property, International Trade and Protection of Traditional Knowledge*, Aldershot, Ashgate

Glowka, L. (2001) 'Towards a certification system for bioprospecting activities', http://www.biodiv.org/doc/meetings/cop/cop-06/other/cop-06-ch-rpt-en.pdf, accessed 28 January 2009

Godt, C. (2003) 'IPRs and environmental protection after Cancún', http://www.ecologic-events.de/Cat-E/en/presentations.htm, accessed 28 January 2009

Godt, C. (2004) 'Von der Biopiraterie zum Biodiversitätsregime – Die sog. Bonner Leitlinien als Zwischenschritt zu einem CBD-Regime über Zugang und Vorteilsausgleich', *Zeitschrift für Umweltrecht* (ZUR), pp202–212

Godt, C. (2007) *Eigentum an Information – Patentschutz und allgemeine Eigentumstheorie am Beispiel genetischer Information*, Tübingen, Mohr Siebeck

Godt, C. and Nde Fru, V. (2008) 'Access and benefit sharing (ABS) zwischen Kamerun und Deutschland: Eine Annäherung an einen grundlegenden

Eigentumskonflikt', in Erdmann, K.-H., Löffler, J. and Roscher, S. (eds) *Naturschutz im Kontext einer nachhaltigen Entwicklung*, pp59–71

Groß, M. (2007) *Der Lizenzvertrag*, Frankfurt, Verl Recht und Wirtschaft

Hahn, A. van (2004) *Traditionelles Wissen indigener und lokaler Gemeinschaften zwischen geistigen Eigentumsrechten und der public domain*, Berlin, Springer

Halfmeier, A. (2009) 'Zur Entwicklung des Internationalen Deliktsrechts im neuen Mittelalter', in Colombi Ciacchi, A. et al (eds) *Haftungsrecht im dritten Millennium / Liability in the Third Millennium*, Baden-Baden, Nomos

Hayward, R. (2006) *Conflict of Laws*, 4th edn, Portland, Cavendish

Hess, B., Pfeiffer, T. and Schlosser, P. (2007) 'Report of the application of regulation Brussels I in the Member States', http://ec.europa.eu/civiljustice/news/docs/study_application_brussels_1_en.pdf, accessed 1 October 2008

Kegel, G. and Schurig, K. (2004) *Internationales Privatrecht*, 9th edn, Munich, Beck

Klüber, R. (2007) *Persönlichkeitsschutz und Kommerzialisierung*, Tübingen, Mohr Siebeck

Kongolo, T. (2008) *Unsettled International Intellectual Property Issues*, Alphen, Kluwer

Lewinski, S. von (2008) *Indigenous Heritage and Intellectual Property – Genetic Resources, Traditional Knowledge and Folklore*, 2nd edn, Alphen, Kluwer

McManis, C. R. (ed) (2007) *Biodiversity and the Law – Intellectual Property, Biotechnology and Traditional Knowledge*, pp241–279

Micklitz, H.-W. and Stadler, A. (2003) *Unrechtsabschöpfung*, Baden-Baden, Nomos Verlag

Morris, J. H. C., McClean, D. and Beevers, K. (2005) *Conflict of Laws*, 2nd edn, London, Sweet & Maxwell

Münchener Kommentar-Seiler (2009), *Münchener Kommentar zum Bürgerlichen Gesetzbuch*, vol §§611–704, München, Beck

North, P. and Fawcett, J. J. (1999) *Cheshire and North's International Private Law*, 13th edn, London, Butterworths

Overwalle, G. van (2008) 'Holder and user perspective in the traditional knowledge debate: A European view', in McManis, C. R. (ed) *Biodiversity and the Law – Intellectual Property, Biotechnology and Traditional Knowledge*, pp241–279, 355–370

Peinze, A. (2002) *Internationales Urheberrecht in Deutschland und England*, Tübingen, Mohr Siebeck

Pfaff, D., Osterrieth, C., Axster, O. (2004) *Lizenzverträge*, 2nd edn, Munich, Beck

Ramsauer, T. (2005) *Geistiges Eigentum und kulturelle Identität*, Munich, Beck

Resta, G. (2005) *Autonomia privata e diritti della personalità*, Napoli, Jovene

Rosenthal, J. (2008) 'Politics, culture and governance in the development of prior informed consent and negotiated agreements with indigenous communities', in McManis, C. R. (ed) pp373–393

Rosskopf, R. (2004) Theorie des Selbstbestimmungsrechts und Minderheitenrechts, Berlin, Berliner Wiss. Verlag

Schulte in den Bäumen, T. (2008) (Oral Presentation with Powerpoint slides) 'Boundaries to information property: IP and data protection – A different world or two complementary legal regimes?', Contribution to the Conference 'Boundaries to information property', 12 September 2008, organized by Godt, C., hosted by the Max-Planck-Institute for Intellectual Property in Munich

Staudinger-Hoffmann, B. von (2001) *Staudingers Kommentar zum Bürgerlichen Gesetzbuch*, vol Article 38–42 EGBGB, de Gruyter, Berlin

Staudinger-Schiemann (2005) *Staudingers Kommentar zum Bürgerlichen Gesetzbuch*, §§249–254

Stoll, C. (2000) *Die dreifache Schadensberechnung in Wettbewerbs– und Markenrecht als Anwendungsfall des allgemeinen Schadensrechts*, Tübingen (Diss)

Stoll, P.-T. (2004) 'Genetische Ressourcen, Zugang und Vorteilsausgleich', in Wolf, N. and Köck, W. (eds) *10 Jahre Übereinkommen über die biologische Vielfalt*, pp73–88

Straus, J. (1998) 'Abhängigkeit bei Patenten auf genetische Information – ein Sonderfall?', *Gewerblicher Rechtsschutz und Urheberrecht (GRUR)*, pp314–320

Stuhlmann, C. (2001) *Der zivilrechtliche Persönlichkeitsschutz bei Ehrverletzung und kommerzieller Vermarktung in Deutschland*, Norderstedt, Books on Demand

Teubner, G. and Fischer-Lescano, A. (2008) 'Cannabalizing epistemes: Will modern law protect traditional cultural expressions?', in Graber, C. B. and Burri-Nenova, M. (eds) *Intellectual Property and Traditional Cultural Expressions in a Digital Environment*, pp17–45

Tilmann, W. (2003) 'Gewinnherausgabe im gewerblichen Rechtsschutz', *Gewerblicher Rechtsschutz und Urheberrecht (GRUR)*, pp647–653

Tvedt, M. W. and Young, T. (2007) *Beyond Access: Exploring Implementation of the Fair and Equitable Sharing Commitment in the CBD*, http://www.fni.no/doc&pdf/beyond_access.pdf, accessed 28 January 2009

Wachs, A. (2007) *Entschädigungszahlungen bei Persönlichkeitsrechtsverletzungen*, Hamburg, Kovac

Wild, T. (2007) *Die Anknüpfung an den Handlungsort im derzeitigen internationalen Deliktsrecht: Article 40 Absatz1 Satz 1 EGBGB und 'Rom II'*, Berlin, Wiss Verlag Berlin

Xanthaki, A. (2007) *Indigenous Rights, UN Standards, Self-Determination, Culture & Land*, New York, Cambridge University Press

Chapter 23

Enforcement of ABS Agreements in User States

Hiroji Isozaki

Introduction

It is recognized that laws and regulations on ABS and issues of utilization of GRs are very complicated. They closely relate to policies and laws on biodiversity, agriculture, intellectual property rights (IPRs), trade, commerce, development and poverty alleviation. Even in the field of law, complicated and layered relations are seen ranging from the global level to the local level and from the public level to the private level. They include the CBD and other global treaties, regional agreements such as the Andean Agreement, bilateral agreements which set out mutually agreed terms and national laws and regulations of both provider and user countries, local laws and ordinances, and also private contracts, which serve as material transfer agreements (MTAs).

For a desirable ABS regime, both paragraphs 3 and 7 of Article 15 CBD should be taken into consideration in order to balance the provider countries' and the user countries' obligations.

Sovereign right and domestic law enforcement

Although Article 15(1) of the CBD stipulates that 'the authority to determine access to GR rests with the national governments and is subject to national legislation', only a few countries have adopted ABS legislations. Even among countries which adopted such legislations, those laws and regulations are diverse. It is the legal consequence of the provision of CBD cited above.

Under international law all countries have the sovereignty to regulate and enforce matters within their own territorial jurisdiction. As a result, countries are not compelled to apply laws and policies of other countries or, in other words, countries cannot enforce their laws within the jurisdiction of another sovereign state. Many countries may not wish to make a concession on matters of sovereignty so as to change their domestic ABS laws and regulations adopted in accordance with the provisions of the CBD to adapt to an internationally agreed approach.

Looking from the other side, the Lacey Act[1] of the United States provides an interesting example of how a national law and/or regulation can incorporate relevant foreign laws so as to control activities that are illegal under the law of that foreign country. Also in Japan, in relation to the implementation of the Convention on International Trade in Endangered Species of Wild Fauna and Flora (CITES), the Management Authority of Japan is required to reconfirm directly to the Management Authority of the country of export, prior to the authorization of the import of the specimen, which export is prohibited under the law of that exporting country.[2] User countries are encouraged to take such legal measures as accepting the result of laws of other countries. For taking such measures in ABS issues, it would be necessary for a provider country to set up a clear category of GRs by a negative listing to which a strict control may be applied.

Administrative and criminal procedures are central to the national sovereignty. It would be unusual in administrative and criminal matters to agree to enforce the unexamined laws of any other state around the world. The same can be said about the recognition and enforcement of foreign judgements in administrative and criminal matters.[3] In international society today, which is based on national sovereignty, international cooperation on administrative and criminal matters is carried out through a bilateral treaty on extradition and mutual legal cooperation in administrative and criminal matters. All countries need to further explore desirable mutual legal cooperation agreements applicable on ABS issues.

However, without a bilateral treaty on extradition and mutual legal cooperation, there is another legal framework where domestic courts of a country can issue the orders that punish nationals as well as foreigners who, outside of that country, committed a crime under the criminal code of that country. It can be found in a traditional criminal law provision on an offence committed while outside the territory of a country. Because a violation of the foreign ABS law may involve such specified offences, this provision could work out a better remedial measure. In Japan, the Penal Code has provisions on an offence committed while outside Japan,[4] and a person, who committed one of the specified crimes while in a provider country, is punishable in Japan under Japan's Penal Code.

To take a further step for utilization of the user country law, Article 15(7) of the CBD should be recalled. All countries are obliged to take legislative, administrative or policy measures in order to share, in a fair and equitable way, the results of research and development and the benefits arising from the commercial and other utilization of GR with the provider country. Where provisions in accordance with Article 15(7) are in place, benefit-sharing clauses in MTAs are enforceable by the user country laws including court procedure law. Also benefits arising from genetic material obtained without the prior informed consent (PIC) of a provider country may be controlled. Thus user countries especially are highly responsible to improve their relevant laws in accordance with Article 15(7).

As for an international public law approach, it will be useful to develop agreements on the definitions of misappropriation or misuse, especially in cases in which the user does not have an MTA or other compliance with the ABS requirements of a provider country. Based on such definitions, illegal taking of GRs is not only a wrongful act under the law of a provider country, but also a wrongful act under international law, and that constitutes an illegality in the user country. This could avoid difficulties faced when domestic laws are applied. In this context, a series of conventions on human rights and the UNESCO Convention on the Means of Prohibiting and Preventing the Illicit Import, Export and Transfer of Ownership of Cultural Property offer valuable examples for developing a framework for an international public law approach. In addition, the Convention against Transnational Organized Crime is noteworthy because it provides a basis for cooperation in addressing international crime and could be applied in relation to organized illegal taking of GRs. This Convention contains provisions regarding extradition and mutual legal assistance, enforcement cooperation and confiscation. It applies to defined crimes that are transnational and involve a group of three or more persons acting in concert with an aim of obtaining a financial or other material benefit.

Compliance with MTAs

In case one of the parties to a commercial contract does not comply with the contract, it is general and effective for the other party, the plaintiff, to institute a lawsuit at the court, which has a jurisdiction over the place where the defendant is. In this case, the plaintiff has to bear the burden of participating in foreign judicial procedure using foreign laws and foreign language. On the other hand, the case may be filed to the court within the plaintiff's country if the plaintiff so decides. In this case, concern arises that the defendant's appearance is not mandatory and that the ruling cannot surely be enforced. These are, however, common challenges always faced in international transactions of goods and services. Various approaches have been made so far to address these difficulties, including the recognition and enforcement of civil and commercial judgements rendered by a foreign country's court (Young, T. R., 2007). These available mechanisms can contribute to the removal of the above-mentioned difficulty when the court of the plaintiff's country is used, and they are useful for ensuring compliance with ABS provisions in MTAs.

The Bonn Guidelines have provisions on dispute settlement and require a Party to take effective measures. Under 'C.' ('Legal provisions') of Appendix I ('Suggested Elements for Material Transfer Agreements') to the Bonn Guidelines, basic items for dispute settlement are set out in paragraphs 7 ('Dispute settlement arrangements'), 10 ('Choice of law') and 11 ('Confidentiality clause'). In line with these provisions, it is necessary to make certain that the MTA is sufficiently clear and specific to enable a court to come to an unambiguous decision.

Although, in many countries, a foreign citizen has an equal access to domestic court, as the nationality is not a prerequisite for litigation, it needs to develop further supporting measures for foreign plaintiffs. Both parties to an MTA, especially a provider, may need to have access to special supporting measures in order to utilize national remedial laws and processes of the other party, considering the fact that many providers lack the funds, expertise and ability to engage in a protracted action in another country seeking redress from a user who is probably better funded and more familiar with the relevant legal system. In order to cope with these obstacles, judicial cooperation among countries is key in facilitating the judicial process for foreign applicants. The Convention on Access to Information and Public Participation in Environmental Matters (Aarhus Convention) is a well-known example in this respect.

At international level, in order to minimize the obstacles and impediments faced by plaintiffs in seeking remedies in the defendant's country, it would be useful to adopt an agreement to provide international standards of procedure, evidence and interpretation. Such efforts have been taken for harmonization of legal systems on civil and commercial law in the international forum, including the Hague Conference on Private International Law, UNCITRAL (United Nations Commission on International Trade Law) and UNIDROIT (International Institute for the Unification of Private Law).

Among them the Hague Conference has achieved appreciable results. A series of conventions for international legal cooperation and litigation, and for jurisdiction and enforcement of judgements, have been adopted.[5] The most recent Convention on the Choice of Court sets rules for when a court shall have jurisdiction to decide a dispute where commercial parties have entered into an exclusive choice of court agreement. It also provides for the recognition and enforcement of resulting judgements, with an option for

States Parties to agree on a reciprocal basis to recognize judgements based on a choice of court agreement that is not exclusive.

However, the negotiation on the Draft Hague Convention on Jurisdiction and the Recognition and Enforcement of Foreign Judgments in Civil and Commercial Matters is not yet agreed, mainly because of differences among opinions of countries on conditions of recognition and enforcement in cases related to IPRs. It is worth recalling that such difficulty is very similar to the negotiation of the international regime on ABS. In other words, consideration on ABS compliance mechanisms cannot be separate from the negotiation on harmonization of private laws.

Recognition and enforcement of foreign judgements in Japan

In general, the recognition and enforcement of foreign judgements in civil and commercial matters is not so difficult to obtain in Japan. As mentioned above, a foreign citizen has equal access to domestic courts in Japan. Nevertheless, there is a need to develop further supporting measures for foreign plaintiffs and judicial cooperation with other countries in order to facilitate the judicial process for foreign applicants.

With regard to foreign judgements, the Japanese courts recognize and enforce not only money judgements, but also foreign judgements awarding other forms of relief (Iwasaki, 1999). However, the relief must be of a type that can be awarded by a Japanese court. Under the Code of Civil Procedure (CCP), Japanese courts can grant three forms of relief. Firstly, an order for the payment of a certain sum of money or the transfer of property, or the performance or non-performance of a certain act. Secondly, an order for confirmation of the existence or non-existence of certain rights or legal relationships. Thirdly, an order for declaration on the creation or change of certain rights or legal relationships. Thus, foreign judgements ordering the payment of taxes or penalties in relation to commercial contracts are not recognized under the CCP since the foreign court that rendered such a judgement is not considered to be a court exercising civil jurisdiction.

Article 118 of the CCP,[6] which sets out conditions for recognition, requires that the foreign judgement is final. In Japan a foreign judgement is considered final when the time for appealing has expired under the laws of that foreign country. Its paragraph 1 requires that the competence of the foreign court is not denied in laws and orders or treaty. If the foreign court is competent under Japanese concepts, its jurisdiction is to be recognized even if the Japanese courts possess concurrent adjudicatory competence over the same case. Respecting the right to refute, paragraph 2 stipulates that the defeated defendant has received service of summons or other necessary process to commence the proceedings. And paragraph 3 requires that the judgement is not contrary to public policy, in other words, public order or good morals, in Japan. Although not clearly defined, a punitive or disciplinary punishment, or exemplary damages, are considered to be contrary to the public policy.

Reciprocity or mutual guarantee is required by paragraph 4. Again, there is not a statutory definition of this requirement. It is considered to be sufficient if the respective requirements in Japan and the foreign country for recognition of foreign judgements do not differ so as to lose balance, and both countries' requirements are identical with each other in important aspects. In addition, theoretically, there is a need to consider whether a foreign judgement that is sought to be recognized conflicts with a prior judgement of a domestic or foreign court, as well as a matter that is currently the subject of litigation. The Japanese courts have considered this question in the context of the public policy requirement mentioned above.

Even if a foreign judgement meets the requirements for its recognition under Article 118 of the CCP, the foreign judgement is not directly enforceable in Japan. That foreign judgement must be transformed to a judgement of execution under Article 24 of the Civil Execution Act (CEA).[7] Then it may be enforced by all means available under CEA. Therefore, a method of enforcement, which is not permitted under CEA, is not enforceable. For example, a specific enforcement by putting the debtor in jail is not available in Japan. According to paragraph 2, Article 24 of CEA, a Japanese court cannot review the facts which were found by the foreign court, and the parties cannot allege any new fact which had existed but was not raised before the foreign judgement was rendered.[8]

Avoidance of dispute and alternative dispute resolution

Many environmental treaties encourage, promote or require an effective, enhanced and positive implementation of their provisions. Such implementation means that a Party should take not only measures clearly obliged by provisions of the treaty, but also every possible measure necessary to tackle the core issues within the purpose of the treaty. This approach plays a very important role for implementation of the treaty and for avoidance of future dispute.

There exist various measures for the positive implementation of international environmental agreements. Those measures include those on public awareness, guidelines, synergies, surveillance and assistance, as well as voluntary measures including codes of conduct and best practice, incentives and voluntary certification schemes. For a positive implementation, every effort should also be made in order to ensure accountability, transparency, information dissemination and participation.

Avoidance of dispute

Since the cost of a dispute-avoidance process is evidently much lower than that of a dispute-settlement process, more attention should be paid to dispute avoidance and to voluntary measures. Firstly, one of the major causes of failure to comply with ABS regulation is the general lack of awareness of the CBD and its third objective, as well as ABS requirements in different countries. So awareness-raising and communication tools on ABS are probably one of the most cost-effective ways of improving compliance of users of GR. In this context, a reference should be made to Communication, Education and Public Awareness (CEPA) developed and implemented by the CBD and the Ramsar Convention.[9] CEPA activities range from a small local meeting to a large international symposium, from a brief pamphlet to a thick scientific report, and from an introductory lecture to a specialized and technical training. These activities encourage providers and users of GRs to agree with appropriate contracts and inform them of the appropriate formulation of MTA by presenting best practices of MTA or internationally recognized components of MTA, and also raise awareness among other stakeholders, including administrative officers and lawyers. Accordingly, countries need to ensure that a sufficient level of information on CBD and its third objective, as well as ABS requirements in major countries, is readily available to both provider and user.

Secondly, consideration should also be given to voluntary measures that address compliance issues and complement the Bonn Guidelines. For example, sectoral menus, guidelines, model MTA clauses, codes of conduct for users and identification of best practices have been demonstrated to be practical and effective mechanisms. Some of these measures have been developed by governments and some by user groups.[10] Voluntary certification schemes have been equipped in the field of forestry, fishery and other commercial activities.[11] Such certification schemes vary in their concept, scope, method, authority, tracing or administration. The scheme developed with a consideration to practicability, flexibility and cost-effectiveness may have the advantage of providing flexibility for their implementation. With regard to certification schemes, an individual identification of GRs by a codified unique identifier has been proposed in order to facilitate international recognition (Barber et al, 2003; CBD, 2007;

Tobin et al, 2008). However, due attention must be paid to the fact that even clearly identified goods have been traded illegally. As pointed out in relation to prevention of laundering activities under CITES, the key to a credible scheme is how to prevent any separate movement of the certificate from the certified GR.

Alternative dispute resolution

The actual implementation of measures to ensure the legality of GRs could be further promoted by an effective dispute settlement system. The dispute settlement system to be examined includes both judicial and non-judicial mechanisms. Mediation, conciliation and arbitration have been the most fundamental and traditional mechanisms for dispute resolution. Those mechanisms are of a non-judicial and voluntary nature, with high flexibility and simplified processes, and require lower costs compared with judicial mechanism. In addition to the traditional mechanisms, the non-compliance procedures (NCP) have been utilized and became popular as an informal mechanism in many environmental treaties.[12] Nowadays, the non-judicial mechanism and the non-compliance procedures have been given a high value in dispute resolution. Such a non-judicial mechanism is also referred to as an alternative dispute settlement (ADR). Thus, ADR is a well-known and established dispute settlement mechanism having advantages – compared with other judicial mechanism – of commonality, universality, less time and less cost. Such an ADR could also be highly suitable for the disputes concerning GRs. It is recommended that MTAs should stipulate the use of ADR as a dispute settlement mechanism.

Although the arbitration is a selective and flexible procedure, its process is clear and its decision is binding. It has been used in various cases of civil and commercial affairs and considered to be very effective and useful. Actually, arbitration clauses are included in many environmental treaties.[13] One of the most recent ADR procedures was established by the International Treaty on Plant Genetic Resources for Food and Agriculture (ITPGRFA) (Article 22) and its Standard Material Transfer Agreement (SMTA) (Article 8). In addition, major international organizations have adopted specific rules on arbitration and some of them established the arbitration court. They include the International Chamber of Commerce (ICC), the International Centre for Settlement of Investment Disputes (ICSID) and the UNCITRAL.[14]

The United Nations Convention on the Recognition and Enforcement of Foreign Arbitral Awards (New York Convention) is widely recognized as a foundation instrument of international arbitration. It requires courts of contracting States to give effect to an arbitration agreement and also to recognize and enforce awards made in other States, subject to specific limited exceptions.[15]

Since most obligations arising under MTAs will be between providers and users, disputes arising in the MTA should be solved in accordance with the relevant contractual arrangements on ABS and the applicable law and practices. In an ABS context, many MTAs already include settlement of dispute clauses based on ADR.

Permanent Court of Arbitration Optional Rules

One of the new developments in the arbitration mechanism was made by the Permanent Court of Arbitration (PCA) in 2001. PCA Optional Rules for Arbitration of Disputes Relating to Natural Resources and/or the Environment were designed to fill the principal lacunae in environmental dispute resolution and adopted on 19 June 2001 and entered into force on the same day.[16] They are based primarily on the PCA and UNCITRAL Conciliation Rules and applicable for all parties, including States, Inter-Governmental Organizations (IGOs), Non-Governmental Organizations (NGOs) and private entities when they seek resolution of controversies concerning environmental protection or conservation of natural resources. The procedures are optional with flexibility and respect the party autonomy. The parties to the submitted dispute have complete freedom to agree to any individual or institution making appointments of arbitrators. As stated in Article 1 para 1, the characterization of the dispute as relating to the environment or natural resources is not necessary for jurisdiction. The Rules are applied if all the parties have agreed to settle a specific dispute under these Rules.

Because time may be a matter of primary concern in disputes related to natural resources and the environment, the Rules provide for arbitration in a shorter period of time than under previous PCA Optional Rules or the UNCITRAL Rules. Thus, the Rules are equipped with a fail-safe mechanism. In order to prevent frustration or delay of the arbitration, the Rules provide that the Secretary-General will act as the appointing authority if the parties do not agree upon the authority or if the authority does not act. Similarly, as for the applicable laws, the parties shall designate the applicable law. If a party fails in such designation, the arbitral tribunal shall decide the applicable law.

Issues on ABS and intellectual property involve sensitive commercial information and commercial secrets. In these cases, the PCA Rules also have a clause on confidentiality. Measures to protect the confidentiality of information provided by the parties are specifically described in Article 15 paras 4–6, respectively on 'application for confidentiality of any information', 'determination by the arbitral tribunal' and 'appointment of a confidentiality advisor'. These provisions also intended to save the time required in the process. In case arbitrations deal with highly technical questions, the parties may submit a document summarizing and providing background to any scientific or technical issues, which the parties may wish to raise in their memorials or at oral hearings. The Rules also have a provision that authorizes the arbitral tribunal, unless the parties chose otherwise in their compromise, to order within the subject matter of the dispute before the tribunal any interim measures necessary to prevent serious harm to the environment.

In addition, the PCA Rules provide a legal and technical assistance to parties to the dispute by establishing a panel of arbitrators with experience and expertise in environmental or conservation of natural resources law (Article 8(3)) and a panel of environmental scientists who can provide expert scientific assistance (Article 27(5)), as well as the place and facility for an arbitral tribunal and necessary secretariat, registrar and translation services. Thus the PCA rules are recommended as a reliable dispute settlement mechanism in MTAs.

International Treaty on Plant Genetic Resources for Food and Agriculture

Another existing ADR mechanism can be found in the regime for plant GRs for food and agriculture. The ITPGRFA under the UN's Food and Agriculture Organization (FAO) was adopted in 2001 and entered into force in 2004. The aim of the ITPGRFA is promotion of agriculture and development of agricultural research activities. For that end, the multilateral system (MLS), composed of a limited category of plant genetic resources (PGRs) listed in Annex I, is established and all transaction of the resources in MLS must be conducted in the form of the SMTA. The aim of the MLS is the facilitation of access to and free use of PGRs within the MLS. When a product – that is a PGR for food and agriculture and that incorporates material accessed from the MLS – is not available without restriction to others for further research and breeding, a recipient who commercializes such a product shall pay to the Trust Account an equitable share of the benefits arising from the commercialization of that product. The amount of payment from such a recipient does not flow directly to the provider. The accumulated fund from the Account is to be allocated, directly or indirectly, primarily to farmers in all countries, especially in developing countries and countries with economies in transition, who conserve and sustainably utilize PGR for food and agriculture.

Although the ITPGR covers a very limited category of PGRs clearly identified, it sets out a mechanism for ABS. It also promotes cooperative and effective operational mechanisms to enhance compliance with its SMTA. These include 'monitoring, offering advice or assistance, including legal advice or legal assistance, when needed, in particular to developing countries and countries with economies in transition'. Article 7 of the SMTA provides for applicable law and Article 8 provides for dispute settlement, including ADR. It is worth noticing here that under Article 8, any party to the dispute may submit the dispute for a final settlement under the ICC Rules of Arbitration, when the dispute has not been settled by negotiation or mediation and the parties to the dispute have failed to agree on arbitration rules of an international body.

Non-commercial activities and the change of intent

Among the disputes on the use of GRs and ABS, issues of change of intent and/or purpose of the GR use are involved. The treatment of activities with non-commercial intent and the subsequent change of the intent needs to be examined here. Most in situ GRs are also accessed by non-commercial research oriented organizations; for example, university professors and students, biologists working for public research institutes and taxonomists. However, it is impossible to predict at the outset of such research whether subsequent study and analysis of the collected GRs will result in commercial applications. Sometimes commercial applications may occur after many years and a number of transfers of the GRs to third parties from the original accessor. Considering these circumstances, it would be better to distinguish activities based on the intent of use rather than the type of activities.

In that context, especially for academic and scientific research activities, special consideration might be taken. For example, Article 247 of the UN Convention on the Law of the Sea (UNCLOS) covers marine scientific research projects undertaken by or under the auspices of international organizations.[17] It stipulates that in the exclusive economic zone (EEZ) or on the continental shelf, sovereign rights of coastal states apply. Scientific research activity in the area, where the sovereign rights apply, logically needs a clear consent from that country. This provision reflected a strong concern raised by the academic and research associations worldwide that an EEZ regime might impede marine scientific research within the EEZ. Recognizing the scientific research activities and the ultimate well-being of humanity, it sets out that an authorized international research organization is deemed to have got an implicit consent from that State. A similar approach could be considered for facilitated procedures for academic and scientific research activities on GRs, planned and conducted by authorized international organizations.

For other scientific research activities with non-commercial intent, compliance is enhanced where there are transparent, non-discriminatory and practical ABS arrangements by MTA in advance. In particular, a clear provision on the change of intent, including procedures for a report to, an application to and a permit by the original provider, as well as a remedy and sanction in the case of non-reported and non-permitted changes of intent, would be effective. Such a change of intent is not specific to ABS issues, and there are a number of commercial practices. Some of the existing national laws and SMTA of ITPGR provide examples on how a change of intent can be addressed. Thus, in order to avoid disputes and also to provide a better resolution of disputes on the use of GRs, it is advisable to include provisions for compliance relating to change of intent or transfer to a third-party in MTAs, as well as national ABS regimes as part of the PIC procedure.

It can be pointed out here that the definition of non-commercial research, if possible, can contribute to tackling this issue. For example, ITPGR provides the definition for 'commercialization' in the SMTA under 'Definitions'. The application of that definition to the more general use of GRs could be pursued in the context of ABS compliance.

Conclusion

In considering the necessary measures for ABS, mutual benefit must be the central principle. For that purpose, such measures should be on a fair and equitable basis with accountability and transparency. In developing such measures and such a system, including in the stage of their operation and among stakeholders, the user company plays the most important role.

As explained above, existing measures for compliance and enforcement, whether international or national, public law or private law, court procedures or ADR, should be utilized at first to the utmost extent of their function. It is recommended that MTAs should stipulate the use of ADR as a reliable dispute settlement mechanism, especially, among others, the PCA Optional Rules for Arbitration of Disputes Relating to Natural Resources and/or the Environment. Since the cost of a dispute-avoidance process is evidently much lower than that of a dispute-settlement process, more attention should be paid to dispute avoidance and to voluntary measures, including CEPA activities. In addition, it needs to further develop financial and legal measures for supporting foreign plaintiffs.

Establishment of centres for ABS at regional, national or local levels and their networking would be most desirable and necessary for promotion of sustainable use of GRs. Such centres could work for promoting and facilitating the sound development of GR activities with the aim of assuring the legality of international transactions in GRs and necessary ABS procedures. Those centres might include a clearing-house mechanism and a databank that included existing laws and regulations, as well as customary rules of indigenous communities. A network of ABS centres could work effectively to cope with the complicated issues of ABS and to assist all providers and users in their negotiation of ABS, and to monitor the status of compliance with and implementation of relevant international laws, domestic laws and private contractual agreements.

The Bonn Guidelines cover necessary measures to be taken for effective implementation of relevant provisions and for dispute settlement. In concrete terms, they include paragraph 50 (Accountability) and paragraphs 54–58 (Monitoring of Compliance, Verification, Certification and Dispute Settlement). Such functions would be carried out efficiently by centres, especially with an informal mechanism for implementation and dispute settlement.

In cooperation with experts in various fields, those centres could provide advice from a biological, ecological, technical, commercial and legal point of view that would be sought by different stakeholders in the process of negotiation, implementation and dispute resolution concerning the MTA on GRs. Those centres would take the role of facilitator or mediator, as well as providing capacity building. Support and promotion of positive and enhanced implementation of ABS regulations could also be given. It is certain that these recommendations are useful and effective in implementing relevant laws in a proper way and in avoiding the occurrence of disputes over GR use. A standardized system for documenting evidence of PIC and tracing flows of GRs required for enhancing transparency, equity and compliance with ABS arrangements could also be operated through the network of ABS centres.

Considering the fact that biological resources including GR as well as TK exist across national borders, for such shared resources and knowledge among several countries, ABS faces ambiguity and difficulty in practice. Since few countries sharing resources have either agreed on an individual or collective right to control access and share benefits, or agreed on allocation percentage of such resources, a regional centre is also effective for equitable sharing of benefits arising from the utilization of common resources.[18]

Notes

1 Public Law 110–246, 18 June 2008. The recent amendments to it extend the statute's reach to encompass plant products, including timber, that derive from illegally harvested trees in foreign countries and are taken into the USA.

2 This obligation was put by the Import Trade Control Order under the Foreign Exchange and Foreign Trade Act in order to introduce a stricter control measure than the CITES provisions.

3 Hence, a judgement ordering the payment of tax or penalty is not considered to be a civil matter. Thus provider countries need further improvement of their legislation so as to maximize the ability of the country, as well as other private providers, to bring action directly in the user country. One of the considerable technical measures is to establish that all ABS permits, licences and other instruments are private contracts, in order to ensure them enforceable in the civil and commercial dispute resolution mechanism. In this context, see Young (2007).

4 Articles 2 and 3 of the Penal Code, Act No. 45 of 24 April 1907, http://www.japaneselawtranslation.go.jp/law/detail/?ft=3&re=01&dn=1&bu=16&x=77&y=35&ky=&page=5, accessed 3 April 2009.

5 The following conventions were adopted by the Hague Conference, http://www.hcch.net/index_en.php?act=text.display&tid=10#litigation, accessed 25 January 2009:
   • Convention of 5 October 1961 on Abolishing the Requirement of Legalization for Foreign Public Documents
   • Convention of 15 November 1965 on the Service Abroad of Judicial and Extrajudicial Documents in Civil or Commercial Matters
   • Convention of 18 March 1970 on the Taking of Evidence Abroad in Civil or Commercial Matters
   • Convention of 25 October 1980 on International Access to Justice
   • Convention of 1 March 1954 on civil procedure
   • Convention of 15 April 1958 on the jurisdiction of the selected forum in the case of international sales of goods
   • Convention of 25 November 1965 on the Choice of Court
   • Convention of 1 February 1971 on the Recognition and Enforcement of Foreign Judgements in Civil and Commercial Matters
   • Supplementary Protocol of 1 February 1971 to the Hague Convention on the Recognition and Enforcement of Foreign Judgements in Civil and Commercial Matters
   • Convention of 30 June 2005 on the Choice of Court Agreements.

6 Act No. 109 of 26 June 1996, http://www.japaneselawtranslation.go.jp/law/detail_main?re=01&ft=2&kn%5B%5D=&page=1&vm=&id=90#en_pt1ch5sc5at5, accessed 3 April 2009.

7 Act No. 4 of 30 March 1979, http://japaneselawtranslation.go.jp/law/detail_main?re=01&ft=3&kn%5B%5D=&page=6&vm=&id=70#en_ch2sc1at3, accessed 3 April 2009.

8 For an explanation of Japan's law on civil procedures, see Iwasaki (1999).

9 About CEPA, see among others, CBD Decision VIII/6 (Global Initiative on Communication, Education and Public Awareness: overview of implementation of the programme of work and options to advance future work; Toolkit CEPA,

CBD/IUCN/CEC), http://www.cbd.int/cepa/toolkit/2008/cepa/index.htm, accessed 3 April 2009 and the Ramsar Convention Resolution VIII.31 (The Convention's Programme on communication, education and public awareness (CEPA) 2003–2008).

10 See, among others, the Guidelines for Access to Genetic Resources for Users in Japan (METI and JBA, 2006), the ABS Management Tool – Best Practice Standard and Handbook for Implementing Genetic Resources Access and Benefit-sharing Activities (Stratos and Swiss Department of Economic Affairs, 2007), Access and Benefit-sharing, Good practice for academic research on GRs (Swiss Academy of Sciences, 2006). See, among others, the Guidelines for BIO Members Engaging in Bioprospecting (Biotechnology Industry Organization, 2005), the Principles on Access to Genetic Resources and Benefit-sharing (Botanical Garden Conservation International, 2000) or the Guidelines for IFPMA Members on Access to Genetic Resources and Equitable Sharing of the Benefits Arising out of their Utilization (International Federation of Pharmaceutical Manufacturers and Associations, 2007). Some industries have developed or publicly committed themselves to respect ABS requirement; for example, Novo Nordisk (Guiding Principles), http://www. novonordisk.com/old/press/environmental/er97/bio/Guidingprinciple.html, accessed 29 May 2009.

11 ISO 14000 and ISO 14001 adopted by the International Organization for Standardization (ISO) in 1996; the Forest Stewardship Council (FSC) founded in 1993; the Marine Stewardship Council (MSC) founded in 1997.

12 See, among others, Article 8 and Appendix IV of the Montreal Protocol and Article 18 of the Kyoto Protocol.

13 For example, the ECE/EIA Convention, UNCLOS, the Basel Convention, CRAMRA, the Antarctic Environmental Protocol, the Antarctic Marine Living Resources Conservation Convention, CBD, the Convention for Combating Desertification (28.2a), Ozone Layer Convention (11.3a), United Nations Framework Convention for Climate Change (UNFCCC) (14.2b), CITES (18.2) and CMS (13.2).

14 ICC adopted the following rules: ICC International Court of Arbitration and ICC 1998 Rules of Arbitration. International Centre for Settlement of Investment Disputes (ICSID) adopted the following rules: Administrative and Financial Regulations; Rules of Procedure for the Institution of Conciliation and Arbitration Proceedings (Institution Rules); Rules of Procedure for Arbitration Proceedings (Arbitration Rules); Rules of Procedure for Conciliation Proceedings (Conciliation Rules); Administrative and Financial Rules (Additional Facility); Conciliation (Additional Facility) Rules; Arbitration (Additional Facility) Rules and Fact-Finding (Additional Facility) Rules. UNCITRAL adopted the following rules: UNCITRAL Arbitration Rules (1976); UNCITRAL Conciliation Rules (1980); UNCITRAL Notes on Organizing Arbitral Proceedings (1996); UNCITRAL Model Law on International Commercial Arbitration; UNCITRAL Model Law on International Commercial Conciliation. Other administered systems include the London Court of International Arbitration, the American Arbitration Association, the China International Economic and the Inter-American Commercial Arbitration Commission.

15 Almost identical provisions to those in the New York Convention and other Arbitration Rules are found in the Japan's Arbitration Law (Law No. 138 of 2003),

http://www.kantei.go.jp/foreign/policy/sihou/arbitrationlaw.pdf, accessed 3 April 2009.

16 These Optional Rules can be viewed at http://www.pca-cpa.org/upload/files/ENVI-RONMENTAL(3).pdf, accessed 25 January 2009. PCA also adopted the Optional Rules for Conciliation of Disputes Relating to Natural Resources and/or the Environment on 16 April 2002, http://www.pca-cpa.org/upload/files/ENV%20CONC.pdf, accessed 25 January 2009.

17 Article 247: Marine scientific research projects undertaken by or under the auspices of international organizations. A coastal state that is a member of or has a bilateral agreement with an international organization, and in whose exclusive economic zone or on whose continental shelf that organization wants to carry out a marine scientific research project, directly or under its auspices, shall be deemed to have authorized the project to be carried out in conformity with the agreed specifications if that State approved the detailed project when the decision was made by the organization for the undertaking of the projector is willing to participate in it, and has not expressed any objection within four months of notification of the project by the organization to the coastal state.

18 For a detailed discussion of this aspect, see Winter ('common pools') in this book.

References

Barber, C. V., Johnston, S. and Tobin, B. (2003) *User Measures: Options for Developing Measures in User Countries to Implement the Access and Benefit-Sharing Provisions of the Convention on Biological Diversity*, UN University/Institute of Advanced Studies (UNU/IAS)

CBD (2007) 'Report of the meeting of the group of technical experts on an internationally recognized certificate of origin/source/legal provenance', UNEP/CBD/WG-ABS/5/7

Iwasaki, K. (1999) 'Recognition and enforcement of foreign judgments, dispute resolution in Japan', http://www.gsid.nagoya-u.ac.jp/project/apec/lawdb/japan/dispute/disp2-en.html, accessed 29 January 2009

Tobin, B., Burton, G. and Fernandez-Ugalde, J. C. (2008) 'Certificates of clarity or confusion: The search for a practical, feasible and cost effective system for certifying compliance with PIC and MAT', UNU-IAS Report

Young, T. R. (2007) 'Analytical study on administrative and judicial remedies available in countries with users under their jurisdiction and in international agreements', UNEP/CBD/WG-ABS/5/INF/3

Australian Government

DEED OF AGREEMENT

BETWEEN

COMMONWEALTH OF AUSTRALIA

AND

INSERT NAME OF ACCESS PARTY

IN RELATION TO

ACCESS TO BIOLOGICAL RESOURCES IN COMMONWEALTH AREAS AND BENEFIT SHARING

**MODEL ACCESS AND BENEFIT SHARING AGREEMENT
AUSTRALIAN GOVERNMENT AND ACCESS PARTY**

CONTENTS

MODEL ACCESS AND BENEFIT SHARING AGREEMENT
AUSTRALIAN GOVERNMENT AND ACCESS PARTY

DEED OF AGREEMENT
ACCESS TO BIOLOGICAL RESOURCES AND BENEFIT SHARING

DATE

This Deed is dated insert date.

PARTIES

This Deed is made between and binds the following parties:

1. **COMMONWEALTH OF AUSTRALIA** (**Commonwealth**), as represented by and acting through the Department of the Environment and Water Resources ABN 34 190 894 983 (**the Department**)

2. NAME OF ACCESS PARTY of address ABN 11 111 111 111 (**Access Party**)

CONTEXT AND PURPOSE

This Deed is made in the following context:

A.   The Convention on Biological Diversity and the Bonn Guidelines under give parties to the Convention the responsibility to manage their biological diversity to ensure, inter alia, fair and equitable sharing of the benefits arising from the use of genetic resources.

B.   Section 301 of the *Environment Protection and Biodiversity Conservation Act 1999* (**EPBC Act**) provides for regulations to be made for the control of access to biological resources in Commonwealth areas, including the equitable sharing of the benefits arising from the use of biological resources in Commonwealth areas.

C.   Part 8A of the *Environment Protection and Biodiversity Conservation Regulations 2000* (**EPBC Regulations**) makes provisions for the purposes of section 301 of the EPBC Act. The regulations require access to biological resources in a Commonwealth area to be in accordance with a permit under the regulations unless the biological resources have been declared exempt. An applicant for a permit to access biological resources for commercial purposes or potential commercial must enter into a benefit-sharing agreement with each access provider for the resources.

D.   The Commonwealth is the access provider for the purposes of Part 8A of the EPBC Regulations for biological resources in Commonwealth areas (as defined in the EPBC Act).

MODEL ACCESS AND BENEFIT SHARING AGREEMENT
AUSTRALIAN GOVERNMENT AND ACCESS PARTY

E.   The Access Party is the applicant for, or intends to apply for, a permit under Part 8A of the EPBC Regulations to access the biological resources, in the Commonwealth area or areas, specified in Schedule 2 to this Deed.

F.   This Deed constitutes a Benefit Sharing Agreement for the purposes of Part 8A of the EPBC Regulations.

G.   In consideration of the Access Party entering into this Deed the Commonwealth grants the Access Party access to the biological resources, in the Commonwealth area or areas, specified in Schedule 2.

H.   In consideration of the Commonwealth granting access the Access Party will access and use the biological resources in accordance with this Deed and will provide the Commonwealth with the benefits specified in Schedules 3 and 4 to this Deed.

I.   This Deed, in conjunction with an access permit issued under Part 8A of the EPBC Regulations, gives the Access Party access to biological resources in the Access Area.

OPERATIVE PROVISIONS

The parties to this Deed agree as follows:

1.   Interpretation

1.1.   Definitions

1.1.1.   In this Deed, unless the context indicates otherwise:

Access Area	means the Commonwealth area or areas specified in Schedule 2 where the Access Party may have access to biological resources;
Access Party	means the person or persons (individual or organisation) named as the Access Party and includes their officers, employees, agents and contractors, or any of them, where the context permits;
access to biological resources	has the meaning given by the EPBC Regulations and means the taking of biological resources of native species for research and development on any genetic resources, or biochemical compounds, comprising or contained in the biological resources, but does not include activities described in regulation 8A.03(3);

2

MODEL ACCESS AND BENEFIT SHARING AGREEMENT
AUSTRALIAN GOVERNMENT AND ACCESS PARTY

access permit	means a permit issued in accordance with Part 17 of the EPBC Regulations, for the purposes of Part 8A of the Regulations, authorising access to biological resources in the Access Area;
biological resources	has the meaning given by the EPBC Act and includes genetic resources, organisms, parts of organisms, populations and any other biotic component of an ecosystem with actual or potential use or value for humanity;
Business Day	in relation to the doing of any action in a place, means a weekday other than a public holiday in that place;
Commencement Date	means the date of this Deed;
Commonwealth area	has the meaning given by section 525 of the EPBC Act;
Confidential Information	means:

    a. any information described as confidential in Schedule 1 to this Deed; and

    b. any information that is agreed between the Parties after the Date of this Deed as constituting Confidential Information for the purposes of this Deed;

Deed	means this Deed, the Schedules to this Deed and any attachments;
Department	means the Department of the Environment and Water Resources and includes any department or agency of the Commonwealth of Australia that succeeds to the functions of the Department;
EPBC Act	means the *Environment Protection and Biodiversity Conservation Act 1999*
EPBC Regulations	means the *Environment Protection and Biodiversity Conservation Regulations 2000*
Exploitation Revenue	means any monies received by the Access Party from third parties arising from the Access Party's use of biological resources, including monies received for:

    a. transferring, delivering, or providing access to Samples or Products; or

    b. assigning or granting rights (including Intellectual Property) in Samples or Products; or

MODEL ACCESS AND BENEFIT SHARING AGREEMENT
AUSTRALIAN GOVERNMENT AND ACCESS PARTY

	c. Sale,
	but not including funds received by the Access Party for the explicit purpose of research.
genetic resources	has the meaning given by the EPBC Act and means any material of plant, animal, microbial or other origin that contains functional units of heredity and that has actual or potential value for humanity;
Intellectual Property	Includes:
	a. copyright
	b. all rights in relation to inventions (including patent rights)
	c. all rights in relation to plant varieties (including plant breeders rights);
	d. registered and unregistered trademarks (including service marks), designs, and circuit layouts, and
	e. all other rights resulting from intellectual activity;
	f. know-how (whether patentable or not);
Material	means any matter or thing the subject of any category of property rights including Intellectual Property;
Product	means Material produced, obtained, extracted or derived through R & D Activity;
R & D Activity	means research or development on a Sample or Product;
Sample	means a sample of biological resources collected from the Access Area under a permit issued in conjunction with this Agreement;
Sale	means a payment received by the Access Party from a third party in consideration of the transfer to the third party of:
	a. Products; or
	b. Material containing a Product,
	by way of retail sale;
Threshold Payment	means the percentage of gross Exploitation Revenue to be paid by the Access Party to the Commonwealth in accordance with this Deed.

**MODEL ACCESS AND BENEFIT SHARING AGREEMENT
AUSTRALIAN GOVERNMENT AND ACCESS PARTY**

1.2.   **Interpretation**

1.2.1.   In this Deed, unless the contrary intention appears:

a.   words importing a gender include any other gender

b.   words in the singular include the plural and words in the plural include the singular

c.   clause headings are inserted for convenient reference only and have no effect in limiting or extending the language of provisions to which they refer

d.   words importing a person include a partnership and a body whether corporate or otherwise

e.   all references to dollars are to Australian dollars

f.   a reference to any legislation or legislative provision includes any statutory modification substitution or re-enactment of such legislation or legislative provision

g.   where any word or phrase is given a defined meaning, any other part of speech or other grammatical form in respect of that word or phrase has a corresponding meaning

h.   reference to an Item is to an Item in a schedule

i.   the schedules and any attachments form part of this Deed

j.   reference to a schedule (or an attachment) is a reference to a schedule (or an attachment) to this Deed, including as amended or replaced from time to time by agreement in writing between the parties and

k.   a reference to writing means any representation of words, figures or symbols, whether or not in a visible form.

1.3.   **Guidance on Construction of this Deed**

1.3.1.   This Deed records the entire agreement between the parties in relation to its subject matter.

1.3.2.   This Deed may only be varied by a formal deed of variation executed by both parties.

1.3.3.   As far as possible all provisions must be construed so as not to be invalid, illegal or unenforceable.

1.3.4.   If anything in this Deed is unenforceable, illegal or void then it is severed and the rest of this Deed remains in force.

MODEL ACCESS AND BENEFIT SHARING AGREEMENT
AUSTRALIAN GOVERNMENT AND ACCESS PARTY

1.3.5. Any reading down or severance of a particular provision does not affect the other provisions of this Deed.

1.3.6. If a provision cannot be read down, that provision will be void and severable and the remaining provisions will not be affected.

1.3.7. A provision of this Deed Lease will not be construed to the disadvantage of a Party solely on the basis that it proposed that provision.

2. Effect, Commencement and Review

2.1. Deed Subject to Issue of Permit

2.1.1. This Deed takes effect only if an access permit is issued to the Access Party for the proposed access to biological resources to which the Deed relates.

2.1.2. This Deed commences on the date a permit is issued to the Access Party to access biological resources to which the Deed relates.

2.2. Review

2.2.1. The operation of this Deed will be reviewed at the request of either party.

2.2.2. The first review may be conducted 2 years after the Commencement Date, and further reviews may be conducted at intervals not less than 2 years.

2.2.3. The timing and form of reviews will be agreed between the parties.

2.2.4. Either party may request that a review be conducted by an independent person agreed by the parties, and the other party will accede to that request.

2.2.5. Where a review is conducted by an independent person:

a. the parties will provide all reasonable assistance to, and respond to all reasonable requests for information and assistance from, the person conducting the review; and

b. the cost of the review will be borne by the party requesting the review unless both parties agree beforehand to share equally the costs of the review.

2.2.6. The requirements of this Deed relating to Confidential Information will apply to the conduct of a review and the parties will take all

MODEL ACCESS AND BENEFIT SHARING AGREEMENT
AUSTRALIAN GOVERNMENT AND ACCESS PARTY

practicable steps to ensure that the person conducting a review complies with those requirements.

2.2.7. The parties will discuss the findings and recommendations of each review and may agree to vary the terms and conditions of this Deed in accordance with clause 1.3.2.

3.   Benefit Sharing

3.1. **Benefits to be Provided**

3.1.1. The Access Party will provide the Commonwealth with the benefits specified in Schedule 3.

3.1.2. The Access Party will provide the Commonwealth with the additional benefits (if any) specified in Schedule 4.

3.1.3. Where the access to biological resources under this Deed leads to the discovery of new taxa, the Access Party must offer voucher specimens for permanent loan to an Australian public institution that is a repository of taxonomic specimens of the same order or genus as those collected.

3.1.4. In offering voucher specimens for permanent loan, the Access Party may set reasonable conditions for use of the loaned specimens.

3.1.5. The operation of this clause survives the expiration or earlier termination of this Deed.

4.   Performance Standards

4.1. **Standards**

4.1.1. In performing this Deed the Access Party will:

a. comply with the conditions specified in Schedule 2;

b. carry on its activities to a high standard and in accordance with relevant best practice, including any policies, codes of practice or guidelines specified in Schedule 1 or notified by the Commonwealth from time to time;

c. comply with the conditions of the Access Party's access permits;

d. comply with all relevant laws of the Commonwealth and any applicable laws of the States, Territories or local government;

e. obtain and hold all necessary approvals and licences;

**MODEL ACCESS AND BENEFIT SHARING AGREEMENT
AUSTRALIAN GOVERNMENT AND ACCESS PARTY**

    f.   liaise with the Department, provide any information the Department may reasonably require and comply with any reasonable request made by the Department; and

4.2.   **Animal Ethics**

4.2.1.   Where any activity under this Deed involves the use and care of living non-human vertebrate animals or tissue for scientific purposes, the Access Party will obtain review of and approval for such scientific purposes from a recognised animal ethics committee operating under the Australian Code of Practice for the Care and Use of Animals for Scientific Purposes or equivalent body.

4.2.2.   The Access Party will comply with all laws, policies, codes of practice and guidelines relating to animal welfare as they apply to the jurisdiction where the research will be undertaken.

5.   **Rights in and Dealings With Samples and Products**

5.1.   **Rights In Samples and Products**

5.1.1.   Subject to this clause, as between the parties the Access Party has the exclusive rights to all Samples and Products.

5.2.   **Intellectual Property**

5.2.1.   As between the Commonwealth and the Access Party (but without affecting the position between the Access Party and a third party) Intellectual Property arising from R & D Activity is vested or will vest in the Access Party.

5.3.   **Dealings with Samples and Products and Intellectual Property**

5.3.1.   Without limiting clause 5.2.1, the Access Party may grant third parties the right to exploit the Intellectual Property arising from R & D Activity.

5.3.2.   The Access Party will not:

    a.   transfer, deliver or provide access to Samples or Products; or

    b.   transfer, assign or grant rights (including Intellectual Property) in Samples or Products,

to a third party unless:

    c.   it does so under an agreement on proper terms, being terms consistent with this Deed so far as practicable and which would normally be contained in a contract, agreement or transaction

MODEL ACCESS AND BENEFIT SHARING AGREEMENT
AUSTRALIAN GOVERNMENT AND ACCESS PARTY

between persons dealing with each other at arms length and from positions of comparable bargaining power; or

d. the third party has entered into an agreement with the Commonwealth, or provided an enforceable undertaking to the Commonwealth, to provide the Commonwealth with the benefits and to comply with the requirements of this clause 5.3.2 in the event of any further dealing;

5.3.3. An agreement under clause 5.3.2.c must ensure the Commonwealth will continue to receive an equitable share of the benefits arising from subsequent use of the Samples or Products, or the rights in those Samples or Products by the third party and any subsequent parties.

5.3.4. An agreement under clause 5.3.2.c relating to use of Samples or Products or associated Intellectual Property by a third party for non-commercial purposes must include an undertaking not to carry out, or allow others to use the Material for commercial purposes unless a benefit-sharing agreement has been entered into with the Access Party

5.3.5. The Access Party must provide the Commonwealth with the name of each third party that an agreement is made with under clause 5.3.2 and details of the terms of the agreement.

5.4.   Exploitation Revenue and Threshold Payments

5.4.1. Exploitation Revenue received by the Access Party is subject to the Threshold Payment requirements under Schedule 3.

6.   Financial Arrangements

6.1.   Payments by the Access Party

6.1.1. Moneys payable by the Access Party to the Commonwealth under this Deed will be paid annually following delivery of Annual Reports in accordance with this Deed and within 28 days following receipt of a correctly rendered tax invoice.

6.2.   Taxes, Duties and Government Charges

6.2.1. Subject to this clause, all taxes, duties and government charges imposed or levied in Australia or overseas in connection with this Deed must be borne by the party liable for them.

6.2.2. Amounts payable by the Access Party to the Commonwealth under this Deed will include an amount to cover any liability of the

MODEL ACCESS AND BENEFIT SHARING AGREEMENT
AUSTRALIAN GOVERNMENT AND ACCESS PARTY

Commonwealth for GST on any supplies made by the Commonwealth under this Deed which are taxable supplies within the meaning of the GST Act.

6.2.3.   In relation to taxable supplies made under this Deed, the Commonwealth will issue the Access Party a tax invoice in accordance with the GST Act.

6.2.4.   In this clause:

a.   GST has the meaning given to it in the GST Act

b.   GST ACT means *A New Tax System (Goods and Services Tax) Act 1999* (Cth).

7.   Acknowledgment and Publicity

7.1.   Acknowledgement and Publicity

7.1.1.   The Access Party will acknowledge the provision of access to biological resources in Commonwealth areas in all dealings with third parties with respect to R & D Activity.

7.1.2.   The Access Party will ensure that an agreement with a third party under clause 5.3.2.c includes a requirement that the third party acknowledges the Commonwealth is the access provider to the source Sample.

7.1.3.   The operation of this clause survives the expiration or earlier termination of the Term of this Deed.

8.   Record Keeping

8.1.   Accounts and Records

8.1.1.   The Access Party will maintain complete, accurate and up to date accounts and records in relation to this Deed that:

a.   include appropriate audit trails for transactions performed;

b.   separately record all receipts;

c.   be kept in such a manner that permits them to be conveniently and properly accessed and audited;

d.   be drawn in accordance with generally accepted accounting practices and standards.

8.1.2.   Without limiting clause 8.1.1 the Access Party accounts and records will enable tracking of Exploitation Revenue to ensure correct delivery of Threshold Payments to the Commonwealth.

MODEL ACCESS AND BENEFIT SHARING AGREEMENT
AUSTRALIAN GOVERNMENT AND ACCESS PARTY

8.1.3.   The Access Party must hold accounts and records in relation to the provision of the Samples for a period of 7 years from the date of expiry or termination of this Deed.

8.1.4.   The operation of this clause survives the expiration or earlier termination of the Term of this Deed.

8.2.   Collection Reports

8.2.1.   Within six (6) months of collecting Samples under this Deed or on or before 31 March first occurring after collection, whichever is the later, the Access Party will provide a report to the Commonwealth containing the following records for each Sample taken:

   a.   for each record about a Sample, a unique identifier for the sample that is also on a label attached to the sample or its container;

   b.   the date the Sample was taken;

   c.   the place from which the Sample was taken and a description of the habitat from which the Sample was collected;

   d.   an appropriate indication of the quantity or size of the Sample;

   e.   the scientific name of, or given to, the Sample;

   f.   the location of the Sample when first entered in the record;

   g.   the details about any subsequent disposition of the Sample, including the names and addresses of others having possession of the Sample or a part of the Sample.

8.2.2.   Where a report under clause 8.2.1 includes a Sample of an undescribed species the Sample must be given a unique identifier, and the Access Party must subsequently advise the Commonwealth the scientific name of, or given to, the Sample when described.

8.2.3.   A Collection Report may be included in an Annual Report under 8.3.

8.3.   Annual Reports

8.3.1.   The Access Party will provide an initial Annual Report to the Commonwealth on activities under this Deed in the period from the date this Deed commences to the end of the calendar year immediately following completion of the collection of Samples. The report will include, but need not be limited to, the following information for the reporting period

   a.   identification of this Deed as the Benefit Sharing Agreement to which the report relates;

MODEL ACCESS AND BENEFIT SHARING AGREEMENT
AUSTRALIAN GOVERNMENT AND ACCESS PARTY

    b.   a summary of all Samples collected under this Deed (including collection locations, summary of taxa collected and isolated);

    c.   results of research on the biology of the taxa and ecology assessments of populations from which the Samples were collected;

    d.   species inventories, ecological data and imagery for sites sampled;

    e.   summary of screening results, or other genetic or biochemical research results;

    f.   summary of structures found;

    g.   publications and conference presentations arising from research into the Samples;

    h.   research opportunities and capacity building opportunities provided in Australia;

    i.   progress in establishing third party agreements as they relate to the Samples;

    j.   Exploitation Revenue received from third parties and the Threshold Payments payable to the Commonwealth; and

    k.   Disposal of Samples and Products.

8.3.2.   Subsequent Annual Reports will report on activities for the preceding calendar year (the reporting period) and will include, but need not be limited to, the following information for the reporting period

    a.   identification of this Deed as the Benefit Sharing Agreement to which the report relates;

    b.   results of research on the biology of the taxa and ecology assessments of populations from which the Samples were collected;

    c.   species inventories, ecological data and imagery for sites sampled not included in previous Annual Reports;

    d.   summary of screening results, or other genetic or biochemical research results;

    e.   summary of structures found;

    f.   publications and conference presentations arising from research into the Samples;

    g.   the progress toward commercialisation of Products;

MODEL ACCESS AND BENEFIT SHARING AGREEMENT
AUSTRALIAN GOVERNMENT AND ACCESS PARTY

h.   research opportunities and capacity building opportunities provided in Australia;

i.   progress in establishing third party agreements as they relate to the Samples and Products;

j.   Exploitation Revenue received from third parties and the Threshold Payments payable to the Commonwealth; and

k.   Disposal of Samples and Products.

8.3.3.   Annual Reports will be provided on or before 31 March in the year following the year to which the report relates.

8.4.   Other Reports

a.   The Access Party will provide such other reports as may reasonably be requested by the Commonwealth from time to time.

8.5.   Form of Reports

8.5.1.   Annual Reports will be provided by the Access Party in two Parts:

a.   the first part will only contain information considered by the Access Party to be non-confidential and may be made available to the public by the Commonwealth without prior consent of the Access Party, and may be published by the Access Party for publicity purposes;

b.   the second part will contain information the Access Party reasonably requires to be treated as commercial-in-confidence for the purpose of protection of Intellectual Property,

and any material identified as commercial-in-confidence will be Confidential Information for the purposes of this Deed.

8.5.2.   All reports provided by the Access Party under this Deed will be provided in hard copy and digital copy.

9.   Confidential Information

9.1.1.   Subject to clause 9.1.5, a Party must not, without the prior written consent of the other Party, use or disclose any Confidential Information of the other Party.

9.1.2.   In giving written consent to use or disclose its Confidential Information, a Party may impose such conditions as it thinks fit, and the other Party agrees to comply with these conditions.

MODEL ACCESS AND BENEFIT SHARING AGREEMENT
AUSTRALIAN GOVERNMENT AND ACCESS PARTY

9.1.3.   A Party may at any time require the other Party to arrange for the other Party's employees, servants or agents to give a written undertaking in the form of a Deed relating to the use and non-disclosure of the first Party's Confidential Information.

9.1.4.   If a Party receives a request under clause 9.1.3, it must promptly arrange for all such undertakings to be given.

9.1.5.   The obligations on a Party under this clause will not be taken to have been breached to the extent that Confidential Information:

   a.   is disclosed by a Party to its employees, servants or agents solely in order to comply with obligations, or to exercise rights, under this Deed;

   b.   is disclosed to a Party's internal management personnel, solely to enable effective management or auditing of activities related to this Deed;

   c.   is disclosed by the Department to the Department's Minister;

   d.   is shared by a Party within its organisation, or in the case of the Department with another Commonwealth department or agency, where this serves the Party's legitimate interests;

   e.   is disclosed by a Party, in response to a request by a House or a Committee of the Parliament of the Commonwealth of Australia;

   f.   is authorised or required by law to be disclosed;

   g.   is disclosed by a Party and is information in a material form in respect of which an interest, whether by licence or otherwise, in the Intellectual Property Rights in relation to that material form, has vested in, or is assigned to, the Party under this Deed or otherwise, and that disclosure is permitted by that licence or otherwise; or

   h.   is in the public domain otherwise than due to a breach of this clause.

9.1.6.   Where a Party discloses Confidential Information to another person:

   a.   pursuant to clauses 9.1.5 (a), (b) or (d) – the disclosing Party must:

      A.   notify the receiving person that the information is Confidential Information; and

      B.   not provide the information unless the receiving person agrees to keep the information confidential; or

**MODEL ACCESS AND BENEFIT SHARING AGREEMENT
AUSTRALIAN GOVERNMENT AND ACCESS PARTY**

      b.   pursuant to clauses 9.1.5 (c) or (e) – the disclosing Party must notify the receiving person that the information is Confidential Information.

9.1.7.   The obligations under this clause continue, notwithstanding the expiry or termination of the Term of this Deed:

      a.   in relation to an item of information described in Schedule 1 – for the period set out in the Schedule in respect of that item; and

      b.   in relation to any information that is agreed between the Parties after the Date of this Deed as constituting Confidential Information for the purposes of this Deed – for the period agreed by the Parties.

9.1.8.   Nothing in this clause derogates from any obligation which the Access Party may have either under the Privacy Act, or under this Deed, in relation to the protection of Personal Information.

10.   Indemnity

10.1.1.   The Access Party indemnifies (and keeps indemnified) the Commonwealth against any:

      a.   loss or liability incurred by the Commonwealth;

      b.   loss of or damage to the Commonwealth's property; or

      c.   loss or expense incurred by the Commonwealth in dealing with any claim against the Commonwealth, including legal costs and expenses on a solicitor/own client basis and the cost of time spent, resources used, or disbursements paid by the Commonwealth;

    arising from:

      d.   any act or omission by the Access Party in connection with this Deed, where there was fault on the part of the person whose conduct gave rise to that liability, loss, damage, or expense;

      e.   any breach by the Access Party of its obligations under this Deed;

10.1.2.   the Access Party's liability to indemnify the Commonwealth under this clause will be reduced proportionally to the extent that any fault on the Commonwealth's part contributed to the relevant loss, damage, expense, or liability.

10.1.3.   The Commonwealth's right to be indemnified under this clause is in addition to, and not exclusive of, any other right, power, or remedy

---

MODEL ACCESS AND BENEFIT SHARING AGREEMENT
AUSTRALIAN GOVERNMENT AND ACCESS PARTY

provided by law, but the Commonwealth is not entitled to be compensated in excess of the amount of the relevant liability, damage, loss, or expense.

10.1.4.   In this clause, "fault" means any negligent or unlawful act or omission or wilful misconduct.

10.1.5.   This operation of this clause survives the expiration or earlier termination of the Term of this Deed.

11.   Insurance

11.1.1.   The Access Party must, for as long as any obligations remain in connection with this Deed, have insurance as specified in Schedule 1.

11.1.2.   Whenever requested, the Access Party must provide the Department, within 10 Business Days of the request, with evidence satisfactory to the Department that the Access Party has complied with its obligation to insure.

11.1.3.   All insurance under this clause is to be taken out with an insurer recognised under Australian law, and whenever requested, the Access Party must provide the Department with evidence satisfactory to the Department that the Access Party has complied with its obligation to insure.

11.1.4.   The operation of this clause survives the expiration or earlier termination of the Term of this Deed.

12.   Access to Premises and Records

12.1.   **Access for Audit Purposes**

12.1.1.   The Access Party will give to the Commonwealth, or to any persons authorised in writing by the Commonwealth's, access to premises occupied by the Access Party and permit those persons to participate in audits, inspect and take copies of any Material relevant to this Deed.

12.1.2.   The rights referred to in clause 12.1.1 are subject to:

a.   the provision of reasonable prior notice by the Commonwealth;

b.   the Access Party's reasonable security procedures;

c.   if appropriate, execution of a deed of confidentiality relating to non-disclosure of the Access Party's Confidential Information; and

MODEL ACCESS AND BENEFIT SHARING AGREEMENT
AUSTRALIAN GOVERNMENT AND ACCESS PARTY

    d.   the Commonwealth not unreasonably interfering with the Access Party's performance under this Deed in any material respect.

12.1.3.   Without in any way affecting the statutory powers of the Auditor-General under the *Auditor-General Act 1997 (Cth)* and subject to the provisions of that Act, the Auditor-General is a person authorised for the purposes of this clause.

12.1.4.   This clause applies for the Term of this Deed and for a period of 7 years from the date of termination of this Deed.

13.   Termination

13.1.   Termination by Agreement

13.1.1.   This Deed may be terminated at any time by mutual agreement in writing.

13.2.   Termination on Cancellation of Permit

13.2.1.   If the permit issued to the Access Party to access biological resources to which the Deed relates is cancelled the Commonwealth may immediately terminate this Deed by written notice to the Access Party.

13.3.   Termination for Default

13.3.1.   If:

    a.   the Access Party fails to satisfy any of its obligations under this Deed;

    b.   the Access Party breaches any law of the Commonwealth, or of a State or Territory in relation to the subject matter of this Deed;

    c.   the Commonwealth is satisfied that any statement made, or document provided, to the Commonwealth by the Access Party in connection with this Deed is defective by reason of being incorrect, incomplete, false or misleading,

the Commonwealth may immediately terminate this Deed by giving written notice to the Access Party of the termination provided:

    d.   the Commonwealth has given notice to the Access Party; and

    e.   the Access Party fails within the period specified in the notice (being not less than 20 Business Days) to rectify or explain to the satisfaction of the Commonwealth the failure, breach or defect.

MODEL ACCESS AND BENEFIT SHARING AGREEMENT
AUSTRALIAN GOVERNMENT AND ACCESS PARTY

13.4. **Consequences of Termination**

13.4.1. If this Deed is terminated under clause 13.2.2 or 13.2.3:

a. the Access Party will not thereafter use, or cause, permit or allow to be used:

A. any Samples or Products;

B. Intellectual Property arising from R & D Activity;

b. the Access Party will deliver to the Commonwealth or destroy, at the Commonwealth's discretion, all Samples and Products that are the subject of this Agreement; and

c. the Access Party's rights in all third party agreements referred to in clause 5.3.2, are assigned to the Commonwealth and the Access Party will do all things, and sign all documents, necessary to effect the assignment of those rights,

and the operation of this clause 13.2.2 survives the termination of this Deed.

13.4.2. Termination under this clause will not affect the right of the Access Party to sell Products or material containing a Product, by way of retail sale under commercial arrangements existing at the date of termination, and the Access Party's obligation to provide Exploitation Revenue in accordance with clause 4 and Schedule 3 will survive the termination.

14. **Dispute Resolution**

14.1. **No Legal Proceedings**

14.1.1. Subject to clause 14.2.2, both Parties agree not to commence any legal proceedings in respect of any dispute arising under this Deed, which cannot be resolved by informal discussion, until the procedure provided by this clause has been utilised.

14.2. **Dispute Resolution Procedure**

14.2.1. Both Parties agree that any dispute arising during the course of this Deed is dealt with as follows:

a. the Party claiming that there is a dispute will send the other a written notice setting out the nature of the dispute;

b. the Parties will try to resolve the dispute though direct negotiation by persons who they have given authority to resolve the dispute;

MODEL ACCESS AND BENEFIT SHARING AGREEMENT
AUSTRALIAN GOVERNMENT AND ACCESS PARTY

      c.   the Parties have 10 Business Days from the receipt of the notice to reach a resolution or to agree that the dispute is to be submitted to mediation or some alternative dispute resolution procedure; and

      d.   if:

          A.   there is no resolution of the dispute;

          B.   there is no agreement on submission of the dispute to mediation or some alternative dispute resolution procedure; or

          C.   there is a submission to mediation or some other form of alternative dispute resolution procedure, but there is no resolution within 15 Business Days of the submission, or such extended time as the Parties may agree in writing before the expiration of the 15 Business Days,

          D.   then, either Party may commence legal proceedings.

14.2.2.   This clause does not apply to the following circumstances:

      a.   either Party commences legal proceedings for urgent interlocutory relief;

      b.   termination for default under clause 13 ;or

      c.   an authority of the Commonwealth, a State or Territory is investigating a breach or suspected breach of the law by the Access Party.

14.2.3.   Despite the existence of a dispute, both Parties must (unless requested in writing by the other Party not to do so) continue to perform their respective obligations in accordance with this Deed.

14.2.4.   The operation of this clause survives the expiration or earlier termination of the Term of this Deed.

15.   General Provisions

15.1.   Negation of Employment, Partnership and Agency

15.1.1.   The Access Party agrees not to represent itself, and to use its best endeavours to ensure that its Personnel do not represent themselves, as being an officer, employee, partner or agent of the Commonwealth, or as otherwise able to bind or represent the Commonwealth.

15.1.2.   The Access Party is not by virtue of this Deed an officer, employee, partner or agent of the Commonwealth, nor does the Access Party have any power or authority to bind or represent the Commonwealth.

15.2. Waiver

15.2.1.   A failure or delay by a party to exercise any right it holds under this Deed will not operate as a waiver of that right.

15.2.2.   A single or partial exercise by a party of any right it holds under this Deed will not prevent that party from exercising that right again or exercising that right to the extent it has not already been exercised.

15.2.3.   In this clause, the word "right" means a right or remedy provided by this Deed or at law.

15.3.   Assignment and novation

15.3.1.   Except as otherwise provided by this Deed, the Access Party cannot novate its obligations and must not assign its rights, under this Deed without, in either case, prior approval in writing from the Commonwealth, which will not be unreasonably withheld.

15.3.2.   The Access Party must not consult with any other person for the purposes of entering into an arrangement that will require novation of this Deed without first consulting the Commonwealth.

15.4.   Notices

15.4.1.   Any notice, request or other communication to be given or served pursuant to this Deed will be in writing and dealt with as follows:

a.   if given by the Access Party to the Commonwealth - addressed as specified in Item D [Commonwealth Address for Notices] of Schedule 1 or

b.   if given by the Commonwealth to the Access Party - addressed as specified in Item E [the Access Party Address for Notices] of Schedule 1.

15.4.2.   Any notice, request or other communication is to be delivered by hand, sent by prepaid post or transmitted electronically. If it is sent or transmitted electronically, a copy is to be sent to the addressee by prepaid post.

15.4.3.   A notice will only be deemed as given and received:

a.   if delivered by hand, upon delivery to the relevant address

**MODEL ACCESS AND BENEFIT SHARING AGREEMENT
AUSTRALIAN GOVERNMENT AND ACCESS PARTY**

b.   if sent by pre-paid ordinary post within Australia, upon the expiration of 2 business days after the date on which it was sent

c.   if transmitted electronically, upon receipt by the sender of an acknowledgment that the communication has been properly transmitted to the recipient and

d.   in any event, if received after 5.00pm (local time in the place of receipt) on a Business Day or on a day that is not a Business Day, on the next Business Day.

15.4.4.   Either party may, by written notice to the other, change its Representative.

15.5.   Governing law

15.5.1.   This Deed is to be construed in accordance with the laws of the Australian Capital Territory.

MODEL ACCESS AND BENEFIT SHARING AGREEMENT
AUSTRALIAN GOVERNMENT AND ACCESS PARTY

SIGNATURE PAGE

EXECUTED as a Deed.

SIGNED on behalf of the   )
COMMONWEALTH OF AUSTRALIA   )
by insert name of signatory   )
insert signatory's position   )

IN THE PRESENCE OF
insert name of witness   )

SIGNED on behalf of   )
INSERT NAME OF ACCESS PARTY   )
by insert name of signatory   )
insert signatory's position   )

IN THE PRESENCE OF
insert name of witness   )

22

MODEL ACCESS AND BENEFIT SHARING AGREEMENT
AUSTRALIAN GOVERNMENT AND ACCESS PARTY

SCHEDULE 1. - PARTICULARS

A.   **Applicable Policies, Codes of Practice and Guidelines (clause 4)**

A.1.   Please list and policies, guidelines, codes of conduct or standards which apply to the collection of Samples and subsequent research on them. The list should include any data standards or protocols that will be used in the preparation of reporting data. Examples of possible policies are: the Bonn Guidelines, the Australian Code for the Responsible Conduct of Research, the data standards for the Atlas of Living Australia (http://www.ala.org.au/datastandards.htm#LSID) protocols for research involving Traditional Knowledge and working with Indigenous people; and university ethics committee requirements (such as the Animal Ethics requirements in clause 5.2.). The Australian Government Bioethics Portal (http://www.bioethics.gov.au/) may be of assistance in identifying relevant material.

B.   **Confidential Information (clause 9)**

B.1.   **The Commonwealth's Confidential Information is:**

Item	Period of Confidentiality

B.2.   **Access Party's Confidential Information is:**

23

**MODEL ACCESS AND BENEFIT SHARING AGREEMENT
AUSTRALIAN GOVERNMENT AND ACCESS PARTY**

Item	Period of Confidentiality
Specify confidential information and the period of confidentiality	

C.   **Insurance (clause 11)**

     C.1.   Public liability insurance to the value of at least $10 million per claim, or occurrence giving rise to a claim, in respect to activities undertaken under this Deed, where occurrence means either a single occurrence or a series of occurrences if these are linked or occur in connection with one another from one original cause, as the case may be.

D.   **Commonwealth Address for Notices (clause 15.4)**

     D.1.   The Director
Genetic Resources Management Policy
Department of the Environment and Water Resources
GPO Box 787
CANBERRA ACT 2601
AUSTRALIA

E.   **Access Party's Address for Notices (clause 15.4)**

     E.1.

MODEL ACCESS AND BENEFIT SHARING AGREEMENT
AUSTRALIAN GOVERNMENT AND ACCESS PARTY

SCHEDULE 2. ACCESS AND USE CONDITIONS

A. **Access Area**

A.1. List the areas from which the Samples will be taken, including latitude and longitude references.

B. **Time and Frequency of Entry to Access Area**

B.1. List the anticipated dates and times of entry to the access area(s).

C. **Samples of Biological Resources to be Collected**

C.1. Include name of the species, or lowest level of taxon, to which the resources belong (if known). If the species composition of Samples is not known, list the sampling method(s) and the types of organisms likely to be collected. Note that clause 8.2.1 requires that a collection report must be provided to the Commonwealth within six months of the Samples being taken, or by 1 March, whichever is the later.

D. **Quantity of Resources to be Collected**

D.1. List the anticipated quantity of each Sample to be collected in the Access Area. Please use metric measurements.

E. **Quantity of Resources to be Removed From Access Area**

E.1. List the quantity of each Sample to be removed from the Access Area. Please use metric measurements.

F. **Purpose of Access**

F.1. Provide a brief description of the purpose(s) of collecting samples and the subsequent research that will be undertaken using them.

G. **Labelling of Samples**

G.1. Include a statement setting out the means of labelling the Samples.

The Regulations require that appropriate labelling of Samples must be undertaken and that records relating to the Samples collected must also be kept.
Records must include:
• a unique identifier for each sample, marked either on a label attached to the sample, or the container holding the sample.
• the date the sample was taken.
• the place from which the sample was taken.
• an appropriate indication of the quantity or size of the sample.
• the scientific name of, or given to, the sample.

**MODEL ACCESS AND BENEFIT SHARING AGREEMENT
AUSTRALIAN GOVERNMENT AND ACCESS PARTY**

• the location of the sample when first entered in the record (ie where the Sample is initially housed).
• the details about any subsequent transfer of the sample, including the names and addresses of others having possession of the sample, or a part of the sample.

H.   **Disposition of Ownership in Samples**

H.1.   Include details of any proposed transmission of samples to third parties. Third parties may include museums, research institutions, individual researchers or commercial organisations.

I.   **Use of Indigenous People's Knowledge**

I.1.   Include details of the source of the knowledge, such as, for example, whether the knowledge was obtained from scientific or other public documents, from the access provider or from another group of indigenous persons

J.   **Benefits or Commitments for Use of Indigenous People's Knowledge**

J.1.   If any indigenous people's knowledge of the access provider, or other group of indigenous persons, is to be used, a copy of the agreement regarding use of the knowledge (if there is a written document), or the terms of any oral agreement, regarding the use of the knowledge

K.   **Proposals to Benefit Biodiversity Conservation in Access Area**

K.1.   Provide a statement of the benefits of the research to the conservation of biodiversity in the access areas. Benefits may include (but are not limited to) improved knowledge of: biodiversity; taxonomy; biological and ecological processes; impacts of environmental change; or data and knowledge that will assist in the conservation and management of the environment.

MODEL ACCESS AND BENEFIT SHARING AGREEMENT
AUSTRALIAN GOVERNMENT AND ACCESS PARTY

SCHEDULE 3. - BENEFITS[1]

THRESHOLD PAYMENTS

A.1.   Where the gross Exploitation Revenue received by the Access Party in a calendar year falls within the relevant threshold range specified in column 1 of the table below the Access Party will pay to the Commonwealth the corresponding percentage of gross Exploitation Revenue specified in column 2 of the table (Threshold Payments).

Purpose of the Product	Gross Exploitation Revenue received in one calendar year ($ Australian Dollars)	Threshold Payment (% of gross Exploitation Revenue)
Pharmaceutical, Nutraceutical or Agricultural	< 500 000	0
	500 000 – 5 000 000	2.5
	> 5 000 000	5.0
Research	> 200 000	2.5
	or	
	< 100 000	0
	100 000 – 3 000 000	1.0
	> 3 000 000	3.0
Industrial, Chemical, Diagnostic or Other	> 200 000	1.5
	or	
	< 100 000	0
	100 000 – 3 000 000	1.0
	> 3 000 000	2.0

A.2.   Threshold payments will be paid annually by the Access Party within 28 days after receipt of a correctly tendered tax invoice.

27

MODEL ACCESS AND BENEFIT SHARING AGREEMENT
AUSTRALIAN GOVERNMENT AND ACCESS PARTY

B.   OFFERS OF SPECIMENS

B.1.   The Access Party will offer a taxonomic duplicate of each Sample taken to an Australian public institution which has a statutory responsibility to maintain biological collections, or another institution approved by the Commonwealth, that is a repository of taxonomic specimens of the same order or genus as those collected for permanent loan.

B.2.   Within 3 months of the date of offer under B1, the Access Party must notify the Commonwealth of the name of the Australian public institution(s) to which the duplicate Sample(s) have been offered, the date of the offer, a list of the Sample(s) offered and indicate which Samples were accepted by that institution.

B.3.   The Access Party agrees that the offer of a taxonomic duplicate of Samples to an institution will include that they may be used for genetic analysis for the International Barcode of Life project.

B.4.   The Access Party may impose reasonable conditions on offers made under B1and B3 including, without limiting the generality of the foregoing that the receiving institution may only use the specimens for non-commercial purposes.

C.   KNOWLEDGE TRANSFER

C.1.   The Access Party agrees that knowledge and information, which is not Confidential Information, contained in reports to the Commonwealth, that is relevant to the taxonomy, conservation or sustainable use of biological diversity may be transferred to Australian research institutions, the Atlas of Living Australia, the Census of Marine Life, managers of Commonwealth areas, or to Indigenous Access Providers for non-commercial purposes.

D.   PUBLICATIONS

D.1.   The Access Party will notify the Commonwealth of publications arising from research involving the Samples and supply an electronic or hard copy of such publications on request.

**MODEL ACCESS AND BENEFIT SHARING AGREEMENT
AUSTRALIAN GOVERNMENT AND ACCESS PARTY**

SCHEDULE 4. – ADDITIONAL BENEFITS

A range of monetary and non-monetary benefits may be provided in return for access to biological resources. A broad range of benefits are outlined in the Convention on Biological Diversity's 'Bonn Guidelines on Access to Genetic Resources and Equitable Sharing of the Benefits Arising out of their Utilization' (http://www.cbd.int/doc/publications/cbd-bonn-gdls-en.pdf). The following clauses are offered by way of example only.

A.   ***AD HOC* RESEARCH**

        A.1.   The Access Party will keep the Commonwealth aware of all field trips that will include access to biological resources in the Access Area.

        A.2.   The Commonwealth has the option to request that additional research be conducted on these field trips.

        A.3.   The Commonwealth will meet the reasonable costs of additional research under clause A.2 with terms and conditions to be negotiated with the Access Party separately to the negotiations around this Deed.

B.   **RESEARCH FUNDING**

        B.1.   The Access Party will provide research funding to a local research institution to conduct research on species collected as Samples or the ecosystem from which they were collected.

C.   **JOINT VENTURES**

        C.1.   The Access Party will enter into a joint venture with
                a. an Australian research institution to conduct research on species collected as Samples or the ecosystem from which they were collected;

                b. an Australian company or research institution to undertake bioactivity screening, preclinical and/or clinical trials or otherwise develop commercial products containing the Sample or a Product.

D.   **CAPACITY-BUILDING**

        D.1.   The Access Party will transfer to an Australian research institution or to Indigenous Access Providers knowledge to make use of genetic resources, including biotechnology, or knowledge that is relevant to the conservation and sustainable use of biological diversity.

MODEL ACCESS AND BENEFIT SHARING AGREEMENT
AUSTRALIAN GOVERNMENT AND ACCESS PARTY

E.   TECHNOLOGY TRANSFER

E.1.   The Access Party will transfer to an Australian research institution
technology to make use of genetic resources, including
biotechnology, or technology that is relevant to the conservation and
sustainable use of biological diversity. The terms of transfer will be
negotiated with the receiving institution, and should be developed
under fair and favourable terms, including concessional and
preferential terms.

F.   SCIENTIFIC RESEARCH AND DEVELOPMENT PROGRAMMES

F.1.   The Access Party will collaborate with Australian research
institutions and contribute to scientific research and development
programmes, particularly biotechnological research activities.

Index

Page numbers with t are tables; b are boxes; f are figures; n show notes with the number after the n showing the note number on the page.